DAVID CHRISTOPHER SHOCK

STRUCTURAL DYNAMICS

THEORY AND COMPUTATION

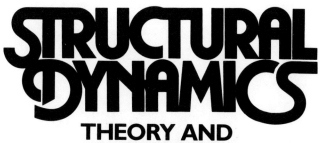

STRUCTURAL DYNAMICS

THEORY AND COMPUTATION

MARIO PAZ

Professor of Civil Engineering
University of Louisville

Van Nostrand Reinhold Environmental Engineering Series

VNR VAN NOSTRAND REINHOLD COMPANY
NEW YORK CINCINNATI TORONTO LONDON MELBOURNE

Van Nostrand Reinhold Company Regional Offices:
New York   Cincinnati

Van Nostrand Reinhold Company International Offices:
London   Toronto   Melbourne

Library of Congress Catalog Card Number: 79-13869
ISBN: 0-442-23019-2

Manufactured in the United States of America

Published by Van Nostrand Reinhold Company Inc.
135 West 50th Street, New York, NY. 10020

Published simultaneously in Canada by Van Nostrand Reinhold Ltd.

15 14 13 12 11 10 9 8 7 6 5 4 3

Library of Congress Cataloging in Publication Data

Paz, Mario.
   Structural dynamics, theory and computation.

   (Van Nostrand Reinhold environmental engineering
series)
   Includes index.
   1. Structural dynamics.   I. Title.
TA654.P39   624'.171   79-13869
ISBN 0-442-23019-2

To Clara, Posthumously

(October 3, 1932–February 11, 1979)

A Tribute

She fought.
  Oh she fought:
  Bravery rode on her smile
  and shone in her eyes.

She fought
  as the enemy
  ate from within;

The Beauty of her
  lived to the end—
  and will live in the hearts
  that knew her.

Feb. 12, 1979   *Sylvia T. Weinberg*

Van Nostrand Reinhold Environmental Engineering Series

THE VAN NOSTRAND REINHOLD ENVIRONMENTAL ENGINEERING SERIES is dedicated to the presentation of current and vital information relative to the engineering aspects of controlling man's physical environment. Systems and subsystems available to exercise control of both the indoor and outdoor environment continue to become more sophisticated and to involve a number of engineering disciplines. The aim of the series is to provide books which, though often concerned with the life cycle—design, installation, and operation and maintenance—of a specific system or subsystem, are complementary when viewed in their relationship to the total environment.

The Van Nostrand Reinhold Environmental Engineering Series includes books concerned with the engineering of mechanical systems designed (1) to control the environment within structures, including those in which manufacturing processes are carried out, and (2) to control the exterior environment through control of waste products expelled by inhabitants of structures and from manufacturing processes. The series includes books on heating, air conditioning and ventilation, control of air and water pollution, control of the acoustic environment, sanitary engineering and waste disposal, illumination, and piping systems for transporting media of all kinds.

To Clara, Posthumously

(October 3, 1932–February 11, 1979)

A Tribute

She fought.
Oh she fought:
Bravery rode on her smile
and shone in her eyes.

She fought
as the enemy
ate from within;

The Beauty of her
lived to the end—
and will live in the hearts
that knew her.

Feb. 12, 1979   *Sylvia T. Weinberg*

Van Nostrand Reinhold Environmental Engineering Series

THE VAN NOSTRAND REINHOLD ENVIRONMENTAL ENGINEERING SERIES is dedicated to the presentation of current and vital information relative to the engineering aspects of controlling man's physical environment. Systems and subsystems available to exercise control of both the indoor and outdoor environment continue to become more sophisticated and to involve a number of engineering disciplines. The aim of the series is to provide books which, though often concerned with the life cycle—design, installation, and operation and maintenance—of a specific system or subsystem, are complementary when viewed in their relationship to the total environment.

The Van Nostrand Reinhold Environmental Engineering Series includes books concerned with the engineering of mechanical systems designed (1) to control the environment within structures, including those in which manufacturing processes are carried out, and (2) to control the exterior environment through control of waste products expelled by inhabitants of structures and from manufacturing processes. The series includes books on heating, air conditioning and ventilation, control of air and water pollution, control of the acoustic environment, sanitary engineering and waste disposal, illumination, and piping systems for transporting media of all kinds.

Van Nostrand Reinhold Environmental Engineering Series

ADVANCED WASTEWATER TREATMENT, by Russell L. Culp and Gordon L. Culp

ARCHITECTURAL INTERIOR SYSTEMS—Lighting, Air Conditioning, Acoustics, John E. Flynn and Arthur W. Segil

SOLID WASTE MANAGEMENT, by D. Joseph Hagerty, Joseph L. Pavoni and John E. Heer, Jr.

THERMAL INSULATION, by John F. Malloy

AIR POLLUTION AND INDUSTRY, edited by Richard D. Ross

INDUSTRIAL WASTE DISPOSAL, edited by Richard D. Ross

MICROBIAL CONTAMINATION CONTROL FACILITIES, by Robert S. Rurkle and G. Briggs Phillips

SOUND, NOISE, AND VIBRATION CONTROL (Second Edition), by Lyle F. Yerges

NEW CONCEPTS IN WATER PURIFICATION, by Gordon L. Culp and Russell L. Culp

HANDBOOK OF SOLID WASTE DISPOSAL: MATERIALS AND ENERGY RECOVERY, by Joseph L. Pavoni, John E. Heer, Jr., and D. Joseph Hagerty

ENVIRONMENTAL ASSESSMENTS AND STATEMENTS, by John E. Heer, Jr. and D. Joseph Hagerty

ENVIRONMENTAL IMPACT ANALYSIS: A New Dimension in Decision Making, by R. K. Jain, L. V. Urban and G. S. Stacey

CONTROL SYSTEMS FOR HEATING, VENTILATING, AND AIR CONDITIONING (Second Edition), by Roger W. Haines

WATER QUALITY MANAGEMENT PLANNING, edited by Joseph L. Pavoni

HANDBOOK OF ADVANCED WASTEWATER TREATMENT (Second Edition), by Russell L. Culp, Geoge Mack Wesner and Gordon L. Culp

HANDBOOK OF NOISE ASSESSMENT, edited by Daryl N. May

NOISE CONTROL: HANDBOOK OF PRINCIPLES AND PRACTICES, edited by David M. Lipscomb and Arthur C. Taylor

AIR POLLUTION CONTROL TECHNOLOGY, by Robert M. Bethea

POWER PLANT SITING, by John V. Winter and David A. Conner

DISINFECTION OF WASTEWATER AND WATER FOR REUSE, by Geo. Clifford White

LAND USE PLANNING: Techniques of Implementation, by T. William Patterson

BIOLOGICAL PATHS TO SELF-RELIANCE, by Russell E. Anderson

HANDBOOK OF INDUSTRIAL WASTE DISPOSAL, by Richard A. Conway and Richard D. Ross

HANDBOOK OF ORGANIC WASTE CONVERSION, by Michael W. Bewick

LAND APPLICATIONS OF WASTE (Volume 1), by Raymond C. Loehr, William J. Jewell, Joseph D. Novak, William W. Clarkson and Gerald S. Friedman

LAND APPLICATIONS OF WASTE (Volume 2), by Raymond C. Loehr, William J. Jewell, Joseph D. Novak, William W. Clarkson and Gerald S. Friedman

STRUCTURAL DYNAMICS: Theory and Computation, by Mario Paz

HANDBOOK OF MUNICIPAL WASTE MANAGEMENT SYSTEMS: Planning and Practice, by Barbara J. Stevens

INDUSTRIAL POLLUTION CONTROL: Issues and Techniques, by Nancy J. Sell

WASTE RECYCLING AND POLLUTION CONTROL HANDBOOK, by A. V. Bridgwater and C. J. Mumford.

WATER CLARIFICATION PROCESSES: Practical Design and Evaluation, by Herbert E. Hudson, Jr.

ENVIRONMENTAL RISK ANALYSIS FOR CHEMICALS, edited by Richard A. Conway

HANDBOOK OF NON-POINT POLLUTION, by Vladimir Novotny

NATURAL SYSTEMS FOR WATER POLLUTION CONTROL, by Ray Dinges

Foreword

In the Latin world, particularly in France, there is a hallowed tradition by which a young author upon the impendent appearance in print of his first book seeks out an older person, who has worked in the same field a generation earlier, and who (in France) usually has reached a ripe age and the dignity of a "Member of the Academy," with the request to write a foreword or preface, wishing the new book well on the beginning of its career. Such forewords (being a generation out of date by definition) rarely are worth reading, but they are intended to pay a compliment to the older man for his past work.

Although we do not live in the Latin world and we have no tradition, hallowed or otherwise, in this respect, Mario Paz (who originally hails from Chile in the Latin world) has still paid me the compliment with the request that I write a foreword and I am very much pleased of course.

I have known Mario in this country for some 15 years and during that time have worked with him on a number of practical vibration problems; both of us as consultants to the Vibrating Equipment Division of Rexnord Inc., makers of vibrating conveyors. I have formed a high opinion of his abilities.

The book gives a thorough and comprehensive discussion of various "structures," which are the stock in trade of the civil engineer. However, the mechanical engineer in the background has the habit of messing up these nice structures by putting unbalanced rotors and other devilish devices on them which cause "harmonic responses." These harmonic responses in the structure are discussed through the whole gamut of increasing complication: from single degree-of-freedom idealizations, to discrete systems, and to systems with distributed inertia. If the reader is willing to admit to his "structure" the presence of rotating or reciprocating structural elements the whole book also gives the methods of attack for the truly mechanical engineering problems of rotor dynamics.

This text is important and timely because it bridges the gap between the differential (or other) equations and the ultimate practical numerical solution of a computer program. In so doing the author has indeed advanced a generation on my own youthful text on vibration (1934–1956).

The present book of Mario Paz is clearly written; it should and hopefully will be appreciated by engineering students and practicing engineers, civil or mechanical. I wish the author and his new book a good career!

J. P. DEN HARTOG

Preface

Natural phenomena and human activities impose forces of time-dependent variability on structures as simple as a concrete beam or a steel pile, or as complex as a multistory building or a nuclear power plant constructed from different materials. Analysis and design of such structures subjected to dynamic loads involve consideration of time-dependent inertial forces. The resistance to displacement exhibited by a structure may include forces which are functions of the displacement and the velocity. As a consequence, the governing equations of motion of the dynamic system are generally nonlinear partial differential equations which are extremely difficult to solve in mathematical terms. Nevertheless, recent developments in the field of structural dynamics enable such analysis and design to be accomplished in a practical and efficient manner. This work is facilitated through the use of simplifying assumptions and mathematical models, and of matrix methods and modern computational techniques.

In the process of teaching courses on the subject of structural dynamics, the author came to the realization that there was a definite need for a text which would be suitable for the advanced undergraduate or the beginning graduate engineering student being introduced to this subject. The author is familiar with the existence of several excellent texts of an advanced nature but generally these texts are, in his view, beyond the expected comprehension of the student. Consequently, it was his principal aim in writing this book to incorporate modern methods of analysis and techniques adaptable to computer programming in a manner as clear and easy as the subject permits. He felt that computer programs should be included in the book in order to assist the student in the application of modern methods associated with computer usage. In addition, the author hopes that this text will serve the practicing engineer for purposes of self-study and as a reference source.

In writing this text, the author also had in mind the use of the book as a possible source for research topics in structural dynamics for students working towards an advanced degree in engineering who are required to write a thesis. At Speed Scientific School, University of Louisville, most engineering students complete a fifth year of study with a thesis requirement leading to a Master in Engineering degree. The author's experience as a thesis advisor leads him to believe that this book may well serve the students in their search and selection of topics in subjects currently under investigation in structural dynamics.

The subjects in this text are organized in four parts. Part I deals with structures modeled as single degree-of-freedom systems. It introduces basic concepts and presents important methods for the solution of such dynamic systems. Part II introduces important concepts and methodology for multidegree-of-freedom systems through the use of structures modeled as *shear buildings*. Part III describes methods for the dynamic analysis of framed structures modeled as discrete systems with many degrees of freedom. Finally, Part IV presents the mathematical solution for some simple structures modeled as systems with distributed properties, thus having an infinite number of degrees of freedom. Part IV also shows mathematically the relation between the exact method of solution of continuous systems and the approximate method of solution of structures modeled as discrete systems. The listing of 16 computer programs is given in the Appendix. These programs are discussed in the appropriate chapters throughout the text. They are not highly sophisticated programs which can solve a large variety of problems with optimal efficiency nor are they intended to compete with commercially developed programs. These computer programs are intended mainly for instructional purposes. The programs have been tested through simple problems solved by hand calculation. As a final note, the author would like to assure the user that these programs are "bug-free." Unfortunately, experience demonstrates the high risk involved in such assurances. Instead, the author will be grateful that any error found in the programs be called to his attention.

Should the text fulfill the expectations of the author in some measure, particularly the elucidation of this subject, he will then feel rewarded for his efforts in the preparation and development of the material in this book.

MARIO PAZ

Acknowledgments

This book has grown out of many years of teaching and professional engineering experience. It is from the combined interplay with my students and associates that I first contemplated writing the present book. During its preparation, I become indebted to many persons to whom I wish to express my appreciation.

I am grateful to Dean Harry Saxe for continuous encouragement and for sabbatical leave so that I could devote a generous amount of time to complete the text. I am thankful to Dr. John Heer Jr., Associate Dean, for carefully reading the final draft and making many editorial improvements; to Dr. Joseph Hagerty for editing the first chapters; and to Dr. C. Eugene Miller for reviewing with me several troublesome sections of the text. My thanks also go to Mrs. Donna Greenwell for her competent typing of the manuscript and to my student José Carrasco for his assistance in testing the computer programs.

A special acknowledgment of gratitude is extended to my friend and colleague Dr. Manuel Schwartz, of the Department of Physics, who spent many evenings reading the text, commenting on the contents, and making excellent suggestions.

I am most grateful to those who so generously authorized permission for the use in text of some of their material. The response spectral charts of Chapter 8 are reproduced with permission from Dr. Nathan M. Newmark and the publisher Portland Cement Association. Subroutine Jacobi listed in the Appendix is reproduced with permission of its authors, Professors Klaus-Jurgen Bathe and E. L. Wilson. I wish to acknowledge Rexnord Inc. for authorization to include in this book several segments of computer programs which I developed in my professional association with this company.

To a great man of singular kindness and modesty, Jacop P. Den Hartog, I am indebted for his complimentary foreword.

Finally, it is with deep affection and gratitude that I pay tribute to my wife who with complete devotion and infinite patience helped me throughout the preparation of this book. This work is dedicated to her.

Contents

STRUCTURAL DYNAMICS

THEORY AND COMPUTATION

Part I

Structures modeled as a single degree-of-freedom system

1

Undamped single degree-of-freedom system

It is not always possible to obtain rigorous mathematical solutions for engineering problems. In fact, analytical solutions can be obtained only for certain simplified situations. For problems involving complex material properties, loading and boundary conditions, the engineer introduces assumptions and idealizations deemed necessary to make the problem mathematically manageable, but still capable of providing sufficiently approximate solutions and satisfactory results from the point of view of safety and economy. The link between the real physical system and the mathematically feasible solution is provided by the *mathematical model* which is the symbolic designation for the substitute idealized system including all the assumptions imposed on the physical problem.

1.1 DEGREES OF FREEDOM

In structural dynamics the number of independent coordinates necessary to specify the configuration or position of a system

3

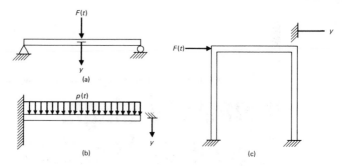

Fig. 1.1   Examples of structures modeled as one-degree-of-freedom systems.

at any time is referred to as the number of degrees of freedom. In general, a continuous structure has an infinite *number of degrees of freedom.* Nevertheless, the process of idealization or selection of an appropriate mathematical model permits the reduction in the number of degrees of freedom to a discrete number and in some cases to just a single degree of freedom. Figure 1.1 shows some examples of structures which may be represented for dynamic analysis as *one-degree-of-freedom systems;* that is, structures modeled as systems with a single displacement coordinate. These one-degree-of-freedom systems may be described conveniently by the mathematical model shown in Fig. 1.2 which has the following elements: (1) a mass element m representing the mass and inertial characteristic of the structure; (2) a spring element k representing the elastic restoring force and potential energy capacity of the structure; (3) a damping element c representing the frictional characteristics and energy losses of the structure; and (4) an excitation force $F(t)$ representing the external forces acting on the structural system. The force $F(t)$ is written this way to indicate that it is a function of time. In adopting the mathematical model shown in Fig. 1.2, it is assumed that each element in the system represents a single property; that is, the mass m represents only the property of inertia and not elasticity or energy dissipation, while the spring k represents exclusively elasticity and not inertia or energy dissipation. Finally, the damper c only dissipates energy. The reader certainly real-

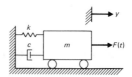

Fig. 1.2   Mathematical model for one-degree-of-freedom systems.

izes that such "pure" elements do not exist in our physical world and that mathematical models are only conceptual idealizations of real structures. As such, mathematical models may provide complete and accurate knowledge of the behavior of the model itself, but only limited or approximate information on the behavior of the real physical system. Nevertheless, from a practical point of view, the information acquired from the analysis of the mathematical model may very well be sufficient for an adequate understanding of the dynamic behavior of the physical system, including design and safety requirements.

1.2 UNDAMPED SYSTEM

We start our study of structural dynamics with the analysis of a fundamental and simple system, the one-degree-of-freedom system in which we disregard or "neglect" frictional forces or damping. In addition, we consider the system, during its motion or vibration to be free from external actions or forces. Under these conditions, the system is in motion governed only by the influence of the so-called *initial conditions;* that is, the given displacement and velocity at time $t = 0$ when the study of the system is initiated. This undamped, one-degree-of-freedom system is often referred to as *the simple undamped oscillator.* It is usually represented as shown in Fig. 1.2 or Fig. 1.3, or any similar arrangements. These two figures represent mathematical models which are dynamically equivalent. It is only a matter of personal preference to adopt one or the other. In these models the mass m is restrained by the spring k and is limited to rectilinear motion along one coordinate axis. The mechanical characteristic of a spring is described by the relation between the magnitude of the force F_s applied to its free end and the resulting end displacement y, as shown graphically in Fig. 1.4 for three different springs.

The curve labeled (a) in Fig. 1.4 represents the behavior of a "hard spring" in which the force required to produce a given displacement becomes increasingly

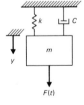

Fig. 1.3   **Alternate representation of mathematical model for one-degree-of-freedom systems.**

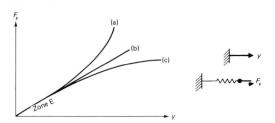

Fig. 1.4   Force   displacement   relation.   (a) Hard   spring.   (b) Linear   spring.
(c) Soft spring.

greater as the spring is deformed. The second spring (b) is designated a *linear*
spring because the deformation is directly proportional to the force and the
graphical representation of its characteristic is a straight line. The constant of
proportionality between the force and displacment [slope of line (b)] of a linear
spring is referred to as the *spring constant*, usually designated by the letter k.
Consequently, we may write the following relation between force and displacement for a linear spring.

$$F_s = ky. \tag{1.1}$$

A spring with characteristics shown by curve (c) in Fig. 1.4 is known as a
"soft spring." For such a spring the incremental force required to produce additional deformation decreases as the spring deformation increases. Undoubtedly,
the reader is aware from his previous exposure to mathematical modeling of
physical systems, that the linear spring is the simplest type to manage analytically. It should not come as a surprise to learn that most of the technical litera-
ture on structural dynamics deals with models using linear springs. In other
words, either because the elastic characteristics of the structural system are, in
fact, essentially linear, or simply because of analytical expediency, it is usually
assumed that the force-deformation properties of the system are linear. In support of this practice, it should be noted that in many cases the displacements
produced in the structure by the action of external forces or disturbances are
small in magnitude (Zone E in Fig. 1.4) thus rendering the linear approximation
close to the actual structural behavior.

1.3   SPRINGS IN PARALLEL OR IN SERIES

Sometimes it is necessary to determine the equivalent spring constant for a system in which two or more springs are arranged in parallel as shown in Fig. 1.5(a)
or in series as in Fig. 1.5(b).

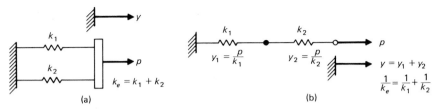

Fig 1.5   Combination of springs. (a) Springs in parallel. (b) Springs in series.

For two springs in parallel the total force required to produce a relative displacement of their ends of one unit is equal to the sum of their spring constants. This total force is by definition the equivalent spring constant and is given by

$$k_e = k_1 + k_2.$$ (1.2)

In general for n springs in parallel

$$k_e = \sum_{i=1}^{n} k_i.$$ (1.3)

For two springs assembled in series as shown in Fig. 1.5(b), the force P produces the relative displacements in the springs

$$y_1 = \frac{P}{k_1}$$

and

$$y_2 = \frac{P}{k_2}.$$

Then, the total displacement y of the free end of the spring assembly is equal to $y = y_1 + y_2$, or

$$y = \frac{P}{k_1} + \frac{P}{k_2}.$$ (1.4)

Consequently, the force necessary to produce one unit displacement (equivalent spring constant) is given by

$$k_e = \frac{P}{y}.$$

Substituting y from this last relation into eq. (1.4), we may conveniently express the reciprocal value of the spring constants as

$$\frac{1}{k_e} = \frac{1}{k_1} + \frac{1}{k_2}.$$ (1.5)

In general for n springs in series the equivalent spring constant may be obtained from

$$\frac{1}{k_e} = \sum_{i=1}^{n} \frac{1}{k_i}. \tag{1.6}$$

1.4   NEWTON'S LAW OF MOTION

We continue now with the study of the simple oscillator depicted in Fig. 1.2. The objective is to describe its motion, that is, to predict the displacement or velocity of the mass m at any time t, for a given set of initial conditions at time $t = 0$. The analytical relation between the displacement, y, and time, t, is given by Newton's Second Law of Motion, which in modern notation may be expressed as

$$F = ma, \tag{1.7}$$

where F is the resultant force acting on a particle of mass m and a is its resultant acceleration. The reader should recognize that eq. (1.7) is a vector relation and as such it can be written in equivalent form in terms of its components along the coordinate axes x, y, and z, namely

$$\sum F_x = ma_x, \tag{1.8a}$$

$$\sum F_y = ma_y, \tag{1.8b}$$

$$\sum F_z = ma_z. \tag{1.8c}$$

The acceleration is defined as the second derivative of the position vector with respect to time; it follows that eqs. (1.8) are indeed differential equations. The reader should also be reminded that these equations as stated by Newton are directly applicable only to bodies idealized as particles, that is, bodies which possess mass but no volume. However, as is proved in elementary mechanics, Newton's Law of Motion is also directly applicable to bodies of finite dimensions undergoing translatory motion.

For plane motion of a rigid body which is symmetric with respect to the reference plane of motion (x-y plane), Newton's Law of Motion yields the following equations:

$$\sum F_x = m(a_G)_x, \tag{1.9a}$$

$$\sum F_y = m(a_G)_y, \tag{1.9b}$$

$$\sum M_G = I_G \alpha. \tag{1.9c}$$

In the above equations $(a_G)_x$ and $(a_G)_y$ are the acceleration components, along the x and y axes, of the center of mass G of the body; α is the angular acceleration; I_G is the mass moment of inertia of the body with respect to an axis through G, the center of mass; and ΣM_G is the sum of the moments of all the forces acting on the body with respect to an axis through G, perpendicular to the x-y plane. Equations (1.9) are certainly also applicable to the motion of a rigid body in pure rotation about a fixed axis. For this particular type of plane motion, alternatively, eq. (1.9c) may be replaced by

$$\sum M_0 = I_0 \alpha \qquad (1.9d)$$

in which the mass moment of inertia I_0 and the moment of the forces M_0 are determined with respect to the fixed axis of rotation. The general motion of a rigid body is described by two vector equations, one expressing the relation between the forces and the acceleration of the mass center, and another relating the moments of the forces and the angular motion of the body. This last equation expressed in its scalar components is rather complicated, but seldom needed in structural dynamics.[1]

1.5 FREE BODY DIAGRAM

At this point, it is advisable to follow a method conducive to an organized and systematic analysis in the solution of dynamics problems. The first and probably the most important practice to follow in any dynamic analysis is to draw a free body diagram of the system, prior to writing a mathematical description of the system.

The free body diagram (FBD), as the student may recall, is a sketch of the body isolated from all other bodies, in which all the forces external to the body are shown. For the case at hand, Fig. 1.6(b) depicts the FBD of the mass m of the oscillator, displaced in the positive direction with reference to coordinate y, and acted upon by the spring force $F_s = ky$ (assuming a linear spring). The weight of the body mg and the normal reaction N of the supporting surface are also shown for completeness, though these forces, acting in the vertical direction, do not enter into the equation of motion written for the y direction. The application of Newton's Law of Motion gives

$$-ky = m\ddot{y}, \qquad (1.10)$$

where the spring force acting in the negative direction has a minus sign, and where the acceleration has been indicated by $\ddot{y}$. In this notation, double overdots denote the second derivative with respect to time and obviously a single overdot denotes the first derivative with respect to time, that is, the velocity.

[1] Meriam, J. L., *Dynamics*, 2nd ed., John Wiley, New York, 1971.

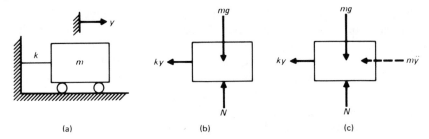

Fig. 1.6   Alternate free body diagrams: (a) Single degree-of-freedom system. (b) Showing only external forces. (c) Showing external and inertial forces.

1.6   D'ALEMBERT'S PRINCIPLE

An alternative approach to obtain eq. (1.10) is to make use of *D'Alembert's Principle* which states that a system may be set in a state of *dynamic equilibrium* by adding to the external forces a fictitious force which is commonly known as the *inertial force*.

Figure 1.6(c) shows the FBD with inclusion of the inertial force $m\ddot{y}$. This force is equal to the mass multiplied by the acceleration, and should always be directed negatively with respect to the corresponding coordinate. The application of D'Alembert's Principle allows us to use equations of equilibrium in obtaining the equation of motion. For example, in Fig. 1.6(c), the summation of forces in the y direction gives directly

$$m\ddot{y} + ky = 0 \tag{1.11}$$

which obviously is equivalent to eq. (1.10).

The use of D'Alembert's Principle in this case appears to be trivial. This will not be the case for a more complex problem, in which the application of D'Alembert's Principle, in conjunction with the *principle of virtual work*, constitutes a powerful tool of analysis. As will be explained later, the Principle of Virtual Work is directly applicable to any system in equilibrium. It follows then that this principle may also be applied to the solution of dynamic problems, provided that D'Alembert's Principle is used to establish the dynamic equilibrium of the system.

Illustrative Example 1.1.   Show that the same differential equation is obtained for a spring-supported body moving vertically as for the same body vibrating along a horizontal axis, as shown in Figs. 1.7(a) and 1.7(b). The FBD's for these two representations of the simple oscillator are shown in Fig. 1.7(c) and 1.7(e) where the inertial forces are included.

Equating to zero the sum of the forces in Fig. 1.7(c), we obtain

$$m\ddot{y} + ky = 0. \tag{a}$$

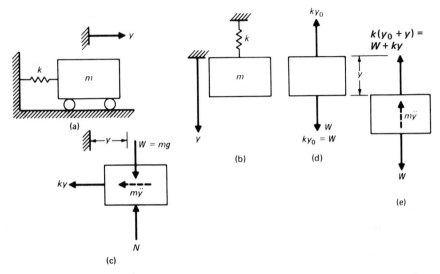

Fig. 1.7   Two representations of the simple oscillator and corresponding free body diagrams.

When the body in Fig. 1.7(d) is in the static equilibrium position, the spring is stretched y_0 units and exerts a force $ky_0 = W$ upward on the body, where W is the weight of the body. When the body is displaced a distance y downward from this position of equilibrium the magnitude of the spring force is given by $F_s = k$ $(y_0 + y)$ or $F_s = W + ky$, since $ky_0 = W$. Using this result and applying it to the body in Fig. 1.7(e) we obtain from Newton's Second Law of Motion

$$-(W + ky) + W = m\ddot{y}$$

or

$$m\ddot{y} + ky = 0 \qquad \text{(b)}$$

which is identical to eq. (a).

1.7   SOLUTION OF THE DIFFERENTIAL EQUATION OF MOTION

The next step toward our objective is to find the solution of the differential equation, eq. (1.11). We should again adopt a systematic approach and proceed to first classify this differential equation. Since the dependent variable y and its second derivative $\ddot{y}$ appear in the first degree in eq. (1.11), this equation is classified as linear and of second order. The fact that the coefficients of y and $\ddot{y}$ (k and m, respectively) are constants, and that the second member (right-hand side) of the equation is zero, further classify the equation as homogeneous with constant coefficients. We should recall, probably with a certain degree of satisfaction, that a general procedure exists for the solution of linear differential equations (homogeneous or nonhomogeneous) of any order. For this simple, second-order differential equation we may proceed directly by assuming a trial solution given by

$$y = A \cos \omega t \tag{1.12}$$

or

$$y = B \sin \omega t, \tag{1.13}$$

where A and B are constants depending on the initiation of the motion while ω is a quantity denoting a physical characteristic of the system as it will be shown next. The substitution of eq. (1.12) into eq. (1.11) gives

$$(-m\omega^2 + k) A \cos \omega t = 0. \tag{1.14}$$

If this equation is to be satisfied at any time, the factor in parentheses must be equal to zero or

$$\omega^2 = \frac{k}{m}. \tag{1.15}$$

The reader should verify that eq. (1.13) is also a solution of the differential equation, eq. (1.11), with ω also satisfying eq. (1.15).

The positive root of eq. (1.15),

$$\omega = \sqrt{k/m}, \tag{1.16}$$

is known as the *natural frequency* of the system for reasons that will soon be apparent.

Since either eq. (1.12) or eq. (1.13) is a solution of eq. (1.11), and since this differential equation is linear, the superposition of these two solutions, indicated by eq. (1.17) below, is also a solution. Furthermore, eq. (1.17) having two constants of integration, A and B, is, in fact, the general solution for this second-order differential equation,

$$y = A \cos \omega t + B \sin \omega t. \tag{1.17}$$

The expression for velocity, $\dot{y}$, is found simply by differentiating eq. (1.17) with respect to time; that is

$$\dot{y} = -A \omega \sin \omega t + B \omega \cos \omega t. \tag{1.18}$$

Next, we should determine the constants of integration A and B. These constants are determined from known values for the motion of the system which almost invariably are the displacement y_0 and the velocity v_0 at the initiation of the motion, that is, at time $t = 0$. These two conditions are referred to as *initial conditions*, and the problem of solving the differential equation for the initial conditions is called an *initial value problem*.

After substituting, for $t = 0$, $y = y_0$, and $\dot{y} = v_0$ into eqs. (1.17) and (1.18), we find that,

$$y_0 = A \qquad \text{(1.19a)}$$

$$v_0 = B\omega. \qquad \text{(1.19b)}$$

Finally, the substitution of A and B from eqs. (1.19) into eq. (1.17) gives:

$$y = y_0 \cos \omega t + \frac{v_0}{\omega} \sin \omega t \qquad \text{(1.20)}$$

which is the expression of the displacement y of the simple oscillator as a function of the time variable t; thus we have accomplished our objective of describing the motion of the simple undamped oscillator modeling structures with a single degree of freedom.

1.8 FREQUENCY AND PERIOD

An examination of eq. (1.20) shows that the motion described by this equation is *harmonic* and, therefore, periodic; that is, it can be expressed by a sine or cosine function of same frequency, ω. The period may easily be found since the functions sine and cosine both have a period of 2π. The period T of the motion is determined from

$$\omega T = 2\pi$$

or

$$T = \frac{2\pi}{\omega}. \qquad \text{(1.21)}$$

The period is usually expressed in seconds per cycle or simply in seconds with the tacit understanding that it is "per cycle." The value reciprocal to the period is the *natural frequency, f.* From eq. (1.21)

$$f = \frac{1}{T} = \frac{\omega}{2\pi}. \qquad \text{(1.22)}$$

The natural frequency f is usually expressed in hertz or cycles per second (cps). Because the quantity ω differs from the natural frequency f only by the constant factor, 2π, ω also is sometimes referred to as the natural frequency. To distinguish between these two expressions for natural frequency, ω may be called the *circular or angular* natural frequency. Most often, the distinction is understood from the context or from the units. The natural frequency f is measured in cps as indicated, while the circular frequency ω should be given in radians per second (rad/sec).

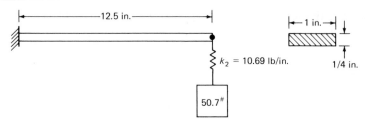

Fig. 1.8   System for Example 1.2.

Illustrative Example 1.2.   Determine the natural frequency of the system shown in Fig. 1.8 consisting of a weight of 50.7 lb attached to a horizontal cantilever beam through the coil spring k_2. The cantilever beam has a thickness $t = \frac{1}{4}$ in, a width $b = 1$ in. modulus of elasticity $E = 30 \times 10^6$ psi, and a length $L = 12.5$ in. The coil spring has a stiffness, $k_2 = 10.69$ (lb/in).

The deflection Δ at the free end of a cantilever beam acted upon by a static force P at the free end is given by

$$\Delta = \frac{PL^3}{3EI}.$$

The corresponding spring constant is then

$$k_1 = \frac{P}{\Delta} = \frac{3EI}{L^3},$$

where $I = \frac{1}{12}bt^3$ (for rectangular section). Now, the cantilever and the coil spring of this system are connected as springs in series. Consequently, the equivalent spring constant as given from eq. (1.5) is

$$\frac{1}{k_e} = \frac{1}{k_1} + \frac{1}{k_2}.$$

Substituting corresponding numerical values, we obtain

$$I = \frac{1}{12} \times 1 \times \left(\frac{1}{4}\right)^3 = \frac{1}{768}(\text{in})^4,$$

$$k_1 = \frac{3 \times 30 \times 10^6}{(12.5)^3 \times 768} = 60 \text{ lb/in}$$

and

$$\frac{1}{k_e} = \frac{1}{60} + \frac{1}{10.69},$$

$$k_e = 9.07 \text{ lb/in}.$$

The natural frequency for this system is then given by

$$\omega = \sqrt{k_e/m}$$
$$\omega = \sqrt{9.07 \times 386/50.7}$$
$$\omega = 8.31 \text{ rad/sec}$$

or

$$f = 1.32 \text{ cps.} \qquad \text{(Ans.)}$$

1.9 AMPLITUDE OF MOTION

Let us now examine in more detail eq. (1.20), the solution describing the free vibratory motion of the undamped oscillator. A simple trigonometric transformation may show us that we can rewrite this equation in the equivalent forms, namely

$$y = C \sin (\omega t + \alpha) \qquad (1.23)$$

or

$$y = C \cos (\omega t - \beta) \qquad (1.24)$$

where

$$C = \sqrt{y_0^2 + (v_0/\omega)^2}, \qquad (1.25)$$

$$\tan \alpha = \frac{y_0}{v_0/\omega}, \qquad (1.26)$$

and

$$\tan \beta = \frac{v_0/\omega}{y}. \qquad (1.27)$$

The simplest way to obtain eq. (1.23) or eq. (1.24) is to multiply and divide eq. (1.20) by the factor C defined in eq. (1.25) and to define α (or β) by eq. (1.26) [or eq. (1.27)] . Thus

$$y = C\left(\frac{y_0}{C} \cos \omega t + \frac{v_0/\omega}{C} \sin \omega t\right). \qquad (1.28)$$

With the assistance of Fig. 1.9, we recognize that

$$\sin \alpha = \frac{y_0}{C} \qquad (1.29)$$

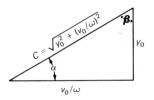

Fig. 1.9   Definition of angle α.

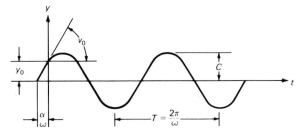

Fig. 1.10   Undamped free-vibration response.

and

$$\cos \alpha = \frac{v_0/\omega}{C}. \tag{1.30}$$

The substitution of eqs. (1.29) and (1.30) into eq. (1.28) gives

$$y = C(\sin \alpha \cos \omega t + \cos \alpha \sin \omega t). \tag{1.31}$$

The expression within the parentheses of eq. (1.31) is identical to $\sin(\omega t + \alpha)$, which yields eq. (1.23). Similarly, the reader should verify without difficulty, the form of solution given by eq. (1.24).

The value of C in eq. (1.23) [or eq. (1.24)] is referred to as the amplitude of motion and the angle α (or β) as the phase angle. The solution for the motion of the simple oscillator is shown graphically in Fig. 1.10.

Illustrative Example 1.3.   Consider the frame shown in Fig. 1.11(a). This is a rigid steel frame to which a horizontal dynamic force is applied at the upper level. As part of the overall structural design it is required to determine the natural frequency of the frame. Two assumptions are made: (1) the masses of the columns and walls are negligible; and (2) the girder is sufficiently rigid to prevent rotation at the tops of the columns. These assumptions are not manda-tory for the solution of the problem, but they serve to simplify the analysis.

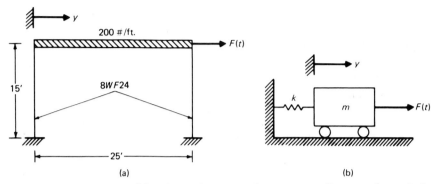

Fig. 1.11   One-degree-of-freedom   frame   and   corresponding   mathematical
model for Example 1.3.

Under these conditions, the frame may be modeled by the spring-mass system
shown in Fig. 1.11(b).

The parameters of this model may be computed as follows:

$$W = 200 \times 25 = 5000 \text{ lb,}$$

$$I = 82.5 \text{ in}^4,$$

$$E = 30 \times 10^6 \text{ psi,}$$

$$k^* = \frac{12E(2I)}{h^3} = \frac{12 \times 30 \times 10^6 \times 165}{(15 \times 12)^3},$$

$$k = 10,185 \text{ lb/in.}$$

Therefore, the natural frequency is

$$f = \frac{1}{2\pi} \sqrt{\frac{kg}{W}} = \frac{1}{2\pi} \sqrt{\frac{10,185 \times 386}{5000}}$$

$$f = 4.46 \text{ cps.} \hspace{3cm} \text{(Ans.)}$$

1.10   SUMMARY

Several basic concepts were introduced in this chapter.

(1) The mathematical model of a structure is an idealized representation for
its analysis.

*A unit displacement of the top of a fixed column creates a resisting force equal to $12EI/L^3$.

(2) The number of degrees of freedom of a system is equal to the number of independent coordinates necessary to describe its position.

(3) The free body diagram (FBD) for dynamic equilibrium (to allow application of D'Alembert's Principle) is a diagram of the system isolated from all other bodies, showing all the external forces on the system, including the inertial force.

(4) The stiffness or spring constant of a linear system is the force necessary to produce a unit displacement.

(5) The differential equation of the undamped simple oscillator in free motion is

$$m\ddot{y} + ky = 0$$

and its general solution is

$$y = A \cos \omega t + B \sin \omega t,$$

where A and B are constants of integration determined from initial conditions:

$$A = y_0,$$

$$B = v_0/\omega,$$

$\omega = \sqrt{k/m}$ is the natural frequency in rad/sec,

$f = \dfrac{\omega}{2\pi}$ is the natural frequency in cps,

$T = \dfrac{1}{f}$ is the natural period in seconds.

(6) The equation of motion may be written in the alternate forms:

$$y = C \sin (\omega t + \alpha),$$

or

$$y = C \cos (\omega t - \beta),$$

where

$$C = \sqrt{y_0^2 + (v_0/\omega)^2},$$

and

$$\tan \alpha = \frac{y_0}{v_0/\omega},$$

$$\tan \beta = \frac{v_0/\omega}{y_0}.$$

PROBLEMS

1.1 Determine the natural period for the system in Fig. P1.1. Assume that the beam and springs supporting the weight W are massless.

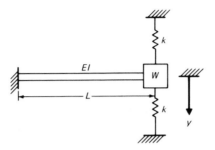

Fig. P1.1

1.2 The following numerical values are given in Problem 1.1: $L = 100$ in, $EI = 10^8$ (lb $\cdot$ in^2), $W = 3000$ lb, and $k = 2000$ lb/in. If the weight W has an initial displacement of $y_0 = 1.0$ in and an initial velocity $v_0 = 20$ in/sec, determine the displacement and the velocity 1 sec later.

1.3 Determine the natural frequency for horizontal motion of the steel frame in Fig. P1.3. Assume the horizontal girder to be infinitely rigid and neglect the mass of the columns.

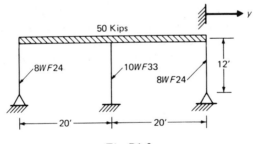

50 Kips

8WF24   10WF33   12'
          8WF24

20'   20'

Fig. P1.3

1.4 Calculate the natural frequency in the horizontal mode of the steel frame in Fig. P1.4 for the following cases: (a) the horizontal member is assumed to be infinitely rigid; (b) the horizontal member is flexible made of steel 18 WF50.

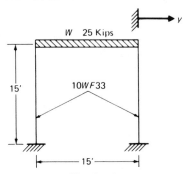

Fig. P1.4

1.5 Determine the natural frequency of the fixed beam in Fig. P1.5 carrying a concentrated weight W at its center. Neglect the mass of the beam.

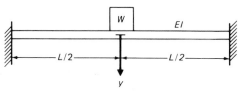

.**Fig. P1.5**

1.6 The numerical values for Problem 1.5 are given as: $L = 120$ in, $EI = 10^9$ (lb · in²), and $W = 5000$ lb. If the initial displacement and the initial velocity of the weight are, respectively, $y_0 = 0.5$ in and $v_0 = 15$ in/sec, determine the displacement, velocity, and acceleration of W when time $t = 2$ sec.

1.7 A vertical pole of length L and flexural rigidity EI carries a mass m at its top, as shown in Fig. P1.7. Neglecting the weight of the pole, derive the differential equation for small horizontal vibrations of the mass, and find the natural frequency. Assume that the effect of gravity is small and nonlinear effects may be neglected.

Fig. P1.7

2

Damped single degree-of-freedom system

We have seen in the preceding chapter that the simple oscillator under idealized conditions of no damping, once excited, will oscillate indefinitely with a constant amplitude at its natural frequency. Experience indicates, however, that it is not possible to have a device which vibrates under these ideal conditions. Forces designated as frictional or damping forces are always present in any physical system undergoing motion. These forces dissipate energy; more precisely, the unavoidable presence of these frictional forces constitutes a mechanism through which the mechanical energy of the system, kinetic or potential energy, is transformed to other forms of energy such as heat. The mechanism of this energy transformation or dissipation is quite complex and is not completely understood at this time. In order to account for these dissipative forces in the analysis of dynamic systems, it is necessary to make some assumptions about these forces, on the basis of experience.

2.1   VISCOUS DAMPING

In considering damping forces in the dynamic analysis of structures, it is usually assumed that these forces are proportional to the magnitude of the velocity, and opposite to the direction of motion. This type of damping is known as viscous damping; it is the type of damping force that could be developed in a body restrained in its motion by a surrounding viscous fluid.

There are situations in which the assumption of viscous damping is realistic and in which the dissipative mechanism is approximately viscous. Nevertheless, the assumption of viscous damping is often made regardless of the actual dissipative characteristics of the system. The primary reason for such wide use of this method is that it leads to a relatively simple mathematical analysis.

2.2   EQUATION OF MOTION

Let us assume that we have modeled a structural system as a simple oscillator with viscous damping, as shown in Fig. 2.1(a). In this figure, m and k are respectively, the mass and spring constant of the oscillator and c is the viscous damping coefficient. We proceed, as in the case of the undamped oscillator, to draw the free body diagram (FBD) and apply Newton's Law to obtain the differential equation of motion. Figure 2.1(b) shows the FBD of the damped oscillator in which the inertial force $m\ddot{y}$ is also shown, so that we can use D'Alembert's Principle. The summation of forces in the y direction gives the differential equation of motion,

$$m\ddot{y} + c\dot{y} + ky = 0. \tag{2.1}$$

The reader may verify that a trial solution $y = A \sin \omega t$ or $y = B \cos \omega t$ will not satisfy eq. (2.1). However, the exponential function $y = Ce^{pt}$, does satisfy

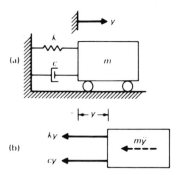

Fig. 2.1   (a) Viscous damped oscillator. (b) Free body diagram.

this equation. Substitution of this function into eq. (2.1) results in the equation

$$mCp^2 e^{pt} + cCp e^{pt} + kC e^{pt} = 0$$

which, after cancellation of the common factors, reduces to an equation called *the characteristic equation* for the system, namely

$$mp^2 + cp + k = 0. \tag{2.2}$$

The roots of this quadratic equation are

$$\frac{p_1}{p_2} = -\frac{c}{2m} \pm \sqrt{\left(\frac{c}{2m}\right)^2 - \frac{k}{m}}. \tag{2.3}$$

Thus the general solution of eq. (2.1) is given by the superposition of the two possible solutions, namely

$$y(t) = C_1 e^{p_1 t} + C_2 e^{p_2 t} \tag{2.4}$$

where C_1 and C_2 are constants of integration to be determined from the initial conditions.

The final form of eq. (2.4) depends on the sign of the expression under the radical in eq. (2.3). Three distinct cases may occur: the quantity under the radical may either be zero, positive, or negative. The limiting case in which the quantity under the radical is zero is treated first. The damping present in this case is called *critical damping*.

2.3   CRITICALLY DAMPED SYSTEM

For a system oscillating with critical damping, as defined above, the expression under the radical in eq. (2.3) is equal to zero; that is,

$$\left(\frac{c_{cr}}{2m}\right)^2 - \frac{k}{m} = 0 \tag{2.5}$$

or

$$c_{cr} = 2\sqrt{km} \tag{2.6}$$

where c_{cr} designates the critical damping value. Since the natural frequency of the undamped system is designated by $\omega = \sqrt{k/m}$, the critical damping coefficient given by eq. (2.6) may also be expressed in alternative notation as

$$c_{cr} = 2m\omega = \frac{2k}{\omega}. \tag{2.7}$$

In a critically damped system the roots of the characteristic equation are equal, and from eq. (2.3), they are

$$p_1 = p_2 = -\frac{c_{cr}}{2m}. \tag{2.8}$$

Since the two roots are equal, the general solution given by eq. (2.4) would provide only one independent constant of integration; hence, one independent solution, namely

$$y_1(t) = C_1 \, e^{-(c_{cr}/2m)t}. \tag{2.9}$$

Another independent solution may be found by using the function

$$y_2(t) = C_2 \, te^{-(c_{cr}/2m)t}. \tag{2.10}$$

This equation, as the reader may verify, also satisfies the differential equation, eq. (2.1). The general solution for a critically damped system is then given by the superposition of these two solutions,

$$y(t) = (C_1 + C_2 t) \, e^{-(c_{cr}/2m)t}. \tag{2.11}$$

2.4 OVERDAMPED SYSTEM

In an overdamped system, the damping coefficient is greater than the value for critical damping, namely

$$c > c_{cr}. \tag{2.12}$$

Therefore, the expression under the radical of eq. (2.3) is positive, thus the two roots of the characteristic equation are real and distinct, and consequently the solution is given directly by eq. (2.4). It should be noted that, for the overdamped or the critically damped system, the resulting motion is not oscillatory; the magnitude of the oscillations decays exponentially with time to zero. Figure 2.2 depicts graphically the response for the simple oscillator with critical damping. The response of the overdamped system is similar to the motion of the crit-

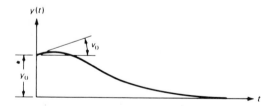

Fig. 2.2   Free-vibration response with critical damping.

ically damped system of Fig. 2.2, but the return toward the neutral position requires more time as the damping ratio is increased.

2.5 UNDERDAMPED SYSTEM

When the value of the damping coefficient is less than the critical value ($c < c_{cr}$), which occurs when the expression under the radical is negative, the roots of the characteristic eq. (2.3) are complex conjugates, so that

$$\frac{p_1}{p_2} = \frac{c}{2m} \pm i \sqrt{\frac{k}{m} - \left(\frac{c}{2m}\right)^2}, \qquad (2.13)$$

where $i = \sqrt{-1}$ is the imaginary unit.

For this case, it is convenient to make use of Euler's equations which relate exponential and trigonometric functions, namely

$$e^{ix} = \cos x + i \sin x,$$
$$e^{-ix} = \cos x - i \sin x. \qquad (2.14)$$

The substitution of the roots p_1 and p_2 from eq. (2.13) into eq. (2.4) together with the use of eq. (2.14) gives the following convenient form for the general solution of the underdamped system:

$$y(t) = e^{-(c/2m)t} (A \cos \omega_D t + B \sin \omega_D t), \qquad (2.15)$$

where A and B are redefined constants of integration and ω_D, the damped frequency of the system, is given by

$$\omega_D = \sqrt{\frac{k}{m} - \left(\frac{c}{2m}\right)^2} \qquad (2.16)$$

or

$$\omega_D = \omega \sqrt{1 - \xi^2}. \qquad (2.17)$$

This last result is obtained after substituting, in eq. (2.16), the expression for the undamped natural frequency

$$\omega = \sqrt{\frac{k}{m}} \qquad (2.18)$$

and defining the *damping ratio* of the system as

$$\xi = \frac{c}{c_{cr}}. \qquad (2.19)$$

Finally, when the initial conditions of displacement and velocity, y_0 and v_0, are introduced, the constants of integration can be evaluated and substituted into eq. (2.15), giving

$$y(t) = e^{-\xi\omega t}\left(y_0 \cos \omega_D t + \frac{v_0 + y_0 \xi\omega}{\omega_D} \sin \omega_D t\right). \tag{2.20}$$

Alternatively, this expression can be written as

$$y(t) = C e^{-\xi\omega t} \cos(\omega_D t - \alpha) \tag{2.21}$$

where

$$C = \sqrt{y_0^2 + \frac{(v_0 + y_0 \xi\omega)^2}{\omega_D^2}} \tag{2.22}$$

and

$$\tan \alpha = \frac{(v_0 + y_0 \xi\omega)}{\omega_D y_0}. \tag{2.23}$$

A graphical record of the response of an underdamped system with initial displacement y_0 but starting with zero velocity ($v_0 = 0$) is shown in Fig. 2.3. It may be seen in this figure that the motion is oscillatory, but not periodic. The amplitude of vibration is not constant during the motion but decreases for successive cycles; nevertheless, the oscillations occur at equal intervals of time. This time interval is designated as the damped period of vibration and is given from eq. (2.17) by

$$T_D = \frac{2\pi}{\omega_D} = \frac{2\pi}{\omega\sqrt{1 - \xi^2}}. \tag{2.24}$$

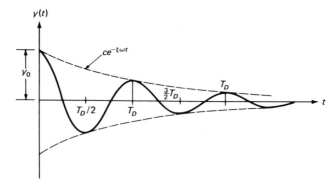

Fig. 2.3   Free vibration response for underdamped system.

The value of the damping coefficient for real structures is much less than the critical damping coefficient and usually ranges between 2 to 20% of the critical damping value. Substituting for the maximum value $\xi = 0.20$ into eq. (2.17),

$$\omega_D = 0.98\omega. \tag{2.25}$$

It can be seen that the frequency of vibration for a system with as much as a 20% damping ratio is essentially equal to the undamped natural frequency. Thus, in practice, the natural frequency for a damped system may be taken to be equal to the undamped natural frequency.

2.6 LOGARITHMIC DECREMENT

A practical method for determining experimentally the damping coefficient of a system is to initiate free vibration, obtain a record of the oscillatory motion, such as the one shown in Fig. 2.4, and measure the rate of decay of the amplitude of motion. The decay may be conveniently expressed by the *logarithmic decrement* δ which is defined as the natural logarithm of the ratio of any two successive peak amplitudes, y_1 and y_2 in free vibration, that is,

$$\delta = \ln \frac{y_1}{y_2}. \tag{2.26}$$

The evaluation of damping from the logarithmic decrement follows. Consider the damped vibration motion represented graphically in Fig. 2.4 and given analytically by eq. (2.21) as

$$y(t) = C e^{-\xi\omega t} \cos (\omega_D t - \alpha).$$

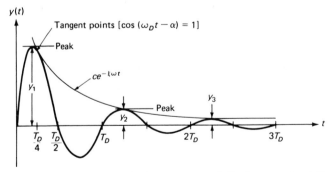

Fig. 2.4   Curve showing peak displacements and displacements at the points of tangency.

We note from this equation that, when the cosine factor is unity, the displacement is on points of the exponential curve $y(t) = Ce^{-\xi\omega t}$ as shown in Fig. 2.4. However, these points are near but not equal to the positions of maximum displacement. The points on the exponential curve appear slightly to the right of the points of maximum amplitude. For most practical problems, the discrepancy is negligible and the displacement curve may be assumed to coincide at the peak amplitude, with the curve $y(t) = Ce^{-\xi\omega t}$ so that we may write, for two consecutive peaks, y_1 at time t_1 and y_2 at T_D seconds later,

$$y_1 = Ce^{-\xi\omega t_1}$$

and

$$y_2 = Ce^{-\xi\omega(t_1+T_D)}.$$

Dividing these two peak amplitudes and taking the natural logarithm, we obtain

$$\delta = \ln\frac{y_1}{y_2} = \xi\omega T_D \tag{2.27}$$

or by substituting, T_D, the damping period, from eq. (2.24),

$$\delta = 2\pi\xi/\sqrt{1-\xi^2}. \tag{2.28}$$

As we can see, the damping ratio ξ can be calculated from eq. (2.28) after determining experimentally the amplitudes of two successive peaks of the system in free vibration. For small values of the damping ratio, eq. (2.28) can be approximated by

$$\delta \simeq 2\pi\xi. \tag{2.29}$$

Illustrative Example 2.1.   A vibrating system consisting of a weight of $W = 10$ lb and a spring with stiffness $k = 20$ lb/in is viscously damped so that the ratio of two consecutive amplitudes is 1.00 to 0.85. Determine: (a) the natural frequency of the undamped system, (b) the logarithmic decrement, (c) the damping ratio, (d) the damping coefficient, and (e) the damped natural frequency.

(a) The undamped natural frequency of the system in radians per second is

$$\omega = \sqrt{k/m} = \sqrt{(20 \text{ lb/in} \times 386 \text{ in/sec}^2)/10 \text{ lb}} = 27.78 \text{ rad/sec}$$

or in cycles per second

$$f = \frac{\omega}{2\pi} = 4.42 \text{ cps.}$$

(b) The logarithmic decrement is given by

$$\delta = \ln\frac{y_1}{y_2} = \ln\frac{1.00}{0.85} = 0.163.$$

(c) The damping ratio from eq. (2.29), is approximately equal to

$$\xi \simeq \frac{\delta}{2\pi} = \frac{0.163}{2\pi} = 0.026.$$

(d) The damping coefficient is obtained from eqs. (2.6) and (2.19) as

$$c = \xi c_{cr} = 2 \times 0.026 \sqrt{(10 \times 20)/386} = 0.037 \ \frac{lb \cdot sec}{in}.$$

(e) The natural frequency of the damped system is given by eq. (2.17), so that

$$\omega_D = \omega \sqrt{1 - \xi^2},$$

$$\omega_D = 27.78 \sqrt{1 - (0.026)^2} = 27.77 \ rad/sec.$$

Illustrative Example 2.2.   A platform of weight $W = 4000$ lb is being sup-ported by four equal columns which are clamped to the foundation as well as to the platform. Experimentally it has been determined that a static force of $F = 1000$ lb applied horizontally to the platform produces a displacement of $\Delta = 0.10$ in. It is estimated that damping in the structures is of the order of 5% of the critical damping. Determine for this structure the following: (a) un-damped natural frequency, (b) absolute damping coefficient, (c) logarithmic decrement, and (d) the number of cycles and the time required for the ampli-tude of motion to be reduced from an initial value of 0.1 in to 0.01 in.

(a) The stiffness coefficient (force per unit displacement) is computed as

$$k = \frac{F}{\Delta} = \frac{1000}{0.1} = 10,000 \ lb/in$$

and the undamped natural frequency

$$\omega = \sqrt{\frac{k}{W/g}} = \sqrt{\frac{10,000 \times 386}{4000}} = 31.06 \ rad/sec.$$

(b) The critical damping is

$$c_{cr} = 2 \sqrt{k \, m} = 2 \sqrt{10,000 \times 4000/386} = 643.8 \ \frac{lb \cdot sec}{in}$$

and the absolute damping

$$c = \xi \, c_{cr} = 0.05 \times 643.8 = 32.19 \ \frac{lb \cdot sec}{in}.$$

(c) Approximately, the logarithmic decrement is

$$\delta = \ln \left(\frac{y_0}{y_1}\right) \simeq 2\pi\xi = 2\pi(0.05) = 0.314$$

and the ratio of two consecutive amplitudes

$$\frac{y_0}{y_1} = 1.37.$$

(d) The ratio between the first amplitude y_0 and the amplitude y_k after k cycles may be expressed as

$$\frac{y_0}{y_k} = \frac{y_0}{y_1} \cdot \frac{y_1}{y_2} \cdots \frac{y_{k-1}}{y_k}.$$

Then taking the natural logarithm, we obtain

$$\ln \frac{y_0}{y_k} = \delta + \delta + \cdots + \delta = k\delta,$$

$$\ln \frac{0.1}{0.01} = 0.314\,k,$$

$$k = \frac{\ln 10}{0.314} = 7.33 \rightarrow 8 \text{ cycles.}$$

The damped frequency ω_D is given by

$$\omega_D = \omega\sqrt{1 - \xi^2} = 31.06\sqrt{1 - (0.05)^2} = 31.02 \text{ rad/sec}$$

and the period T_D by

$$T_D = \frac{2\pi}{\omega_D} = \frac{2\pi}{31.02} = 0.2025 \text{ sec.}$$

Then the time for eight cycles is

$$t(8 \text{ cycles}) = 8\,T_D = 1.62 \text{ sec.}$$

2.7 SUMMARY

Real structures dissipate energy while undergoing vibratory motion. The most common and practical method for considering this dissipation of energy is to assume that it is due to viscous damping forces. These forces are assumed to be proportional to the magnitude of the velocity but acting in the direction opposite to the motion. The factor of proportionality is called the *viscous damping* coefficient. It is expedient to express this coefficient as a fraction of the *critical damping* in the system (the damping ratio, $\xi = c/c_{cr}$). The critical damping may be defined as the least value of the damping coefficient for which the system will not oscillate when disturbed initially, but it simply will return to the equilibrium position.

The differential equation of motion for the damped single degree-of-freedom system is given by

$$m\ddot{y} + c\dot{y} + ky = 0.$$

The analytical expression for the solution of this equation depends on the magnitude of the damping ratio. Three cases are possible: (1) critically damped system ($\xi = 1$), (2) underdamped system ($\xi < 1$), and (3) overdamped system ($\xi > 1$). For the underdamped system ($\xi < 1$) the solution of the differential equation of motion may be written as

$$y(t) = e^{-\xi\omega t}\left[y_0 \cos \omega_D t + \frac{v_0 + y_0 \xi\omega}{\omega_D} \sin \omega_D t\right]$$

in which

$\omega = \sqrt{k/m}$ is the undamped frequency,
$\omega_D = \omega \sqrt{1 - \xi^2}$ is the damped frequency,
$\xi = c/c_{cr}$ is the damping ratio,
$c_{cr} = 2\sqrt{km}$ is the critical damping,

and y_0 and v_0 are, respectively, the initial displacement and velocity.

A common method of determining the damping present in a system is to evaluate experimentally the logarithmic decrement which is defined as the natural logarithm of the ratio of two consecutive peaks in free vibration, that is,

$$\delta = \ln \frac{y_1}{y_2}.$$

The damping ratio in structural systems is usually less than 20% of the critical damping ($\xi < 0.2$). For such systems, the damped frequency is approximately equal to the undamped frequency.

PROBLEMS

2.1 Repeat Problem 1.2 assuming that the system has 15% of critical damping.
2.2 Repeat Problem 1.6 assuming that the system has 10% of critical damping.
2.3 The amplitude of vibration of the system shown in Fig. P2.3 is observed to

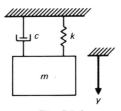

Fig. P2.3

decrease 5% on each consecutive cycle of motion. Determine the damping coefficient c of the system. $k = 200$ lb/in and $m = 10$ lb · sec^2/in.

2.4 It is observed experimentally that the amplitude of free vibration of a certain structure, modeled as a single degree-of-freedom system, decreases from 1 to 0.4 in 10 cycles. What is the percentage of critical damping?

2.5 Show that the displacement for critical and overcritical damped systems with initial displacement y_0 and velocity v_0 may be written as

$$y = e^{-\omega t}[y_0(1 + \omega t) + v_0 t] \quad \text{for } \xi = 1$$

$$y = e^{-\xi \omega t}\left[y_0 \cosh \omega_D' t + \frac{v_0 + y_0 \xi \omega}{\omega_D'} \cosh \omega_D' t\right] \quad \text{for } \xi > 1$$

where $\omega_D' = \omega\sqrt{\xi^2 - 1}$.

2.6 A structure is modeled as a damped oscillator with spring constant $k = 30$ Kips/in and undamped natural frequency $\omega = 25$ rad/sec. Experimentally it was found that a force 1 Kip produced a relative velocity of 1.0 in/sec in the damping element. Find: (a) the damping ratio ξ, (b) the damped period T_D, (c) the logarithmic decrement δ, and (d) the ratio between two consecutive amplitudes.

3

Response of one-degree-of-freedom system to harmonic loading

In this chapter, we will study the motion of structures idealized as single degree-of-freedom systems excited harmonically, that is, structures subjected to forces or displacements whose magnitudes may be represented by a sine or cosine function of time. This type of excitation results in one of the most important motions in the study of mechanical vibrations as well as in applications to structural dynamics. Structures are very often subjected to the dynamic action of rotating machinery which produces harmonic excitations due to the unavoidable presence of mass eccentricities in the rotating parts of such machinery. Furthermore, even in those cases when the excitation is not a harmonic function, the response of the structure may be obtained using the *Fourier Method*, as the superposition of individual responses to the harmonic components of the external excitation. This approach will be dealt with in Chapter 5.

3.1   UNDAMPED HARMONIC EXCITATION

The impressed force $F(t)$ acting on the simple oscillator in Fig. 3.1 is assumed to be harmonic and equal to $F_0 \sin \bar{\omega} t$, where F_0 is the peak amplitude and $\bar{\omega}$ is the frequency of the force in radians per second. The differential equation obtained by summing all the forces in the free body diagram of Fig. 3.1(b) is

$$m\ddot{y} + ky = F_0 \sin \bar{\omega} t. \tag{3.1}$$

The solution of eq. (3.1) can be expressed as

$$y(t) = y_c(t) + y_p(t) \tag{3.2}$$

where $y_c(t)$ is the complementary solution satisfying the homogeneous equation, that is, eq. (3.1) with the left-hand side set equal to zero; and $y_p(t)$ is the particular solution based on the solution satisfying the nonhomogeneous differential equation, eq. (3.1). The complementary solution, $y_c(t)$ is given by

$$y_c(t) = A \cos \omega t + B \sin \omega t, \tag{3.3}$$

where $\omega = \sqrt{k/m}$.

The nature of the forcing function in eq. (3.1) suggests that the particular solution be taken as

$$y_p(t) = Y \sin \bar{\omega} t, \tag{3.4}$$

where Y is the peak value of the particular solution. The substitution of eq. (3.4) into eq. (3.1) followed by cancellation of common factors gives

$$-m\bar{\omega}^2 Y + kY = F_0$$

or

$$Y = \frac{F_0}{k - m\bar{\omega}^2} = \frac{F_0/k}{1 - r^2} \tag{3.5}$$

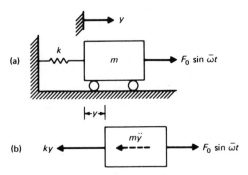

Fig. 3.1   (a) Undamped oscillator harmonically excited. (b) Free body diagram.

in which r represents the ratio (frequency ratio) of the applied forced frequency to the natural frequency of vibration of the system, that is,

$$r = \frac{\overline{\omega}}{\omega}. \tag{3.6}$$

Combining eqs. (3.3) through (3.5) with eq. (3.2) yields

$$y(t) = A \cos \omega t + B \sin \omega t + \frac{F_0/k}{1 - r^2} \sin \overline{\omega}t. \tag{3.7}$$

If the initial conditions at time $t = 0$ are taken as zero ($y_0 = 0$, $v_0 = 0$), the constants of integration determined from eq. (3.7) are

$$A = 0, B = -\frac{rF_0/k}{1 - r^2}$$

which, upon substitution in eq. (3.7) gives

$$y(t) = \frac{F_0/k}{1 - r^2} (\sin \overline{\omega}t - r \sin \omega t). \tag{3.8}$$

As we can see from eq. (3.8), the response is given by the superposition of two harmonic terms of different frequencies. The resulting motion is not harmonic; however, in the practical case, damping forces will always be present in the system and will cause the last term, i.e., the free frequency term in eq. (3.8) to vanish eventually. For this reason, this term is said to represent the *transient response*. The forcing frequency term in eq. (3.8), namely

$$y(t) = \frac{F_0/k}{1 - r^2} \sin \overline{\omega}t, \tag{3.9}$$

is referred to as the *steady-state response*. It is clear from eq. (3.8) that in the case of no damping in the system, the transient will not vanish and the response is then given by eq. (3.8). It can also be seen from eq. (3.8) or eq. (3.9) that when the forcing frequency is equal to the natural frequency ($r = 1.0$), the amplitude of the motion becomes infinitely large. A system acted upon by an external excitation of frequency coinciding with the natural frequency is said to be at *resonance*. In this circumstance, the amplitude will increase gradually to infinite. However, materials that are commonly used in practice are subjected to strength limitations and in actual structures failures occur long before extremely large amplitudes can be attained.

3.2 DAMPED HARMONIC EXCITATION

Now consider the case of the one-degree-of-freedom system in Fig. 3.2 vibrating under the influence of viscous damping. The differential equation of motion is

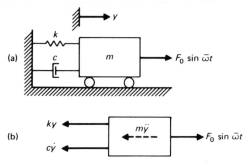

Fig. 3.2 **(a) Damped oscillator harmonically excited. (b) Free body diagram.**

obtained by equating to zero the sum of the forces in the free body diagram of Fig. 3.2(b). Hence

$$m\ddot{y} + c\dot{y} + ky = F_0 \sin \bar{\omega}t. \tag{3.10}$$

The complete solution of this equation again consists of the complementary solution $y_c(t)$ and the particular solution $y_p(t)$. The complementary solution is given for the underdamped case ($c < c_{cr}$) by eq. (2.15) as

$$y_c(t) = e^{-\xi\omega t}(A \cos \omega_D t + B \sin \omega_D t).$$

The particular solution may be found by substituting y_p, in this case, assumed to be of the form

$$y_p(t) = C_1 \sin \bar{\omega}t + C_2 \cos \bar{\omega}t \tag{3.11}$$

into eq. (3.10) and equating the coefficients of the sine and cosine functions. Here we follow a more elegant approach using Euler's relation, namely

$$e^{i\bar{\omega}t} = \cos \bar{\omega}t + i \sin \bar{\omega}t. \tag{3.12}$$

For this purpose, the reader should realize that we can write eq. (3.10) as

$$m\ddot{y} + c\dot{y} + ky = F_0 e^{i\bar{\omega}t} \tag{3.13}$$

with the understanding that only the imaginary component of $F_0 e^{i\bar{\omega}t}$, i.e., the force component of $F_0 \sin \bar{\omega}t$ is acting and, consequently, the response will then consist only of the imaginary part of the total solution of eq. (3.13). In other words, we obtain the solution of eq. (3.13) which has real and imaginary components, and disregard the real component.

It is reasonable to expect that the particular solution of eq. (3.13) will be of the form

$$y_p = C e^{i\bar{\omega}t}. \tag{3.14}$$

Substitution of eq. (3.14) into eq. (3.13) gives

$$-m\bar{\omega}^2 C + ic\bar{\omega}C + kC = F_0$$

or

$$C = \frac{F_0}{k - m\bar{\omega}^2 + ic\bar{\omega}}$$

and

$$y_p = \frac{F_0 e^{i\bar{\omega}t}}{k - m\bar{\omega}^2 + ic\bar{\omega}}. \tag{3.15}$$

By using polar coordinate form, the complex denominator in eq. (3.15) may be written as

$$y_p = \frac{F_0 e^{i\bar{\omega}t}}{\sqrt{(k - m\bar{\omega}^2)^2 + (c\bar{\omega})^2} \, e^{i\theta}}$$

or

$$y_p = \frac{F_0 e^{i(\bar{\omega}t-\theta)}}{\sqrt{(k - m\bar{\omega}^2)^2 + (c\bar{\omega})^2}} \tag{3.16}$$

where

$$\tan \theta = \frac{c\bar{\omega}}{k - m\bar{\omega}^2}. \tag{3.17}$$

The response to the force $F_0 \sin \bar{\omega}t$ (the imaginary component of $F_0 e^{i\bar{\omega}t}$) is then the imaginary component of eq. (3.16), namely,

$$y_p = \frac{F_0 \sin (\bar{\omega}t - \theta)}{\sqrt{(k - m\bar{\omega}^2)^2 + (c\bar{\omega})^2}} \tag{3.18}$$

or

$$y_p = Y \sin (\bar{\omega}t - \theta) \tag{3.19}$$

where

$$Y = \frac{F_0}{\sqrt{(k - m\bar{\omega}^2)^2 + (c\bar{\omega})^2}}$$

is the amplitude of the steady-state motion. Equations (3.18) and (3.17) may conveniently be written in terms of dimensionless ratios as

$$y_p = \frac{y_{st} \sin (\bar{\omega}t - \theta)}{\sqrt{(1 - r^2)^2 + (2\xi r)^2}} \tag{3.20}$$

and

$$\tan \theta = \frac{2\xi r}{1 - r^2}, \tag{3.21}$$

where $y_{st} = F_0/k$ is seen to be the static deflection of the spring acted upon by the force F_0; $\xi = c/c_{cr}$, the damping ratio; and $r = \overline{\omega}/\omega$, the frequency ratio. The total response is then obtained by summing the complementary solution (transient response) from eq. (2.15) and the particular solution (steady-state response) from eq. (3.20), that is,

$$y(t) = e^{-\xi \omega t}(A \cos \omega_D t + B \sin \omega_D t) + \frac{y_{st} \sin (\overline{\omega} t - \theta)}{\sqrt{(1 - r^2)^2 + (2r\xi)^2}}. \qquad (3.22)$$

The reader should be warned that the constants of integration A and B must be evaluated from initial conditions using the total response given by eq. (3.22) and not from just the transient component of the response given in eq. (2.15). By examining the transient component of response, it may be seen that the presence of the exponential factor $e^{-\xi \omega t}$ will cause this component to vanish, leaving only the steady-state motion which is given by eq. (3.20).

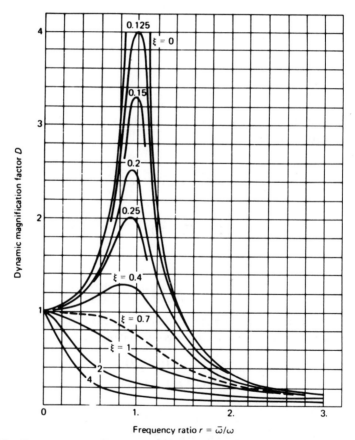

Fig. 3.3   **Dynamic magnification factor as a function of the frequency ratio for various amounts of damping.**

The ratio of the steady-state amplitude of $y_p(t)$ to the static deflection y_{st} defined above is known as the *dynamic magnification factor*, D, and is given from eqs. (3.19) and (3.20) by

$$D = \frac{Y}{y_{st}} = \frac{1}{\sqrt{(1-r^2)^2 + (2r\xi)^2}}. \tag{3.23}$$

It may be seen from eq. (3.23) that the *dynamic magnification factor* varies with the frequency ratio r and the damping ratio ξ. Parametric plots of the dynamic magnification factor are shown in Fig. 3.3. The phase angle θ, given in eq. (3.21), also varies with the same quantities as it is shown in the plots of Fig. 3.4. We note in Fig. 3.3 that for a lightly damped system, the peak amplitude occurs at a frequency ratio very close to one; that is, the dynamic magnification factor has its maximum value virtually at resonance ($r = 1$). It can also be seen from eq. (3.23) that at resonance the dynamic magnification factor is inversely proportional to the damping ratio, that is

$$D(r = 1) = \frac{1}{2\xi}. \tag{3.24}$$

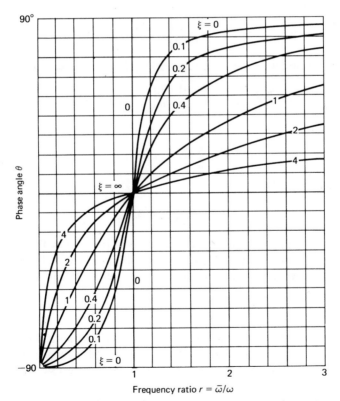

Fig. 3.4 Phase angle θ as a function of the frequency ratio for various amounts of damping.

Although the dynamic magnification factor evaluated at resonance is close to its maximum value, it is not exactly the maximum response for a damped system. However, for moderate amounts of damping, the difference between the approximate value of eq. (3.24) and the exact maximum is negligible.

Illustrative Example 3.1.   A simple beam supports at its center a machine having a weight $W = 16,000$ lb. The beam is made of two standard 8I23 sections with a clear span $L = 12$ ft and total cross-sectional moment of inertia $I = 2 \times 64.2 = 128.4$ in^4. The motor runs at 300 rpm, and its rotor is out of balance to the extent of $W' = 40$ lb at a radius of $e_0 = 10$ in. What will be the amplitude of the steady-state response if the equivalent viscous damping for the system is assumed 10% of the critical?

This dynamic system may be modeled by the damped oscillator. The distributed mass of the beam will be neglected in comparison with the large mass of the machine. Figures 3.5 and 3.6 show, respectively, the schematic diagram of beam-machine system and the adapted model. The force at the center of a simply supported beam necessary to deflect this point one unit (i.e., the stiffness coefficient) is given by the formula

$$k = \frac{48EI}{L^3} = \frac{48 \times 30 \times 10^6 \times 128.4}{(144)^3} = 61,920 \text{ lb/in.}$$

The natural frequency of the system (neglecting the mass of the beam) is

$$\omega = \sqrt{\frac{k}{m}} = \sqrt{\frac{61,920}{16,000/386}} = 38.65 \text{ rad/sec,}$$

the force frequency

$$\bar{\omega} = \frac{300 \times 2\pi}{60} = 31.41 \text{ rad/sec,}$$

and the frequency ratio

$$r = \frac{\bar{\omega}}{\omega} = \frac{31.41}{38.65} = 0.813.$$

Referring to Fig. 3.6, let m be the total mass of the motor and m' the unbalanced rotating mass. Then, if y is the vertical displacement of the nonrotating mass $(m - m')$ from the equilibrium position, the displacement y_1 of m' as shown in Fig. 3.6 is

$$y_1 = y + e_0 \sin \bar{\omega} t. \tag{a}$$

The equation of motion is then obtained by summing forces along the vertical direction in the free body diagram of Fig. 3.6(b) where the inertial forces of both the nonrotating mass and the unbalanced mass are also shown. This summation yields

$$(m - m')\ddot{y} + m'\ddot{y}_1 + c\dot{y} + ky = 0. \tag{b}$$

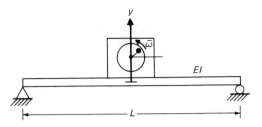

Fig. 3.5   Diagram for beam-machine system of Example 3.1.

Substitution of y_1 from eq. (a) gives

$$(m - m')\ddot{y} + m'(\ddot{y} - e_0\bar{\omega}^2 \sin \bar{\omega}t) + c\dot{y} + ky = 0$$

and with a rearrangement of terms

$$m\ddot{y} + c\dot{y} + ky = m'e_0\bar{\omega}^2 \sin \bar{\omega}t. \qquad (c)$$

This last equation is of the same form as the equation of motion for the damped oscillator excited harmonically by a force of amplitude

$$F_0 = m'e_0\bar{\omega}^2. \qquad (d)$$

Substituting in this equation the numerical values for this example, we obtain

$$F_0 = (40)(10)(31.41)^2/386 = 1022 \text{ lb.}$$

The amplitude of the steady-state resulting motion from eq. (3.20) is then

$$Y = \frac{1022/61,920}{\sqrt{(1 - 0.813^2)^2 + (2 \times 0.813 \times 0.1)^2}},$$

$$Y = 0.044 \text{ in.} \qquad \text{(Ans.)}$$

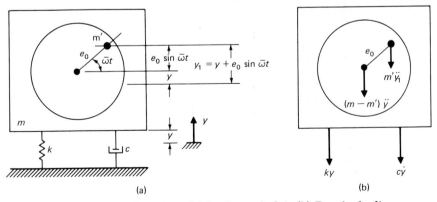

Fig. 3.6   (a) Mathematical model for Example 3.1. (b) Free body diagram.

Illustrative Example 3.2. The steel frame shown in Fig. 3.7 supports a rotating machine which exerts a horizontal force at the girder level, $F(t) = 200 \sin 5.3t$ lb. Assuming 5% of critical damping, determine: (a) the steady-state amplitude of vibration and (b) the maximum dynamic stress in the columns. Assume that the girder is rigid.

This structure may be modeled for dynamic analysis as the damped oscillator shown in Fig. 3.7(b). The parameters in this model are computed as follows:

$$k = \frac{3E(2I)}{L^3} = \frac{3 \times 30 \times 10^6 \times 2 \times 69.2}{(12 \times 15)^3} = 2136 \text{ lb/in},$$

$$\xi = 0.05,$$

$$y_{st} = \frac{F_0}{k} = \frac{200}{2136} = 0.0936 \text{ in},$$

$$\omega = \sqrt{\frac{k}{m}} = \sqrt{\frac{2136 \times 386}{15,000}} = 7.41 \text{ rad/sec},$$

$$r = \frac{\overline{\omega}}{\omega} = \frac{5.3}{7.41} = 0.715.$$

Then the steady-state amplitude from eq. (3.20) is

$$Y = \frac{y_{st}}{\sqrt{(1 - r^2)^2 + (2r\xi)^2}} = 0.189 \text{ in} \qquad \text{(Ans.)}$$

and the maximum shear force in the columns

$$V_{max} = \frac{3EIY}{L^3} = 201.8 \text{ lb}.$$

The maximum bending moment in the columns is

$$M_{max} = V_{max} L = 36,324 \text{ lb} \cdot \text{in}$$

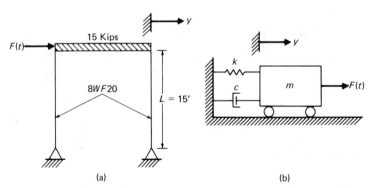

(a)   (b)

Fig. 3.7   (a) Diagram of frame for Example 3.2. (b) Mathematical model.

and the maximum stress

$$\sigma_{max} = \frac{M_{max}}{I/c} = \frac{36,324}{17} = 2136 \text{ psi} \qquad \text{(Ans.)}$$

in which I/c is the section modulus.

3.3   EVALUATION OF DAMPING AT RESONANCE

We have seen in Chapter 2 that the free-vibration decay curve permits the evaluation of damping of a single degree-of-freedom system by simply calculating the logarithm decrement as shown in eq. (2.28). Another technique for determining damping is based on observations of steady-state harmonic response, which requires harmonic excitations of the structure in a range of frequencies in the neighborhood of resonance. With the application of a harmonic force $F_0 \sin \bar{\omega} t$ at closely spaced values of frequencies, the response curve for the structure can be plotted, resulting in displacement amplitudes as a function of the applied frequencies. A typical response curve for such a moderately damped structure is shown in Fig. 3.8. It is seen from eq. (3.24) that, at resonance, the damping ratio is given by

$$\xi = \frac{1}{2D(r = 1)}, \qquad (3.25)$$

where $D(r = 1)$ is the dynamic magnification factor evaluated at resonance. In practice, the damping ratio ξ is determined from the dynamic magnification

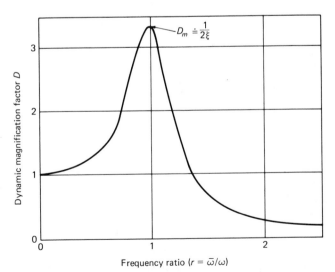

Fig. 3.8   Frequency response curve for moderately damped system.

factor evaluated at the maximum amplitude, namely,

$$\xi = \frac{1}{2D_m},$$ (3.26)

where

$$D_m = \frac{Y_m}{y_{st}}$$

and Y_m is the maximum amplitude. The error involved in evaluating the damping ratio ξ using the approximate eq. (3.26) is not significant in ordinary structures. This method of determining the damping ratio requires only some simple equipment to vibrate the structure in a range of frequencies that span the resonance frequency and a transducer for measuring amplitudes; nevertheless, the evaluation of the static displacement $Y_{st} = F_0/k$ may present a problem since it frequently is difficult to apply a static load to the structure.

3.4 BANDWIDTH METHOD (HALF-POWER) TO EVALUATE DAMPING

An examination of the response curves in Fig. 3.3 shows that the shape of these curves is controlled by the amount of damping present in the system; in particular, the *bandwidth*, that is, the difference between two frequencies correspond-

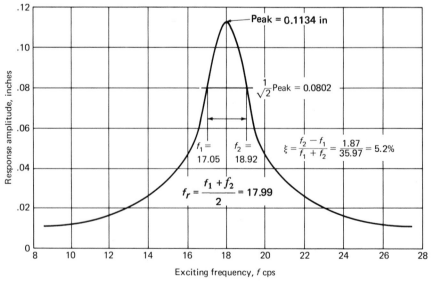

Fig. 3.9   Experimental frequency response curve of Example 3.3.

ing to the same response amplitude, is related to the damping in the system. A typical frequency amplitude curve obtained experimentally for a moderately damped structure is shown in Fig. 3.9. In the evaluation of damping, it is convenient to measure the bandwidth at $1/\sqrt{2}$ of the peak amplitude as shown in this figure. The frequencies corresponding in this bandwidth f_1 and f_2 are also referred to as *half-power* points and are shown in Fig. 3.9. The values of the frequencies for this bandwidth can be determined by setting the response amplitude in eq. (3.20) equal to $1/\sqrt{2}$ times the resonant amplitude given by eq. (3.24), that is,

$$\frac{y_{st}}{\sqrt{(1 - r^2)^2 + (2r\xi)^2}} = \frac{1}{\sqrt{2}} \frac{y_{st}}{2\xi}.$$

Squaring both sides and solving for the frequency ratio results in

$$r^2 = 1 - 2\xi^2 \pm 2\xi\sqrt{1 + \xi^2}$$

or by neglecting ξ^2 in the square root terms

$$r_1^2 \simeq 1 - 2\xi^2 - 2\xi,$$
$$r_2^2 \simeq 1 - 2\xi^2 + 2\xi,$$
$$r_1 \simeq 1 - \xi - \xi^2,$$
$$r_2 \simeq 1 + \xi - \xi^2.$$

Finally, the damping ratio is given approximately by half the difference between these half-power frequency ratios, namely

$$\xi = \frac{1}{2}(r_2 - r_1)$$

or

$$\xi = \frac{1}{2}\frac{\bar{\omega}_2 - \bar{\omega}_1}{\omega} = \frac{f_2 - f_1}{f_2 + f_1} \tag{3.27}$$

since

$$r = \frac{\bar{\omega}}{\omega} = \frac{\bar{f}}{f}, \quad \text{and} \quad f \simeq \frac{f_1 + f_2}{2}.$$

Illustrative Example 3.3.   Experimental data for the frequency response of a single degree-of-freedom system are plotted in Fig. 3.9. Determine the damping ratio of this system. From Fig. 3.9 the peak amplitude is 0.1134 in; hence the amplitude at half-power is equal to

$$0.1134/\sqrt{2} = 0.0802 \text{ in.}$$

The frequencies at this amplitude obtained from Fig. 3.9 are

$$f_1 = 17.05,$$

$$f_2 = 18.92.$$

The damping ratio is then calculated from eq. (3.27) as

$$\xi \simeq \frac{f_2 - f_1}{f_2 + f_1},$$

$$\xi \simeq \frac{18.92 - 17.05}{18.92 + 17.05} = 5.2\%. \qquad \text{(Ans.)}$$

3.5   RESPONSE TO SUPPORT MOTION

There are many actual cases where the foundation or support of a structure is subjected to time varying motion. Structures subjected to ground motion by earthquakes or other excitations such as explosions or dynamic action of machinery are examples in which support motions may have to be considered in the analysis of dynamic response. Let us consider in Fig. 3.10 the case where the support of the simple oscillator modeling the structure is subjected to a harmonic motion given by the expression

$$y_s(t) = y_0 \sin \bar{\omega} t, \qquad (3.28)$$

where y_0 is the maximum amplitude and $\bar{\omega}$ is the frequency of the support motion. The differential equation of motion is obtained by setting equal to zero the sum of the forces (including the inertial force) in the corresponding free body diagram shown in Fig. 3.10(b). The summation of the forces in the hori-

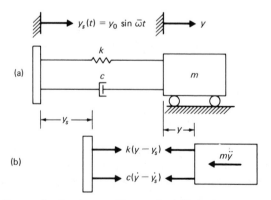

Fig. 3.10   (a) Damped simple oscillator harmonically excited through its support. (b) Free body diagram including inertial force.

zontal direction gives

$$m\ddot{y} + c(\dot{y} - \dot{y}_s) + k(y - y_s) = 0. \tag{3.29}$$

The substitution of eq. (3.28) into eq. (3.29) and the rearrangement of terms result in

$$m\ddot{y} + c\dot{y} + ky = ky_0 \sin \overline{\omega}t + c\overline{\omega}y_0 \cos \overline{\omega}t. \tag{3.30}$$

The two harmonic terms of frequency $\overline{\omega}$ in the right-hand side of this equation may be combined and eq. (3.30) rewritten as

$$m\ddot{y} + c\dot{y} + ky = F_0 \sin (\overline{\omega}t + \beta), \tag{3.31}$$

where

$$F_0 = y_0\sqrt{k^2 + (c\overline{\omega})^2} = y_0 k\sqrt{1 + (2r\xi)^2} \tag{3.32}$$

and

$$\tan \beta = c\overline{\omega}/k = 2r\xi. \tag{3.33}$$

It is apparent that eq. (3.31) is the differential equation for the oscillator excited by the harmonic force $F_0 \sin (\overline{\omega}t + \beta)$ and is of the same form as eq. (3.10). Consequently, the steady-state solution of eq. (3.31) is given as before by eq. (3.20), except for the addition of the angle β in the argument of the sine function, that is,

$$y(t) = \frac{F_0/k \sin (\overline{\omega}t + \beta - \theta)}{\sqrt{(1 - r^2)^2 + (2r\xi)^2}} \tag{3.34}$$

or substituting F_0 from eq. (3.32)

$$\frac{y(t)}{y_0} = \frac{\sqrt{1 + (2r\xi)^2}}{\sqrt{(1 - r^2)^2 + (2r\xi)^2}} \sin (\overline{\omega}t + \beta - \theta). \tag{3.35}$$

Equation (3.35) is the expression for the relative transmission of the support motion to the oscillator. This is an important problem in vibration isolation in which equipment must be protected from harmful vibrations of the supporting structure. The degree of relative isolation is known as *transmissibility* and is defined as the ratio of the amplitude of motion Y of the oscillator to the amplitude y_0, the motion of the support. From eq. (3.35), transmissibility T_r is then given by

$$T_r = \frac{Y}{y_0} = \frac{\sqrt{1 + (2r\xi)^2}}{\sqrt{(1 - r^2)^2 + (2r\xi)^2}}. \tag{3.36}$$

A plot of transmissibility as a function of the frequency ratio and damping ratio is shown in Fig. 3.11. The curves in this figure are similar to curves in Fig. 3.3,

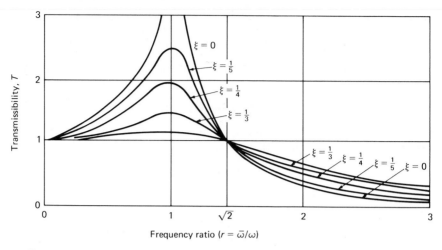

Fig. 3.11   Transmissibility versus frequency ratio for vibration isolation.

representing the frequency response of the damped oscillator. The major difference being that all of the curves in Fig. 3.11 pass through the same point at a frequency ratio $r = \sqrt{2}$. It can be seen in Fig. 3.11 that damping tends to reduce the effectiveness of vibration isolation for frequencies greater than this ratio, that is, for r greater than $\sqrt{2}$.

Equation (3.34) provides the absolute response of the damped oscillator to a harmonic motion of its base. Alternatively, we can solve the differential equation, eq. (3.29), in terms of the relative motion between the mass m and the support given by

$$u = y - y_s \qquad (3.37)$$

which substituted into eq. (3.29) gives

$$m\ddot{u} + c\dot{u} + ku = F_{eff}(t) \qquad (3.38)$$

where $F_{eff}(t) = -m\ddot{y}_s$ may be interpreted as the effective force acting on the mass of the oscillator, and its displacement is indicated by coordinate u. Using eq. (3.28) to obtain $\ddot{y}_s$ and substituting in eq. (3.38) results in

$$m\ddot{u} + c\dot{u} + ku = my_0\bar{\omega}^2 \sin \bar{\omega}t. \qquad (3.39)$$

Again, eq. (3.39) is of the same form as eq. (3.10) with $F_0 = my_0\bar{\omega}^2$. Then, from eq. (3.20), the steady-state response in terms of relative motion is given by

$$u(t) = \frac{my_0\bar{\omega}^2/k \sin (\bar{\omega}t - \theta)}{\sqrt{(1 - r^2)^2 + (2r\xi)^2}} \qquad (3.40)$$

or substituting

$$\frac{\bar{\omega}^2}{k/m} = \frac{\bar{\omega}^2}{\omega^2} = r^2$$

we obtain

$$\frac{u(t)}{y_0} = \frac{r^2 \sin (\bar{\omega}t - \theta)}{\sqrt{(1 - r^2)^2 + (2r\xi)^2}}, \tag{3.41}$$

where θ is given in eq. (3.21).

Illustrative Example 3.4.   If the frame of Example 3.2 (Fig. 3.7) is subjected to a sinusoidal ground motion $y_s(t) = 0.2 \sin 5.3t$, determine: (a) the transmissibility of motion to the girder, (b) the maximum shearing force in the supporting columns, and (c) maximum stresses in the columns.
   The parameters for this system are

$$k = 2136 \text{ lb/in,}$$

$$\xi = 0.05,$$

$$y_0 = 0.2 \text{ in,}$$

$$y_{st} = 0.0936 \text{ in,}$$

$$\omega = 7.41 \text{ rad/sec,}$$

$$\bar{\omega} = 5.3 \text{ rad/sec,}$$

$$r = 0.715.$$

The transmissibility from eq. (3.36) is

$$T_r = \sqrt{\frac{1 + (2r\xi)^2}{(1 - r^2)^2 + (2r\xi)^2}} = 2.1. \tag{Ans.}$$

The maximum relative displacement U from eq. (3.41) is

$$U = \frac{y_0 r^2}{\sqrt{(1 - r^2)^2 + (2r\xi)^2}} = 0.206 \text{ in.}$$

Then the maximum shear force in each column is

$$V_{max} = \frac{kU}{2} = 219.8 \text{ lb.} \tag{Ans.}$$

The maximum bending moment

$$M_{max} = V_{max} L = 39,567 \text{ lb} \cdot \text{in}$$

and the corresponding stress

$$\sigma_{max} = \frac{M_{max}}{I/c} = \frac{39{,}567}{17} = 2327 \text{ psi} \qquad \text{(Ans.)}$$

in which I/c is the section modulus.

3.6   FORCE TRANSMITTED TO THE FOUNDATION

In the preceding section, we determined the response of the structure to a harmonic motion of its foundation. In this section we shall consider a similar problem of vibration isolation; the problem now, however, is to find the force transmitted to the foundation. Consider again the damped oscillator with a harmonic force $F(t) = F_0 \sin \overline{\omega}t$ acting on its mass as shown in Fig. 3.2. The differential equation of motion is

$$m\ddot{y} + c\dot{y} + ky = F_0 \sin \overline{\omega}t$$

with the steady-state solution

$$y = Y \sin (\overline{\omega}t - \theta),$$

where

$$Y = \frac{F_0/k}{\sqrt{(1 - r^2)^2 + (2r\xi)^2}} \qquad (3.42)$$

and

$$\tan \theta = \frac{2\xi r}{1 - r^2}.$$

The force transmitted to the support through the spring is ky and through the damping element is $c\dot{y}$. Hence the total force transmitted F_T is

$$F_T = ky + c\dot{y}. \qquad (3.43)$$

Differentiating eq. (3.19) and substituting in eq. (3.43) yield

$$F_T = Y[k \sin (\overline{\omega}t - \theta) + c\overline{\omega} \cos (\overline{\omega}t - \theta)]$$

or

$$F_T = Y\sqrt{k^2 + c^2\overline{\omega}^2} \sin (\overline{\omega}t - \theta + \beta), \qquad (3.44)$$

$$F_T = Y\sqrt{k^2 + c^2\overline{\omega}^2} \sin (\overline{\omega}t - \phi) \qquad (3.45)$$

in which

$$\tan \beta = \frac{c\overline{\omega}}{k} = 2\xi r \qquad (3.46)$$

and

$$\phi = \theta - \beta. \tag{3.47}$$

Then from eqs. (3.42) and (3.45), the maximum force A_T transmitted to the foundation is

$$A_T = F_0 \sqrt{\frac{1 + (2\xi r)^2}{(1 - r^2)^2 + (2\xi r)^2}}. \tag{3.48}$$

The transmissibility T_r is defined as the ratio between the amplitude of the force transmitted to the foundation and the amplitude of the applied force. Hence from eq. (3.48)

$$T_r = \frac{A_T}{F_0} = \sqrt{\frac{1 + (2\xi r)^2}{(1 - r^2)^2 + (2r\xi)^2}}. \tag{3.49}$$

It is interesting to note that both the transmissibility of motion from the foundation to the structure, eq. (3.36), and the transmissibility of force from the structure to the foundation, eq. (3.49), are given by exactly the same function. Hence the curves of transmissibility in Fig. 3.11 represent either type of transmissibility. An expression for the total phase angle ϕ in eq. (3.45) may be determined by taking the tangent function to both members of eq. (3.47), so that

$$\tan \phi = \frac{\tan \theta - \tan \beta}{1 + \tan \theta \tan \beta}.$$

Then substituting $\tan \theta$ and $\tan \beta$, respectively, from eqs. (3.21) and (3.46), we obtain

$$\tan \phi = \frac{2\xi r^3}{1 - r^2 + 4\xi^2 r^2}. \tag{3.50}$$

Illustrative Example 3.5.   A machine of weight $W = 3860$ lb is mounted on a simple supported steel beam as shown in Fig. 3.12. A piston that moves up and

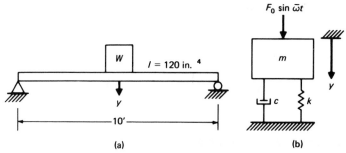

(a)   (b)

Fig. 3.12   (a) Beam-machine system for Example 3.5. (b) Mathematical model.

down in the machine produces a harmonic force of magnitude $F_0 = 7000$ lb and frequency $\overline{\omega} = 60$ rad/sec. Neglecting the weight of the beam and assuming 10% of the critical damping, determine: (a) the amplitude of the motion of the machine, (b) the force transmitted to the beam supports, and (c) the corresponding phase angle.

The damped oscillator in Fig. 3.12(b) is used to model the system. The following parameters are calculated:

$$k = \frac{48EI}{L^3} = 10^5 \text{ lb/in,}$$

$$\omega = \sqrt{\frac{k}{m}} = 100 \text{ rad/sec,}$$

$$\xi = 0.1,$$

$$r = \frac{\overline{\omega}}{\omega} = 0.6,$$

$$y_{st} = \frac{F_0}{k} = 0.07 \text{ in.}$$

(a) From eq. (3.20), the amplitude of motion is

$$Y = \frac{y_{st}}{\sqrt{(1 - r^2)^2 + (2r\xi)^2}} = 0.1075 \text{ in} \qquad \text{(Ans.)}$$

with a phase angle from eq. (3.21)

$$\theta = \tan^{-1} \frac{2r\xi}{1 - r^2} = 10.6°.$$

(b) From eq. (3.48), the transmissibility is

$$T_r = \frac{A_T}{F_0} = \sqrt{\frac{1 + (2r\xi)^2}{(1 - r^2)^2 + (2r\xi)^2}} = 1.547.$$

Hence the amplitude of the force transmitted to the foundation is

$$A_T = F_0 \, T_r = 10{,}827 \text{ lb} \qquad \text{(Ans.)}$$

(c) The corresponding phase angle from eq. (3.50) is

$$\phi = \tan^{-1} \frac{2\xi r^3}{1 - r^2 + (2r\xi)^2} = 3.78°. \qquad \text{(Ans.)}$$

3.7   SEISMIC INSTRUMENTS

When a system of the type shown in Fig. 3.13 is used for the purpose of vibration measurement, the relative displacement between the mass and the base is

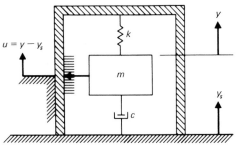

Fig. 3.13   Model of seismograph.

ordinarily recorded. Such an instrument is called a *seismograph* and it can be designed to measure either the displacement or the acceleration of the base. The peak relative response, U/y_0 of the seismograph depicted in Fig. 3.13, for harmonic motion of the base is given from eq. (3.41) by

$$\frac{U}{y_0} = \frac{r^2}{\sqrt{(1 - r^2)^2 + (2r\xi)^2}}. \tag{3.51}$$

A plot of this equation as a function of the frequency ratio and damping ratio is shown in Fig. 3.14. It may be seen from this figure that the response is essentially constant for frequency ratios $r > 1$ and damping ratio $\xi = 0.5$. Consequently, the response of a properly damped instrument of this type is essentially proportional to the base-displacement amplitude for high frequencies of motion of the base. The instrument will thus serve as a displacement meter for measuring such motions. The range of applicability of the instrument is in-

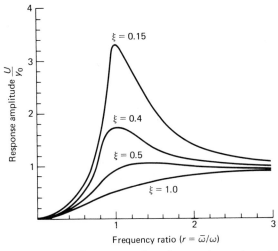

Fig. 3.14   Response of seismograph to harmonic motion of the base.

creased by reducing the natural frequency, i.e., by reducing the spring stiffness or increasing the mass.

Now consider the response of the same instrument to a harmonic acceleration of the base $\ddot{y}_s = \ddot{y}_0 \sin \bar{\omega}t$. The equation of motion of this system is obtained from eq. (3.38) as

$$m\ddot{u} + c\dot{u} + ku = -m\ddot{y}_0 \sin \bar{\omega}t. \qquad (3.52)$$

The steady-state response of this system expressed as the dynamic magnification factor is then given from eq. (3.23) by

$$D = \frac{U}{m\ddot{y}_0/k} = \frac{1}{\sqrt{(1 - r^2)^2 + (2r\xi)^2}}. \qquad (3.53)$$

This equation is represented graphically in Fig. 3.3. In this case, it can be seen from this figure that for a damping ratio $\xi = 0.7$, the value of the response is nearly constant in the frequency range $0 < r < 0.6$. Thus it is clear from eq. (3.53) that the response indicated by this instrument will be directly proportional to the base-acceleration amplitude for frequencies up to about six-tenths of the natural frequency. Its range of applicability will be increased by increasing the natural frequency, that is, by increasing the stiffness of the spring or by decreasing the mass of the oscillator.

3.8   SUMMARY

In this chapter, we have determined the response of a single degree-of-freedom system subjected to harmonic loading. This type of loading is expressed as a sine, cosine, or exponential function and can be handled mathematically with minimum difficulty for the undamped or damped structure. The differential equation of motion for a linear single degree-of-freedom system is the second-order differential equation

$$m\ddot{y} + c\dot{y} + ky = F_0 \sin \bar{\omega}t \qquad (3.10)$$

or

$$\ddot{y} + 2\xi\omega\dot{y} + \omega^2 y = \frac{F_0}{m} \sin \bar{\omega}t$$

in which $\bar{\omega}$ is the forced frequency,

$$\xi = \frac{c}{c_{cr}} \quad \text{is the damping ratio,}$$

and

$$\omega = \sqrt{\frac{k}{m}} \quad \text{is the natural frequency.}$$

The general solution of eq. (3.10) is obtained as the summation of the complementary (transient) and the particular (steady-state) solutions, namely

$$y = e^{-\xi\omega t}\underbrace{(A \cos \omega_d t + B \sin \omega_d t)}_{\text{transient solution}} + \underbrace{\frac{F_0/k \sin (\bar{\omega}t - \theta)}{\sqrt{(1 - r^2)^2 + (2r\xi)^2}}}_{\text{steady-state solution}}$$

in which A and B are constants of integration,

$$r = \frac{\bar{\omega}}{\omega} \quad \text{is the frequency ratio,}$$

$$\omega_d = \omega\sqrt{1 - r^2} \quad \text{is the damped natural frequency,}$$

and

$$\theta = \tan^{-1}\frac{2r\xi}{1 - r^2} \quad \text{is the phase angle.}$$

The transient part of the solution vanishes rapidly to zero because of the negative exponential factor leaving only the steady-state solution. Of particular significance is the condition of resonance $(r = \bar{\omega}/\omega = 1)$ for which the amplitudes of motion become very large for the damped system and tend to become infinity for the undamped system.

The response of the structure to support or foundation motion can be obtained in terms of the absolute motion of the mass or of its relative motion with respect to the support. In this latter case, the equation assumes a much simpler and more convenient form, namely

$$m\ddot{u} + c\dot{u} + ku = F_{\text{eff}}(t) \tag{3.38}$$

in which

$$F_{\text{eff}}(t) = -m\ddot{y}_s(t) \quad \text{is the effective force}$$

and

$$u = y - y_s \quad \text{is the relative displacement.}$$

For harmonic excitation of the foundation, the solution of eq. (3.38) in terms of the relative motion is of the same form as the solution of eq. (3.10) in which the force is acting on the mass.

In this chapter, we have also shown that the damping in the system may be evaluated experimentally either from the peak amplitude or from the bandwidth obtained from a plot of the amplitude-frequency curve when the system is forced to harmonic vibration. Two related problems of vibrating isolation were discussed in this chapter: (1) the motion transmissibility, that is, the relative motion transmitted from the foundation to the structure; and (2) the force transmissibility which is the relative magnitude of the force transmitted from the structure to the foundation. For both of these problems, the transmissibility is given by

$$T_r = \sqrt{\frac{1 + (2r\xi)^2}{(1 - r^2)^2 + (2r\xi)^2}}.$$

PROBLEMS

3.1   An electric motor of total weight $W = 1000$ lb is mounted at the center of a simply supported beam as shown in Fig. P3.1. The unbalance in the rotor is $W'e = 1$ lb · in. Determine the steady-state amplitude of vertical motion of the motor for a speed of 900 rpm. Assume that the damping in the system is 10% of the critical damping. Neglect the mass of the supporting beam.

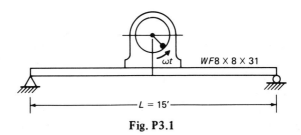

Fig. P3.1

3.2   Determine the maximum force transmitted to the supports of the beam in Problem 3.1.
3.3   Determine the steady-state amplitude for the horizontal motion of the steel frame in Fig. P3.3. Assume the horizontal girder to be infinitely rigid and neglect both the mass of the columns and damping.

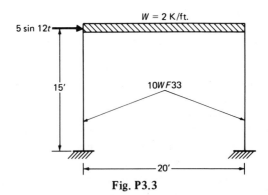

Fig. P3.3

3.4 Solve for Problem P3.3 assuming that the damping in the system is 8% of the critical damping.

3.5 For Problem 3.4 determine: (a) the maximum force transmitted to the foundation and (b) the transmissibility.

3.6 A delicate instrument is to be spring mounted to the floor of a test laboratory where it has been determined that the floor vibrates vertically with harmonic motion of amplitude 0.1 in at 10 cps. If the instrument weighs 100 lb determine the stiffness of the isolation springs required to reduce the vertical motion amplitude of the instrument to 0.001 in. Neglect damping.

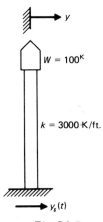

Fig. P3.7

3.7 Consider the water tower shown in Fig. P3.7 which is subjected to ground motion produced by a passing train in the vicinity of the tower. The ground motion is idealized as a harmonic acceleration of the foundation of the tower with an amplitude of $0.1g$ at a frequency of 25 cps. Determine the motion of the tower relative to the motion of its foundation. Assume an effective damping coefficient of 10% of the critical damping in the system.

3.8 Determine the transmissibility in Problem 3.7.

4

Response to general dynamic loading

In the preceding chapter we studied the response of a single degree-of-freedom system with harmonic loading. Though this type of loading is important, real structures are often subjected to loads which are not harmonic. In the present chapter we shall study the response of the single degree-of-freedom system to a general type of force. We shall see that the response can be obtained in terms of an integral which for many simple load functions can be evaluated analytically. For the general case, however, it will be necessary to resort to a numerical integration procedure.

4.1 IMPULSIVE LOADING AND DUHAMEL'S INTEGRAL

An impulsive loading is a load which is applied during a short duration of time. The corresponding impulse of this type of load is defined as the product of the force and the time of its duration. For example, the impulse of the force $F(\tau)$ depicted

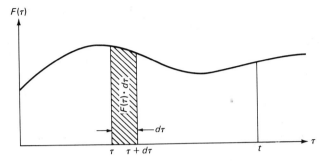

Fig. 4.1   General load history as impulsive loading.

in Fig. 4.1 at time τ during the interval $d\tau$ is represented by the shaded area and it is equal to $F(\tau)\,d\tau$. This impulse acting on a body of mass m, produces a change in velocity which can be determined from Newton's Law of Motion, namely

$$m\frac{dv}{d\tau} = F(\tau).$$

Rearrangement yields

$$dv = \frac{F(\tau)\,d\tau}{m}, \tag{4.1}$$

where $F(\tau)\,d\tau$ is the impulse and dv is the incremental velocity. This incremental velocity may be considered to be an initial velocity of the mass at time τ. Now let us consider this impulse $F(\tau)\,d\tau$ acting on the structure represented by the undamped oscillator. At the time τ the oscillator will experience a change of velocity given by eq. (4.1). This change in velocity is then introduced in eq. (1.20) as the initial velocity v_0 together with the initial displacement $y_0 = 0$ at time τ producing a displacement at a later time t given by

$$dy(t) = \frac{F(\tau)\,d\tau}{m\omega}\sin\omega(t - \tau). \tag{4.2}$$

The loading history may then be regarded as a series of short impulses at successive incremental times $d\tau$, each producing its own differential response at time t of the form given by eq. (4.2). Therefore, we conclude that the total displacement at time t due to the continuous action of the force $F(\tau)$ is given by the summation or integral of the differential displacements $dy(t)$ from time $t = 0$ to time t, that is,

$$y(t) = \frac{1}{m\omega}\int_0^t F(\tau)\sin\omega(t - \tau)\,d\tau. \tag{4.3}$$

The integral in this equation is known as Duhamel's integral. Equation (4.3) represents the total displacement produced by the exciting force $F(\tau)$ acting on the undamped oscillator; it includes both the steady-state and the transient components of the motion. If the function $F(\tau)$ cannot be expressed analytically, the integral of eq. (4.3) can always be evaluated approximately by suitable numerical methods. To include the effect of initial displacement y_0 and initial velocity v_0 at time $t = 0$, it is only necessary to add to eq. (4.3) the solution given by eq. (1.20) for the effects due to the initial conditions. Thus the total displacement of an undamped single degree-of-freedom system with an arbitrary load is given by

$$y(t) = y_0 \cos \omega t + \frac{v_0}{\omega} \sin \omega t + \frac{1}{m\omega} \int_0^t F(\tau) \sin \omega(t - \tau)\, d\tau. \qquad (4.4)$$

Applications of eq. (4.4) for some simple forcing functions for which it is possible to obtain the explicit integration of eq. (4.4) are presented below.

4.1.1   Constant Force

Consider the case of a constant force of magnitude F_0 applied suddenly to the undamped oscillator at time $t = 0$ as shown in Fig. 4.2. For both initial displacement and initial velocity equal to zero, the application of eq. (4.4) to this case gives

$$y(t) = \frac{1}{m\omega} \int_0^t F_0 \sin \omega(t - \tau)\, d\tau$$

and integration yields

$$y(t) = \frac{F_0}{m\omega^2} \left| \cos \omega(t - \tau) \right|_0^t,$$

$$y(t) = \frac{F_0}{k} (1 - \cos \omega t) = y_{st}(1 - \cos \omega t), \qquad (4.5)$$

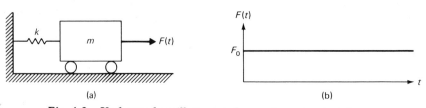

Fig. 4.2   Undamped oscillator acted upon by a constant force.

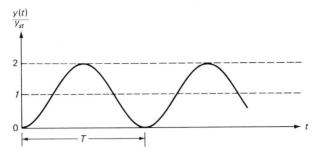

Fig. 4.3 Response of an undamped single degree-of-freedom system to a suddenly applied constant force.

where $y_{st} = F_0/k$. The response for such a suddenly applied constant load is shown in Fig. 4.3. It will be observed that this solution is very similar to the solution for the free vibration of the undamped oscillator. The major difference is that the coordinate axis t has been shifted by an amount equal to $y_{st} = F_0/k$. Also, it should be noted that the maximum displacement $2y_{st}$ is exactly twice the displacement which the force F_0 would produce if it were applied statically. We have found an elementary but important result: the maximum displacement of a linear elastic system for a constant force applied suddenly is twice the displacement caused by the same force applied statically (slowly). This result for displacement is also true for the internal forces and stresses in the structure.

4.1.2 Rectangular Load

Let us consider a second case, that of a constant force F_0 suddenly applied but only during a limited time duration t_d as shown in Fig. 4.4. Up to the time t_d, eq. (4.5) applies and at that time the displacement and velocity are

$$y_d = \frac{F_0}{k}(1 - \cos \omega t_d)$$

and

$$v_d = \frac{F_0}{k}\omega \sin \omega t_d.$$

For the response after time t_d we apply eq. (1.20) for free vibration, taking as the initial conditions the displacement and velocity at t_d. After replacing t by $t - t_d$, and y_0 and v_0 by y_d and v_d, respectively, we obtain

$$y(t) = \frac{F_0}{k}(1 - \cos \omega t_d) \cos \omega(t - t_d) + \frac{F_0}{k} \sin \omega t_d \sin \omega(t - t_d)$$

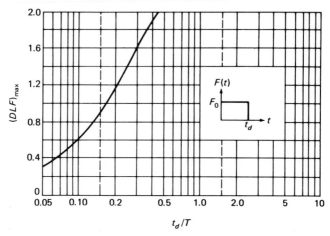

Fig. 4.4   Maximum dynamic load factor for the undamped oscillator acted upon by a rectangular force.

which can be reduced to

$$y(t) = \frac{F_0}{k} \{\cos \omega(t - t_d) - \cos \omega t\}. \qquad (4.6)$$

If the dynamic load factor (DLF) is defined as the displacement at any time t divided by the static displacement $y_{st} = F_0/k$, we may write eqs. (4.5) and (4.6) as

$$\text{DLF} = 1 - \cos \omega t, \qquad t \leqslant t_d$$

and

$$\text{DLF} = \cos \omega(t - t_d) - \cos \omega t, \qquad t \geqslant t_d. \qquad (4.7)$$

It is often convenient to express time as a dimensionless parameter by simply using the natural period instead of the natural frequency ($\omega = 2\pi/T$). Hence eq. (4.7) may be written as

$$\text{DLF} = 1 - \cos 2\pi \frac{t}{T}, \qquad t \leqslant t_d$$

and

$$\text{DLF} = \cos 2\pi \left(\frac{t}{T} - \frac{t_d}{T} \right) - \cos 2\pi \frac{t}{T}, \qquad t \geqslant t_d. \qquad (4.8)$$

The use of dimensionless parameters in eq. (4.8) serves to emphasize the fact that the ratio of duration of the time the constant force is applied to the natural period rather than the actual value of either quantity is the important parameter. The maximum dynamic load factor $(DLF)_{max}$, obtained by maximizing eq. (4.8) is plotted in Fig. 4.4. It is observed from this figure that the maximum dynamic load factor for loads of duration $t_d/T \geq 0.5$ is the same as if the load duration had been infinite.

Charts, as shown in Fig. 4.4, which give the maximum response of a single degree-of-freedom system for a given loading function are called *response spectral* charts. These charts are extremely useful for design purposes, as will be discussed in Chapter 8. Response spectral charts for impulsive loads of short duration are often presented for the undamped system. For short duration of the load, damping does not have a significant effect on the response of the system. The maximum dynamic load factor usually corresponds to the first peak of response and the amount of damping normally found in structures is not sufficient to appreciably decrease this value.

4.1.3 Triangular Load

We consider now a system represented by the undamped oscillator, initially at rest and subjected to a force $F(t)$ which has an initial value F_0 and which decreases linearly to zero at time t_d (Fig. 4.5). The response may be computed by eq. (4.4) in two intervals. For the first interval, $\tau \leq t_d$, the force is given by

$$F(\tau) = F_0 \left(1 - \frac{\tau}{t_d}\right)$$

and the initial conditions by

$$y_0 = 0, \quad v_0 = 0.$$

The substitution of these values in eq. (4.4) and integration gives

$$y = \frac{F_0}{k}(1 - \cos \omega t) + \frac{F_0}{kt_d}\left(\frac{\sin \omega t}{\omega} - t\right) \tag{4.9}$$

or in terms of the dynamic load factor

$$DLF = \frac{y}{y_{st}} = 1 - \cos \omega t + \frac{\sin \omega t}{\omega t_d} - \frac{t}{t_d}, \tag{4.10}$$

which defines the response before time t_d. For the second interval $(t \geq t_d)$, we obtain from eq. (4.10) the displacement and velocity at time t_d as

$$y_d = \frac{F_0}{k}\left(\frac{\sin \omega t_d}{\omega t_d} - \cos \omega t_d\right)$$

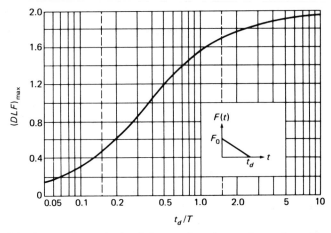

Fig. 4.5   **Maximum dynamic load factor for the undamped oscillator acted upon by a triangular force.**

and

$$v_d = \frac{F_0}{k}\left(\omega \sin \omega t_d + \frac{\cos {}^{\omega}t_d}{t_d} - \frac{1}{t_d}\right). \tag{4.11}$$

These values may be considered as the initial conditions at time $t = t_d$ for this second interval. Replacing in eq. (1.20) t by $t - t_d$ and y_0 and v_0, respectively, by y_d and v_d and noting that $F(\tau) = 0$ in this interval, we obtain the response as

$$y = \frac{F_0}{k\omega t_d}\{\sin \omega t - \sin \omega(t - t_d)\} - \frac{F_0}{k}\cos \omega t$$

and upon dividing by $y_{st} = F_0/k$ gives

$$\text{DLF} = \frac{1}{\omega t_d}\{\sin \omega t - \sin \omega(t - t_d)\} - \cos \omega t. \tag{4.12}$$

In terms of the dimensionless time parameter, this last equation may be written as

$$\text{DLF} = \frac{1}{2\pi t_d/T}\left\{\sin 2\pi \frac{t}{T} - \sin 2\pi\left(\frac{t}{T} - \frac{t_d}{T}\right)\right\} - \cos 2\pi \frac{t}{T}. \tag{4.13}$$

The plot of the maximum dynamic load factor as a function of the relative time duration t_d/T for the undamped oscillator is given in Fig. 4.5. As it would be expected, the maximum value of the dynamic load factor approaches 2 as t_d/T becomes large; that is, the effect of the decay of the force is negligible for the time required for the system to reach the first peak.

We have studied the response of the undamped oscillator for two simple impulse loadings: the rectangular pulse and the triangular pulse. Extensive charts have been prepared by the U.S. Army Corps of Engineers[1] and are available for a variety of other loading pulses. We shall now determine the response for forcing functions which do not permit an analytical solution of Duhamel's integral. In these cases it is necessary to resort to a numerical evaluation of Duhamel's integral in order to obtain the response of the system.

4.2   NUMERICAL EVALUATION OF DUHAMEL'S INTEGRAL—UNDAMPED SYSTEM

In many practical cases the applied loading function is known only from experimental data as in the case of seismic motion and the response must be evaluated by a numerical method. For this purpose we use the trigonometric identity $\sin \omega(t - \tau) = \sin \omega t \cos \omega \tau - \cos \omega t \sin \omega \tau$, in Duhamel's integral. Then, assuming zero initial conditions, we obtain Duhamel's integral, eq. (4.4), in the form

$$y(t) = \sin \omega t \; \frac{1}{m\omega} \int_0^t F(\tau) \cos \omega \tau \, d\tau - \cos \omega t \; \frac{1}{m\omega} \int_0^t F(\tau) \sin \omega \tau \, d\tau$$

or

$$y(t) = \{A(t) \sin \omega t - B(t) \cos \omega t\}/m\omega \tag{4.14}$$

where

$$A(t) = \int_0^t F(\tau) \cos \omega \tau \, d\tau$$

$$B(t) = \int_0^t F(\tau) \sin \omega \tau \, d\tau. \tag{4.15}$$

The calculation of Duhamel's integral thus requires the evaluation of the integrals $A(t)$ and $B(t)$ numerically.

Several numerical integration techniques have been used for this evaluation. In these techniques the integrals are replaced by a suitable summation of the function under the integral and evaluated for convenience at n equal time increments, $\Delta\tau$. The most popular of these methods are the trapezoidal rule and the

---

[1]U.S. Army Corps of Engineers, *Design of Structures to Resist the Effects of Atomic Weapons*, Manuals 415, 415, and 416, March 15, 1957; Manuals 417 and 419, January 15. 1958; Manuals 418, 420, 421, January 15, 1960.

Simpson's rule. Consider the integration of a general function $I(\tau)$

$$A(t) = \int_0^t I(\tau)\, d\tau.$$

The elementary operation required for the trapezoidal rule is

$$A(t) = \Delta\tau \tfrac{1}{2}(I_0 + 2I_1 + 2I_2 + \cdots + 2I_{n-1} + I_n), \qquad (4.16)$$

and for Simpson's rule

$$A(t) = \Delta\tau \tfrac{1}{3}(I_0 + 4I_1 + 2I_2 + \cdots + 4I_{n-1} + I_n), \qquad (4.17)$$

where $n = t/\Delta\tau$ must be an even number for Simpson's rule. The implementation of these rules is straightforward. The response obtained will be approximate since these rules are based on the substitution of the function $I(\tau)$ for a piecewise linear function for the trapezoidal rule, or piecewise parabolic function for Simpson's rule. An alternative approach to the evaluation of Duhamel's integral is based on obtaining the exact analytical solution of the integral for the loading function assumed to be given by a succession of linear segments. This method does not introduce numerical approximations for the integration other than those inherent in the roundoff error, so in this sense it is an exact method.

In using this method, it is assumed that $F(\tau)$, the forcing function may be approximated by a segmentally linear function as shown in Fig. 4.6. To provide a complete response history, it is more convenient to express the integrations in eq. (4.15) in incremental form, namely

$$A(t_i) = A(t_{i-1}) + \int_{t_{i-1}}^{t_i} F(\tau) \cos \omega\tau\, d\tau, \qquad (4.18)$$

$$B(t_i) = B(t_{i-1}) + \int_{t_{i-1}}^{t_i} F(\tau) \sin \omega\tau\, d\tau, \qquad (4.19)$$

where $A(t_i)$ and $B(t_i)$ represent the values of the integrals in eq. (4.15) at time t_i. Assuming that the forcing function $F(\tau)$ is approximated by a piecewise linear function as shown in Fig. 4.6, we may write

$$F(\tau) = F(t_{i-1}) + \frac{\Delta F_i}{\Delta t_i}(\tau - t_{i-1}), \qquad t_{i-1} \leqslant \tau \leqslant t_i \qquad (4.20)$$

where

$$\Delta F_i = F(t_i) - F(t_{i-1})$$

and

$$\Delta t_i = t_i - t_{i-1}.$$

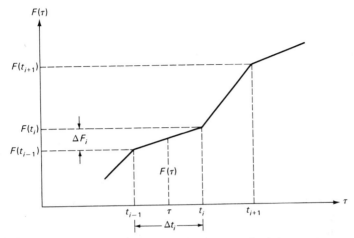

Fig. 4.6   Segmentally linear loading function.

The substitution of eq. (4.20) into eq. (4.18) and integration yield

$$A(t_i) = A(t_{i-1}) + \left(F(t_{i-1}) - t_{i-1}\frac{\Delta F_i}{\Delta t_i}\right)(\sin \omega t_i - \sin \omega t_{i-1})/\omega$$

$$+ \frac{\Delta F_i}{\omega^2 \Delta t_i}\{\cos \omega t_i - \cos \omega t_{i-1} + \omega(t_i \sin \omega t_i - t_{i-1} \sin \omega t_{i-1})\}. \quad (4.21)$$

Analogously from eq. (4.19),

$$B(t_i) = B(t_{i-1}) + \left(F(t_{i-1}) - t_{i-1}\frac{\Delta F_i}{\Delta t_i}\right)(\cos \omega t_{i-1} - \cos \omega t_i)/\omega$$

$$+ \frac{\Delta F_i}{\omega^2 \Delta t_i}\{\sin \omega t_i - \sin \omega t_{i-1} - \omega(t_i \cos \omega t_i - t_{i-1} \cos \omega t_{i-1})\}. \quad (4.22)$$

Equations (4.21) and (4.22) are recurrent formulas for the evaluation of the integrals in eq. (4.15) at any time $t = t_i$.

Illustrative Example 4.1.   Determine the dynamic response of a tower subjected to a blast loading. The idealization of the structure and the blast loading are shown in Fig. 4.7. Neglect damping.

For this system, the natural frequency is

$$\omega = \sqrt{k/m} = \sqrt{100,000/100} = 31.62 \text{ rad/sec.}$$

Since the loading is given as a segmented linear function, the response obtained using Duhamel's integral, eq. (4.14), with the coefficients $A(t)$ and $B(t)$ determined from eqs. (4.21) and (4.22) will be exact. The necessary calculations are

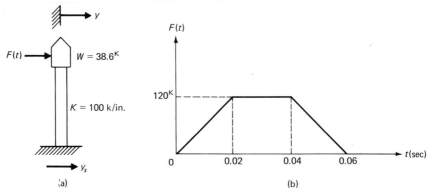

Fig. 4.7   Idealized structure and loading for Example 4.1.

presented in a convenient tabular format in Table 4.1 for a few time steps. The integrals in eqs. (4.18) and (4.19) are labeled ΔA and ΔB in this table, since

$$\Delta A = A(t_i) - A(t_{i-1}) = \int_{t_{i-1}}^{t_i} F(\tau) \cos \omega\tau \, d\tau$$

and

$$\Delta B = B(t_i) - B(t_{i-1}) = \int_{t_{i-1}}^{t_i} F(\tau) \sin \omega\tau \, d\tau.$$

Since the blast terminates at $t = 0.060$ sec, the values of A and B remain constant after this time. Consequently, the free vibration which follows is obtained by substituting these values of A and B evaluated at $t = 0.060$ sec into eq. (4.14), that is,

$$y(t) = (2571 \sin 31.62t - 3585 \cos 31.62t)/3162$$

TABLE 4.1   Numerical Calculation of the Response for Example 4.1.

t (sec)	$F(\tau)$	ωt	$\Delta A(t)$	$A(t)$	$\Delta B(t)$	$B(t)$	$y(t)$ (in)
0.000	0	0	0	0	0	0	0
0.020	120,000	0.6324	1082	1082	486	486	0.078
0.040	120,000	1.2649	1376	2458	1918	2404	0.512
0.060	0	1.8974	113	2571	1181	3585	1.134
0.080	0	2.5298	0	2571	0	3585	1.395
0.100	0	3.1623	0	2571	0	3585	1.117

or

$$y(t) = 0.8130 \sin 31.62t - 1.1338 \cos 31.62t$$

for $t \geqslant 0.060$ sec.

4.3   NUMERICAL EVALUATION OF DUHAMEL'S INTEGRAL—DAMPED SYSTEM

The response of a damped system expressed by the Duhamel's integral is obtained in a manner entirely equivalent to the undamped analysis except that the impulse $F(\tau) \, d\tau$ producing an initial velocity $dv = F(\tau) \, d\tau/m$ is substituted into the corresponding damped free-vibration equation. Setting $y_0 = 0$, $v_0 = F(\tau) \, d\tau/m$, and substituting t for $t - \tau$ in eq. (2.20), we obtain the differential displacement at time t as

$$dy(t) = e^{-\xi\omega(t-\tau)} \frac{F(\tau) \, d\tau}{m\omega_D} \sin \omega_D(t - \tau). \tag{4.23}$$

Summing these differential response terms over the entire loading interval results in

$$y(t) = \frac{1}{m\omega_D} \int_0^t F(\tau) \, e^{-\xi\omega(t-\tau)} \sin \omega_D(t - \tau) \, d\tau, \tag{4.24}$$

which is the response for a damped system in terms of the Duhamel's integral. For numerical evaluation, we proceed as in the undamped case and obtain from eq. (4.24)

$$y(t) = \{A_D(t) \sin \omega_D t - B_D(t) \cos \omega_D t\} \frac{e^{-\xi\omega t}}{m\omega_D}, \tag{4.25}$$

where

$$A_D(t_i) = A_D(t_{i-1}) + \int_{t_{i-1}}^{t_i} F(\tau) \, e^{\xi\omega\tau} \cos \omega_D\tau \, d\tau, \tag{4.26}$$

$$B_D(t_i) = B_D(t_{i-1}) + \int_{t_{i-1}}^{t_i} F(\tau) \, e^{\xi\omega\tau} \sin \omega_D\tau \, d\tau. \tag{4.27}$$

For a linear piecewise loading function, $F(\tau)$ given by eq. (4.20) is substituted into eqs. (4.26) and (4.27) which requires the evaluation of the following integrals:

$$I_1 = \int_{t_{i-1}}^{t_i} e^{\xi\omega\tau} \cos \omega_D\tau \, d\tau = \frac{e^{\xi\omega\tau}}{(\xi\omega)^2 + \omega_D^2} (\xi\omega \cos \omega_D\tau + \omega_D \sin \omega_D\tau) \Big|_{t_{i-1}}^{t_i} ,$$

$$(4.28)$$

$$I_2 = \int_{t_{i-1}}^{t_i} e^{\xi\omega\tau} \sin \omega_D\tau \, d\tau = \frac{e^{\xi\omega\tau}}{(\xi\omega)^2 + \omega_D^2} (\xi\omega \sin \omega_D\tau - \omega_D \cos \omega_D\tau) \Big|_{t_{i-1}}^{t_i} ,$$

$$(4.29)$$

$$I_3 = \int_{t_{i-1}}^{t_i} \tau e^{\xi\omega\tau} \sin \omega_D\tau \, d\tau = \tau - \frac{\xi\omega}{(\xi\omega)^2 + \omega_D^2} I_2 + \frac{\omega_D}{(\xi\omega)^2 + \omega_D^2} I_1 \Big|_{t_{i-1}}^{t_i} ,$$

$$(4.30)$$

$$I_4 = \int_{t_{i-1}}^{t_i} \tau e^{\xi\omega\tau} \cos \omega_D\tau \, d\tau = \left(\tau - \frac{\xi\omega}{(\xi\omega)^2 + \omega_D^2}\right) I_1 - \frac{\omega_D}{(\xi\omega)^2 + \omega_D^2} I_2 \Big|_{t_{i-1}}^{t_i} .$$

$$(4.31)$$

In terms of these integrals, $A_D(t_i)$ and $B_D(t_i)$ may be evaluated from

$$A_D(t_i) = A_D(t_{i-1}) + \left(F(t_{i-1}) - t_{i-1} \frac{\Delta F_i}{\Delta t_i}\right) I_1 + \frac{\Delta F_i}{\Delta t_i} I_4, \qquad (4.32)$$

$$B_D(t_i) = B_D(t_{i-1}) + \left(F(t_{i-1}) - t_{i-1} \frac{\Delta F_i}{\Delta t_i}\right) I_2 + \frac{\Delta F_i}{\Delta t_i} I_3. \qquad (4.33)$$

Finally, the substitution of eqs. (4.32) and (4.33) into eq. (4.25) gives the displacement at time t_i as

$$y(t_i) = \frac{e^{-\xi\omega t_i}}{m\omega_D} \{A_D(t_i) \sin \omega_D t_i - B_D(t_i) \cos \omega_D t_i\} \qquad (4.34)$$

4.4   PROGRAM 1—DYNAMIC RESPONSE USING DUHAMEL'S INTEGRAL (DUHAMEL)

The computer program described in this section calculates the response of the simple damped oscillator excited by a time-dependent external force or by an acceleration acting at the support. The excitation is assumed to be piecewise linear between defining points. The response consists of the displacement,

TABLE 4.2   Description of Input Variables for Program 1.

Variable	Symbol in Text	Description
NTYPE		Excitation index: 0 → force at mass, 1 → acceleration at support
N		Number of points defining the excitation function
M	m	Mass
K	k	Linear spring constant
XI	ξ	Damping ratio
TMAX		Maximum time of integration
DT	dt	Time step of integration
INT		Interpolation index: 0 → no interpolation, 1 → interpolate
GR	g	Acceleration of gravity (GR = 0 when NTYPE = 0)
T(I)	t_i	Time at point i
F(I)	$F(t_i)$	Force or acceleration at time t_i (acceleration in g's)

velocity, acceleration, and the support reaction. It is printed at specified time increments. The program is written in the Fortran computer language. This language is well known and its essentials are readily available to the reader. A short list of the principal input variables and symbols used in the program is given in Table 4.2. The corresponding algebraic symbols that were used in the equations are also given. Input data cards and corresponding formats are given in Table 4.3.

This program is relatively simple, containing one basic loop within which the time interval is incremented and the response is computed at each discrete time interval. The program permits the option of calling the short subroutine INTER which serves to calculate by linear interpolation the value of the forcing function at equal time intervals. The computer program is listed in the Appendix as Program 1–DUHAMEL. Comments are interspersed throughout the program to facilitate the users' understanding of what is being done at each segment of

TABLE 4.3   Input Data Symbols and Format for Program 1.

Format	Variables
(2I5, 5F10.3, I5, F10.0)	NTYPE N M K XI TMAX DT INT GR
(8F10.4)	T(1) F(I), (I = 1,N) (as many cards as needed)

TABLE 4.4   Input Data for Example 4.2

Data Listing								
0	4	100.	100000.	0.05	0.12	0.005	1	0.
0.	0.	0.020	120000.	0.04	120000.	0.06	0.0	

TABLE 4.5   Computer Output for Example 4.2.

RESPONSE OF SINGLE DEGREE SYSTEM USING DUHAMEL INTEGRAL

MASS	=	100.0000
SPRING CONSTANT	=	100000.00
NATURAL FREQUENCY	=	31.623 RADIANS/SEC
DAMPED FREQUENCY	=	31.583 RADIANS/SEC
DAMPING CONSTANT	=	316.228
RELATIVE DAMPING	=	0.0500

TIME	FORCE	DISPL.	VELOCITY	ACC.	SUP. REAC.
0.0000	0.00	0.000	0.000	0.00	0.00
0.0050	30000.00	0.001	0.745	296.40	266.26
0.0100	60000.00	0.010	2.944	580.82	1356.95
0.0150	90000.00	0.033	6.521	846.40	3889.82
0.0200	120000.00	0.077	11.366	1086.86	8515.61
0.0250	120000.00	0.147	16.594	1000.24	15634.93
0.0300	120000.00	0.242	21.329	890.23	25153.09
0.0350	120000.00	0.360	25.463	759.91	36847.58
0.0400	120000.00	0.496	28.901	612.82	50413.83
0.0450	90000.00	0.646	30.825	156.47	65336.24
0.0500	60000.00	0.800	30.471	−296.59	80601.41
0.0550	30000.00	0.947	27.883	−735.21	95112.86
0.0600	0.00	1.076	23.161	−1148.74	107799.47
0.0650	0.00	1.177	17.198	−1230.96	117782.91
0.0700	0.00	1.247	10.904	−1281.42	124741.06
0.0750	0.00	1.285	4.438	−1299.36	128540.42
0.0800	0.00	1.291	−2.035	−1284.87	129132.08
0.0850	0.00	1.265	−8.358	−1238.80	126550.22
0.0900	0.00	1.208	−14.374	−1162.79	120909.47
0.0950	0.00	1.122	−19.939	−1059.19	112401.18
0.1000	0.00	1.010	−24.924	−931.00	101288.64
0.1050	0.00	0.874	−29.214	−781.77	87902.56
0.1100	0.00	0.719	−32.714	−615.54	72639.51
0.1150	0.00	0.548	−35.349	−436.68	55973.74

MAX DISPL.	=	1.29
MAX VELOC.	=	−35.35
MAX ACC.	=	−1299.36
MAX SUP. FORCE	=	129132.08

TABLE 4.6   Input Data for Example 4.3.

1	24	100.00	100000.000	0.050	0.120	0.000	0	386
0.0000	0.0108	0.0420	0.0010	0.9700	0.0159	0.1610	−0.0001	
0.2210	0.0189	0.2630	0.0001	0.2910	0.0059	0.3320	−0.0012	
0.3740	0.0200	0.4290	−0.0237	0.4710	0.0076	0.5810	0.0425	
0.6230	0.0094	0.6650	0.0138	0.7200	−0.0088	0.7250	−0.0256	
0.7890	−0.0387	0.7940	−0.0568	0.8720	−0.0232	0.8770	−0.0343	
0.9410	−0.0402	0.9460	−0.0603	0.9970	−0.0789	1.0660	−0.0666	

the program. It should be noted that this program is independent of the system of units. Any consistent system of units may be used. For instance, pounds for force, seconds for time, inches for length. In this system mass should be given in units of weight divided by $g = 386.0$ in/sec^2, that is, units of lb · sec^2/in. In the SI unit system (Systéme International d'Unités), the units are mass in kilograms, time in seconds, length in meters, and force in newtons.

Illustrative Example 4.2. As the first application of the use of the program on Duhamel's integration, let us find the response of the tower of Example 4.1 assuming that damping is 5% of the critical damping. The required input data are listed in Table 4.4 and the output obtained using Program 1 is given in Table 4.5.

The displacements given in Table 4.5 for the 5% damped structure are, as expected, slightly lower than the displacements $y(t)$ calculated in Table 4.1 for the undamped structure.

Illustrative Example 4.3. As a second example in using Program 1, consider the tower of Examples 4.1 and 4.2 but now subjected to excitation at the foundation of the tower. For this example, the excitation data were obtained from the acceleration record of the first second of the well-documented El Centro earthquake of 1940. Assume that the damping in this system is 5% of the critical value. Input data for this example are listed in Table 4.6 and the computer output from Program 1 is given in Table 4.7. For this example, the control index for interpolation has been set equal to zero (INT = 0) indicating that interpolation between data points of the excitation is not desired. In this case data for the variables DT and TMAX are not needed by the program for processing; hence the corresponding spaces for these data could be left blank or set to zero as it has been done in Table 4.6.

4.5 SUMMARY

We have shown in this chapter that the differential equation of motion for a linear system can be solved in general for any forcing function in terms of

TABLE 4.7   Computer Results for Example 4.3.

RESPONSE OF SINGLE DEGREE SYSTEM USING DUHAMEL INTEGRAL

MASS	=	100.0000
SPRING CONSTANT	=	100000.00
NATURAL FREQUENCY	=	31.623 RADIANS/SEC
DAMPED FREQUENCY	=	31.583 RADIANS/SEC
DAMPING CONSTANT	=	316.228
RELATIVE DAMPING	=	0.0500

TIME	FORCE	DISPL.	VELOCITY	ACC.	SUP. REAC.
0.0000	−416.88	0.000	0.000	−4.17	0.00
0.0420	−38.60	−0.002	−0.054	1.84	250.68
0.9700	−613.74	−0.006	−0.012	−0.50	560.14
0.1610	3.86	0.002	0.003	−1.86	188.90
0.2210	−729.54	−0.004	−0.204	−2.79	391.12
0.2630	−3.86	−0.010	−0.010	10.32	1032.83
0.2910	−227.74	−0.007	0.207	4.23	719.53
0.3320	46.32	0.003	0.230	−3.11	293.85
0.3740	−772.00	0.006	−0.169	−12.77	561.04
0.4290	914.82	−0.007	−0.008	16.13	695.39
0.4710	−293.36	0.001	0.252	−5.12	159.80
0.5810	−1640.50	−0.022	−0.381	7.25	2247.92
0.6230	−362.84	−0.026	0.326	21.00	2567.79
0.6650	−532.68	−0.001	0.663	−6.75	220.24
0.7200	339.68	0.017	−0.088	−13.11	1678.35
0.7250	988.16	0.016	−0.136	−5.90	1621.19
0.7890	1493.82	0.007	0.003	8.09	683.60
0.7940	2192.48	0.007	0.060	14.75	698.25
0.8720	895.52	0.024	−0.057	−14.54	2367.90
0.8770	1323.98	0.023	−0.117	−9.62	2323.55
0.9410	1551.72	0.008	−0.158	8.12	791.56
0.9460	2327.58	0.007	−0.097	16.34	725.09
0.9970	3045.54	0.023	0.615	5.58	2300.83
1.0660	2570.76	0.046	−0.264	−19.01	4555.68

MAX DISPL.	=	0.05
MAX VELOC.	=	0.66
MAX ACC.	=	21.00
MAX SUP. FORCE	=	4555.68

Duhamel's integral. The numerical evaluation of this integral can be accomplished by any standard method such as the trapezoidal or the Simpson's rule. We have preferred the use of a numerical integration by simply assuming that the forcing function is a linear function between defining points and on this basis, we have obtained the exact response for each time increment. The computer program described in this chapter allows us to obtain the response in terms of displacement, velocity, and acceleration as a function of time for any single degree-of-freedom system of linear-elastic behavior when subjected to a general force function of time applied to the mass or to an acceleration applied to the support.

PROBLEMS

4.1   The steel frame shown in Fig. P4.1 is subjected to a horizontal force applied at the girder level. The force decreases linearly from 5 kip at time $t = 0$ to zero at $t = 0.6$ sec. Determine: (a) the horizontal deflection at $t = 0.5$ sec and (b) the maximum horizontal deflection. Assume the columns massless and the girder rigid. Neglect damping.

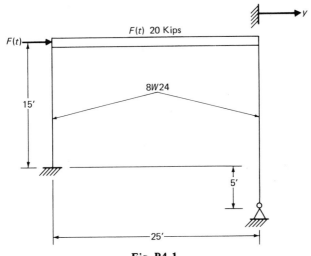

Fig. P4.1

4.2   Repeat Problem 4.1 for 10% of critical damping.
4.3   For the load-time function in Fig. P4.3, derive the expression for the dynamic load factor for the undamped simple oscillator as a function of t, ω, and t_d.

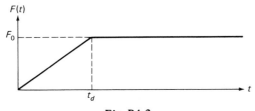

Fig. P4.3

4.4   The frame shown in Fig. P4.1 is subjected to a sudden acceleration of 0.5g applied to its foundation. Determine the maximum shear force in the columns. Neglect damping.

4.5   Repeat Problem 4.4 for 10% of critical damping.

4.6   For the dynamic system shown in Fig. P4.6, determine and plot the displacement as a function of time for the interval $0 \leqslant t \leqslant 0.5$ sec. Neglect damping.

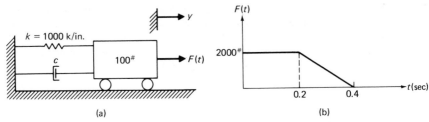

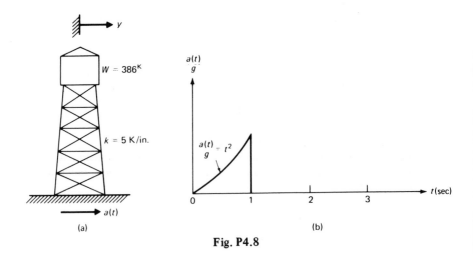

Fig. P4.6

4.7   Repeat Problem 4.6 for 10% of critical damping.

4.8   The tower of Fig. P4.8(a) is subjected to horizontal ground acceleration $a(t)$ shown in Fig. P4.8(b). Determine the relative displacement at the top of the tower at time $t = 1.0$ sec. Neglect damping.

Fig. P4.8

4.9   Repeat Problem 4.8 for 20% of critical damping.

4.10 Determine for the tower of Problem 4.9, the maximum displacement at the top of the tower relative to the ground displacement.

5

Fourier analysis and response in the frequency domain

This chapter presents the application of Fourier series to determine: (1) the response of a system to periodic forces, and (2) the response of a system to nonperiodic forces in the frequency domain as an alternate approach to the usual analysis in the time domain. In either case, the calculations require the evaluation of integrals which, except for some relatively simple loading functions, employ numerical methods for their computation. Thus, in general, to make practical use of the Fourier method, it is necessary to replace the integrations with finite sums.

5.1 FOURIER SERIES

The subject of Fourier series and Fourier analysis has extensive ramifications in its application to many fields of science and mathematics. We begin by considering single degree-of-freedom system under the action of a periodic loading, that is, a forcing function that repeats itself at equal intervals of time (the pe-

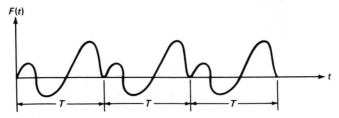

Fig. 5.1   Arbitrary periodic function.

riod of the function). Fourier has shown that a periodic function may be expressed as the summation of an infinite number of sine and cosine terms. Such a sum is known as a Fourier series.

For a periodic function, such as the one shown in Fig. 5.1, the Fourier series may be written as

$$F(t) = a_0 + a_1 \cos \overline{\omega}t + a_2 \cos 2\overline{\omega}t + a_3 \cos 3\overline{\omega}t + \ldots a_n \cos n\overline{\omega}t + \ldots$$

$$+ b_1 \sin \overline{\omega}t + b_2 \sin 2\overline{\omega}t + b_3 \sin 3\overline{\omega}t + \ldots b_n \sin n\overline{\omega}t + \ldots \quad (5.1)$$

or

$$F(t) = a_0 + \sum_{n=1}^{\infty} \{a_n \cos n\overline{\omega}t + b_n \sin n\overline{\omega}t\}, \quad (5.2)$$

where $\overline{\omega} = 2\pi/T$ is the frequency and T the period of the function. The evaluation of the coefficients a_0, a_n, and b_n to fit a given function $F(t)$ is determined from the following expressions:

$$a_0 = \frac{1}{T} \int_{t_1}^{t_1 + T} F(t)\, dt,$$

$$a_n = \frac{2}{T} \int_{t_1}^{t_1 + T} F(t) \cos n\overline{\omega}t\, dt,$$

$$b_n = \frac{2}{T} \int_{t_1}^{t_1 + T} F(t) \sin n\overline{\omega}t\, dt, \quad (5.3)$$

where t_1 in the limits of the integrals may be any value of time, but is usually equal to either $-T/2$ or zero. The constant a_0 represents the average of the periodic function $F(t)$.

5.2   RESPONSE TO A LOADING REPRESENTED BY FOURIER SERIES

The response of a single degree-of-freedom system to a periodic force represented by its Fourier series is found as a superposition of the response to each component of the series. When the transient is omitted, the response of an undamped system to any sine term of the series is given by eq. (3.9) as

$$y_n(t) = \frac{b_n/k}{1 - r_n^2} \sin n\bar{\omega}t. \tag{5.4}$$

Similarly, the response to any cosine term is

$$y_n(t) = \frac{a_n/k}{1 - r_n^2} \cos n\bar{\omega}t, \tag{5.5}$$

where $r_n = n\bar{\omega}/\omega$ and $\omega = \sqrt{k/m}$. The total response of an undamped single degree-of-freedom system may then be expressed as the superposition of the responses to all the force terms of the series, including the response a_0/k (steady-state response) to the constant force a_0. Hence we have

$$y(t) = \frac{a_0}{k} + \sum_{n=1}^{\infty} \frac{1}{1 - r_n^2} \left(\frac{a_n}{k} \cos n\bar{\omega}t + \frac{b_n}{k} \sin n\bar{\omega}t \right). \tag{5.6}$$

When the damping in the system is considered, the steady-state response for the general sine term of the series is given from eq. (3.20) as

$$y_n(t) = \frac{b_n/k \sin (n\bar{\omega}t - \theta)}{\sqrt{(1 - r_n^2)^2 + (2r_n\xi)^2}} \tag{5.7}$$

or

$$y_n(t) = \frac{b_n}{k} \cdot \frac{\sin n\bar{\omega}t \cos \theta - \cos n\bar{\omega}t \sin \theta}{\sqrt{(1 - r_n^2)^2 + (2r_n\xi)^2}}.$$

The substitution of $\sin \theta$ and $\cos \theta$ from eq. (3.21) gives

$$y_n(t) = \frac{b_n}{k} \cdot \frac{(1 - r_n^2) \sin n\bar{\omega}t - 2r_n\xi \cos n\bar{\omega}t}{(1 - r_n^2)^2 + (2r_n\xi)^2}. \tag{5.8}$$

Similarly for a cosine term of the series, we obtain

$$y_n(t) = \frac{a_n}{k} \cdot \frac{(1 - r_n^2) \cos n\bar{\omega}t + 2r_n\xi \sin n\bar{\omega}t}{(1 - r_n^2)^2 + (2r_n\xi)^2}. \tag{5.9}$$

Finally, the total response is then given by the superposition of the terms expressed by eqs. (5.8) and (5.9) in addition to the response to the constant term of the series. Therefore, the total response of a damped single degree-of-freedom system may be expressed as

$$y(t) = \frac{a_0}{k} + \frac{1}{k} \sum_{n=1}^{\infty} \left\{ \frac{a_n 2 r_n \xi + b_n(1 - r_n^2)}{(1 - r_n^2)^2 + (2 r_n \xi)^2} \sin n\bar{\omega}t \right.$$

$$\left. + \frac{a_n(1 - r_n^2) - b_n 2 r_n \xi}{(1 - r_n^2)^2 + (2 r_n \xi)^2} \cos n\bar{\omega}t \right\}. \quad (5.10)$$

Illustrative Example 5.1.   As an application of the use of Fourier series in determining the response of a system to a periodic loading, consider the undamped simple oscillator in Fig. 5.2(a) which is acted upon by the periodic force shown in Fig. 5.2(b). The first step is to determine the Fourier series expansion of $F(t)$. The corresponding coefficients are determined from eq. (5.3) as follows:

$$a_0 = \frac{1}{T} \int_0^T \frac{F_0}{T} t \, dt = \frac{F_0}{2},$$

$$a_n = \frac{2}{T} \int_0^T \frac{F_0}{T} t \cos n\bar{\omega}t \, dt = 0,$$

$$b_n = \frac{2}{T} \int_0^T \frac{F_0}{T} t \sin n\bar{\omega}t \, dt = -\frac{F_0}{n\pi}.$$

The response of the undamped system is then given from eq. (5.6) as

$$y(t) = \frac{F_0}{2k} - \sum_{n=1}^{\infty} \frac{F_0 \sin n\bar{\omega}t}{n\pi k(1 - r_n^2)}$$

or in expanded form as

$$y(t) = \frac{F_0}{2k} - \frac{F_0 \sin \bar{\omega}t}{\pi k(1 - r_1^2)} - \frac{F_0 \sin 2\bar{\omega}t}{2\pi k(1 - 4r_1^2)} - \frac{F_0 \sin 3\bar{\omega}t}{3\pi k(1 - 9r_1^2)} - \cdots$$

where $r_1 = \bar{\omega}/\omega$, $\omega = \sqrt{k/m}$, and $\bar{\omega} = 2\pi/T$.

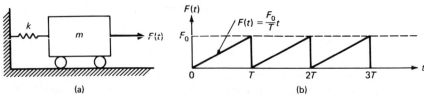

Fig. 5.2   Undamped oscillator acted upon by a periodic force.

5.3 FOURIER COEFFICIENTS FOR PIECEWISE LINEAR FUNCTIONS

As we proceeded before, in the evaluation of Duhamel's integral, we can represent the forcing function by a piecewise linear function as shown in Fig. 5.3. The calculation of Fourier coefficients, eq. (5.3), is then obtained as a summation of the integrals evaluated for each linear segment of the forcing function, that is,

$$a_0 = \frac{1}{T} \sum_{i=1}^{N} \int_{t_{i-1}}^{t_i} F(t)\, dt, \tag{5.11}$$

$$a_n = \frac{2}{T} \sum_{i=1}^{N} \int_{t_{i-1}}^{t_i} F(t) \cos n\bar{\omega} t\, dt, \tag{5.12}$$

$$b_n = \frac{2}{T} \sum_{i=1}^{N} \int_{t_{i-1}}^{t_i} F(t) \sin n\bar{\omega} t\, dt, \tag{5.13}$$

where N is the number of segments of the piecewise forcing function. The forcing function in any interval $t_{i-1} \leqslant t \leqslant t_i$ is expressed by eq. (4.20) as

$$F(t) = F(t_{i-1}) + \frac{\Delta F_i}{\Delta t_i} (t - t_{i-1}) \tag{5.14}$$

in which $\Delta F_i = F(t_i) - F(t_{i-1})$ and $\Delta t_i = t_i - t_{i-1}$. The integrals required in the expressions of a_n and b_n have been evaluated in eqs. (4.21) and (4.22) and designated as $A(t_i)$ and $B(t_i)$ in the recurrence expressions eqs. (4.18) and (4.19). The

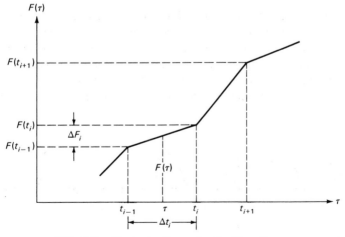

Fig. 5.3   Piecewise linear forcing function.

use of eqs. (4.18) through (4.22) to evaluate the coefficients a_n and b_n yields

$$a_n = \frac{2}{T} \sum_{i=1}^{N} \left\{ \frac{1}{n\bar{\omega}} (F(t_{i-1}) - t_{i-1} \frac{\Delta F_i}{\Delta t_i} (\sin n\bar{\omega}t_i - \sin n\bar{\omega}t_{i-1}) \right.$$

$$\left. + \frac{\Delta F_i}{n^2 \bar{\omega}^2 \Delta t_i} ((\cos n\bar{\omega}t_i - \cos n\bar{\omega}t_{i-1}) + n\bar{\omega}(t_i \sin n\bar{\omega}t_i - t_{i-1} \sin n\bar{\omega}t_{i-1})) \right\}$$

$$(5.15)$$

$$b_n = \frac{2}{T} \sum_{i=1}^{N} \left\{ \frac{1}{n\bar{\omega}} (F(t_{i-1}) - t_{i-1} \frac{\Delta F_i}{\Delta t_i} (\cos n\bar{\omega}t_i - \cos n\bar{\omega}t_{i-1}) \right.$$

$$\left. + \frac{\Delta F_i}{n^2 \omega^2 \Delta t_i} ((\sin n\bar{\omega}t_i - \sin n\bar{\omega}t_{i-1}) + n\bar{\omega}(t_i \cos n\bar{\omega}t_i - t_{i-1} \cos n\bar{\omega}t_{i-1})) \right\} .$$

$$(5.16)$$

The integral appearing in the coefficient a_0 is readily evaluated after substituting $F(t)$ from eq. (5.14) into eq. (5.11). This evaluation yields,

$$a_0 = \frac{1}{T} \sum_{i=1}^{n} \{\Delta t_i (F_i + F_{i-1})/2\}.$$

$$(5.17)$$

5.4   PROGRAM 2–DYNAMIC RESPONSE OF A SINGLE DEGREE-OF-FREEDOM SYSTEM USING FOURIER SERIES (FOURIER)

A computer program is presented in this section to calculate the response of a damped oscillator excited either by a periodic time-dependent external force applied to the mass or by a periodic time-dependent acceleration applied to the support. Any force, acting during a finite interval of time, may be assumed to be periodic by simply extending its interval to include a final portion in which the force remains zero as shown in Fig. 5.4. The force is assumed to be piecewise lin-

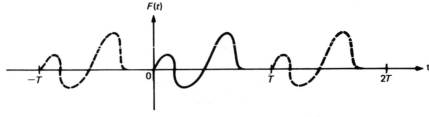

Fig. 5.4   Arbitrary loading represented by a periodic function.

TABLE 5.1   Description of Input Variables for Program 2.

Variable	Symbol in Text	Description
NTYPE		Excitation index: $0 \to$ force at mass, $1 \to$ acceleration at support
N		Number of points defining the excitation
NT		Number of terms in the Fourier series
DT	dt	Printing time interval
TP		Period of the excitation
TT		Total printing time
AK	k	Spring constant
AM	m	Mass
XSI	ξ	Damping ratio
INT		Interpolation index: $0 \to$ do not interpolate, $1 \to$ interpolate
NPRT		Printing index: $0 \to$ do not print history response, $1 \to$ print
T(I)	t_i	Time at point i
F(I)	$F(t_i)$	Excitation at time t_i

ear between defining points. The computer output consists of a listing of any desired number of Fourier coefficients (for the force and for the response) and of the response of the oscillator in terms of displacement, velocity, acceleration, and support reaction. The response is printed at specified equal time intervals.

The program is written in the Fortran language. A list of the principal input variables and symbols used in the program is given in Table 5.1. To assist the reader, corresponding algebraic symbols that were used in the text are also given. Input data and corresponding formats are given in Table 5.2. The listing of Program 2 is given in the Appendix where all the programs of the text are presented. This program contains two basic loops. The first loop computes the Fourier coefficients for the series expansion of the forcing function and the second loop computes the response of the oscillator to the harmonic excitation of each term of the series. The computer program comments are interspersed throughout the program to facilitate the understanding of what is being done in each segment. It

TABLE 5.2   Input Data Symbols and Format for Program 2.

Format	Variables
F (3I5, 6F10.2, 2I2)	NTYPE N NT DT TP TT AK AM XSI INT NPRT
F (8F10.2)	T(1) F(1) T(2) F(2) $\cdots$ T(N) F(N)

should be noted that this program may be used for any consistent system of units.

Illustrative Example 5.2.   As the first example of the use of the computer Program 2, consider the water tower shown in Fig. 5.5(a) subjected to the force history in Fig. 5.5(b). Determine the first 20 Fourier coefficients of the forcing function and of the steady-state displacement response of this structure. Assume 10% of the critical damping.

Data input for this problem expressed in units of pounds, inches, and seconds are given in Table 5.3.

The computer output for this problem giving the first 20 Fourier coefficients of both the applied force and the steady-state response is shown in Table 5.4.

TABLE 5.3   Data Input for Example 5.2.

Data Listing										
0	4	20	0.01	0.12	0.12	100000.	100.0	0.10	1	1
0.0		0.0	0.025		120000.	0.050	0.0	0.12		0.0

TABLE 5.4   Force and Response Fourier Coefficients for Example 5.2.

	FREQUENCY		FOURIER FORCE COEFF.		FOURIER REPS. COEFF.	
N	RAD/SEC	CPS	A(N)	B(N)	A(N)	B(N)
0	0	0	0.2400D 05		0.2400D 00	
1	52.36	8.33	0.1068D 05	0.3986D 05	-0.1012D 00	-0.2096D 00
2	104.72	16.67	-0.2189D 05	0.1264D 05	0.2102D-01	-0.1408D-01
3	157.08	25.00	-0.6485D 04	-0.6485D 04	0.2849D-02	0.2620D-02
4	209.44	33.33	0.2836D-06	-0.4913D-06	-0.6257D-13	0.1165D-12
5	261.80	41.67	-0.1594D 04	-0.4272D 03	0.2375D-03	0.5744D-04
6	314.16	50.00	0.1048D-11	-0.2911D-06	0.6057D-15	0.2979D-13
7	366.52	58.33	-0.8135D 03	0.2180D 03	0.6071D-04	-0.1740D-04
8	418.88	66.67	-0.1418D-06	-0.2456D-06	0.8341D-14	0.1395D-13
9	471.24	75.00	-0.7205D 03	0.7205D 03	0.3215D-04	-0.3303D-04
10	523.60	83.33	-0.8754D 03	-0.5054D 03	0.3227D-04	0.1811D-04
11	575.96	91.67	0.8827D 02	-0.3294D 03	-0.2559D-05	0.9989D-05
12	628.32	100.00	-0.9456D-22	-0.8769D-13	0.2247D-22	0.2227D-20
13	680.68	108.33	0.6320D 02	0.2359D 03	-0.1414D-05	-0.5089D-05
14	733.04	116.67	-0.4466D 03	0.2579D 03	0.8285D-05	-0.4879D-05
15	785.40	125.00	-0.2594D 03	-0.2594D 03	0.4245D-05	0.4178D-05
16	837.76	133.33	0.7091D-07	-0.1228D-06	-0.9985D-15	0.1760D-14
17	890.12	141.67	-0.1379D 03	-0.3696D 02	0.1746D-05	0.4546D-06
18	942.48	150.00	0.1156D-12	-0.9705D-07	0.7346D-17	0.1094D-14
19	994.84	158.33	-0.1104D 03	0.2959D 02	0.1115D-05	-0.3063D-06
20	1047.20	166.67	-0.5673D-07	-0.9826D-07	0.5232D-15	0.8937D-15

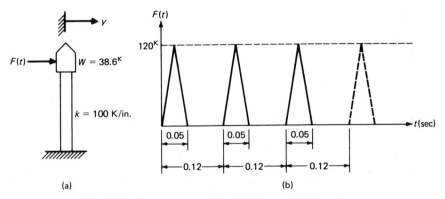

Fig. 5.5   Water tower subjected to a period impulse.

TABLE 5.5   Steady-State Response for Example 5.2.

TIME	FORCE	DISPL.	VELOC.	ACC.	FOUND. FORCE
0.0000	0.1504D 04	0.1631D 00	−0.1203D 02	−0.7190D 02	0.1799D 05
0.0100	0.4807D 05	0.4829D−01	−0.9950D 01	0.4954D 03	0.7932D 04
0.0200	0.9439D 05	−0.1771D−01	−0.2422D 01	0.9769D 03	0.2342D 04
0.0300	0.9439D 05	0.6697D−02	0.7161D 01	0.8919D 03	0.4578D 04
0.0400	0.4807D 05	0.1134D 00	0.1314D 02	0.2843D 03	0.1406D 05
0.0500	0.1504D 04	0.2485D 00	0.1287D 02	−0.3149D 03	0.2615D 05
0.0600	0.8103D 02	0.3591D 00	0.9098D 01	−0.4158D 03	0.3637D 05
0.0700	−0.7443D 02	0.4284D 00	0.4687D 01	−0.4587D 03	0.4294D 05
0.0800	0.2814D 02	0.4522D 00	0.9575D−01	−0.4525D 03	0.4522D 05
0.0900	0.2814D 02	0.4312D 00	−0.4225D 01	−0.4042D 03	0.4320D 05
0.1000	−0.7443D 02	0.3700D 00	−0.7874D 01	−0.3209D 03	0.3733D 05
0.1100	0.8103D 02	0.2770D 00	−0.1054D 02	−0.2095D 03	0.2849D 05
0.1200	0.1504D 04	0.1631D 00	−0.1203D 02	−0.7190D 02	0.1799D 05

The time history response in terms of displacements, velocity, acceleration, and foundation force is shown in Table 5.5. Note that the force given in Table 5.5 has been computed using Fourier series and serves an an indication of the precision of the calculations.

Illustrative Example 5.3.   As a second example illustrating the use of Program 2, consider the water tower structure of Example 5.2 shown in Fig. 5.5(a), subjected to the earthquake acceleration record shown in Fig. 5.6 applied to its foundation. Determine the following: (a) the first 20 Fourier coefficients of the applied effective force, $F_{eff} = -m\ddot{y}_s(t)$; (b) the first 20 Fourier coefficients of the steady-state relative displacement motion of the mass; and (c) the time history response (relative displacement, relative velocity, relative acceleration, and force transmitted to the foundation).

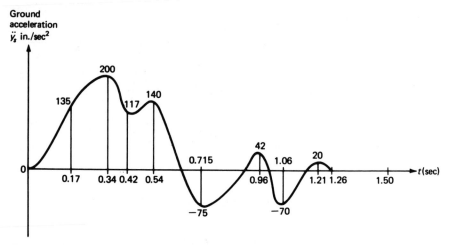

Fig. 5.6 Acceleration history applied to the foundation of the water tower in Fig. 5.5.

Figure 5.7 shows the damped oscillator used to model this system. The equation of motion obtained by equating to zero the sum of the forces in the free body diagram of Fig. 5.7(b) is

$$m\ddot{y} + c(\dot{y} - \dot{y}_s) + k(y - y_s) = 0. \tag{a}$$

In terms of the relative motion

$$u = y - y_s,$$
$$\dot{u} = \dot{y} - \dot{y}_s,$$
$$\ddot{u} = \ddot{y} - \ddot{y}_s.$$

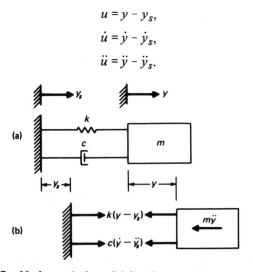

Fig. 5.7 Mathematical model for the water tower structure.

TABLE 5.6   Input Data for Example 5.3.

Data Listing									
1	10	20	0.05	1.5	1.5	100000.	100	0.1	0	1
0.	0.	0.17	135.		0.34	200.		0.42	117.
0.54	140.	0.71	−75.		1.06	−70.		1.21	20.
1.26	0.	1.5	0.						

Equation (a) becomes

$$m\ddot{u} + c\dot{u} + ku = F_{eff}(t) \tag{b}$$

in which $F_{eff}(t) = -m\ddot{y}_s(t)$ is defined as the effective force. With the excitation index, NTYPE set equal to 1, the program considers $-m\ddot{y}_s(t)$ to be the effective force. The response is then calculated in terms of the relative motion. A listing of the input data for this example is shown in Table 5.6.

Table 5.7 gives the computer calculated values of the first 20 Fourier coeffi-

TABLE 5.7   Force and Response Fourier Coefficients for Example 5.3.

OUTPUT RESULTS

N	FREQUENCY RAD/SEC	CPS	FOURIER FORCE COEFF. A(N)	B(N)	FOURIER REPS. COEFF. A(N)	B(N)
0	0	0	−0.2997D 04		−0.2997D−01	
1	4.19	0.67	−0.4406D 04	−0.1018D 05	−0.4202D−01	−0.1048D 00
2	8.38	1.33	0.4781D 04	0.6358D 03	0.5087D−01	0.9736D−02
3	12.57	2.00	0.6648D 03	−0.2678D 03	0.8122D−02	−0.2413D−02
4	16.76	2.67	0.1486D 04	−0.4857D 03	0.2119D−01	−0.3630D−02
5	20.94	3.33	−0.1066D 04	0.6751D 03	−0.2068D−01	0.7147D−02
6	25.13	4.00	0.2416D 03	−0.5519D 03	0.1098D−01	−0.1025D−01
7	29.32	4.67	0.8509D 03	0.3905D 03	0.8680D−02	0.3932D−01
8	33.51	5.33	−0.2890D 02	0.1606D 03	−0.5077D−02	−0.4309D−02
9	37.70	6.00	−0.2874D 03	−0.1341D 03	0.6532D−02	−0.5131D−03
10	41.89	6.67	0.1056D 03	−0.2459D 03	−0.2269D−03	0.3339D−02
11	46.08	7.33	0.1021D 03	0.1413D 03	−0.1157D−02	−0.9575D−03
12	50.27	8.00	0.4274D 02	0.3802D 02	−0.3180D−03	−0.1828D−03
13	54.45	8.67	0.8992D 02	0.9365D 02	−0.5249D−03	−0.3845D−03
14	58.64	9.33	0.5124D 00	−0.1799D 03	0.1076D−03	0.7211D−03
15	62.83	10.00	0.1032D 03	−0.1842D 02	−0.3356D−03	0.1077D−03
16	67.02	10.67	−0.6642D 02	0.1093D 03	0.1500D−03	−0.3313D−03
17	71.21	11.33	0.1430D 02	0.2302D 02	−0.4088D−04	−0.5203D−04
18	75.40	12.00	0.3967D 02	−0.2345D 02	−0.7877D−04	0.5807D−04
19	79.59	12.67	0.2851D 02	−0.2229D 01	−0.5258D−04	0.9141D−05
20	83.78	13.33	−0.1184D 02	−0.5023D 02	0.2681D−04	0.8111D−04

cients for the effective force and for the response which is obtained in terms of the relative displacement between the mass and the support.

The time history response also given in the output of Program 2 is shown in Table 5.8. The response is obtained by the superposition of the steady-state solution of the equation of motion for each Fourier component of the effective force. Consequently, the response does not include the contribution of the transient. The response is based on the assumption that the excitation is a periodic function.

TABLE 5.8   History Response for Example 5.3.

STEADY-STATE RESPONSE

TIME	FORCE	DISPL.	VELOC.	ACC.	FOUND. FORCE
0.0000	-0.3125D 03	0.6177D-02	0.5337D 00	-0.1268D 02	0.7039D 03
0.0500	-0.3973D 04	0.1340D-02	-0.9297D 00	-0.3519D 02	0.6030D 03
0.1000	-0.7914D 04	-0.7530D-01	-0.1764D 01	0.7313D 01	0.7613D 04
0.1500	-0.1198D 05	-0.1403D 00	-0.6898D 00	0.2482D 02	0.1404D 05
0.2000	-0.1466D 05	-0.1514D 00	0.7156D-01	0.4299D 01	0.1514D 05
0.2500	-0.1658D 05	-0.1515D 00	-0.2276D 00	-0.1283D 02	0.1515D 05
0.3000	-0.1853D 05	-0.1769D 00	-0.7068D 00	-0.3953D 01	0.1769D 05
0.3500	-0.1899D 05	-0.2097D 00.	-0.3921D 00	0.2222D 02	0.2097D 05
0.4000	-0.1349D 05	-0.1855D 00	0.1545D 01	0.4079D 02	0.1857D 05
0.4500	-0.1222D 05	-0.8753D-01	0.1613D 01	-0.4484D 02	0.8813D 04
0.5000	-0.1329D 05	-0.7893D-01	-0.1267D 01	-0.4593D 02	0.7934D 04
0.5500	-0.1273D 05	-0.1656D 00	-0.1405D 01	0.4721D 02	0.1658D 05
0.6000	-0.6283D 04	-0.1482D 00	0.2317D 01	0.7068D 02	0.1489D 05
0.6500	-0.1805D 03	0.1355D-01	0.3219D 01	-0.3571D 02	0.2445D 04
0.7000	0.6175D 04	0.1097D 00	0.5128D 00	-0.5122D 02	0.1098D 05
0.7500	0.7499D 04	0.8914D-01	-0.9722D 00	-0.7995D 01	0.8935D 04
0.8000	0.7440D 04	0.4848D-01	-0.3370D 00	0.2805D 02	0.4852D 04
0.8500	0.7232D 04	0.6256D-01	0.6943D 00	0.5368D 01	0.6272D 04
0.9000	0.7235D 04	0.9081D-01	0.2202D 00	-0.1985D 02	0.9082D 04
0.9500	0.7208D 04	0.7925D-01	-0.5383D 00	-0.3767D 01	0.7933D 04
1.0000	0.7024D 04	0.5756D-01	-0.1825D 00	0.1383D 02	0.5757D 04
1.0500	0.6983D 04	0.6478D-01	0.3795D 00	0.2651D 01	0.6482D 04
1.1000	0.4645D 04	0.7386D-01	-0.2659D 00	-0.2573D 02	0.7388D 04
1.1500	0.1661D 04	0.3055D-01	-0.1289D 01	-0.5790D 01	0.3162D 04
1.2000	-0.1439D 04	-0.2785D-01	-0.8484D 00	0.1882D 02	0.2836D 04
1.2500	-0.3826D 03	-0.3875D-01	0.5415D 00	0.3150D 02	0.3890D 04
1.3000	-0.5724D 02	0.1104D-01	0.1006D 01	-0.1797D 02	0.1274D 04
1.3500	0.1818D 02	0.2814D-01	-0.3791D 00	-0.2557D 02	0.2825D 04
1.4000	0.5702D 02	-0.8202D-02	-0.7332D 00	0.1341D 02	0.9422D 03
1.4500	-0.6482D 02	-0.2043D-01	0.2749D 00	0.1805D 02	0.2051D 04
1.5000	-0.3125D 03	0.6177D-02	0.5337D 00	-0.1268D 02	0.7039D 03

5.5 EXPONENTIAL FORM OF FOURIER SERIES AND DISCRETE FOURIER INTEGRAL

The Fourier series expression, eq. (5.2), may also be written in exponential form by substituting the trigonometric functions in terms as given by Euler's relations

$$\sin n\bar{\omega} = \frac{e^{in\bar{\omega}} - e^{-in\bar{\omega}}}{2i},$$

$$\cos n\bar{\omega} = \frac{e^{in\bar{\omega}} + e^{-in\bar{\omega}}}{2}. \tag{5.18}$$

The result of this substitution may be written as

$$F(t) = \sum_{n=-\infty}^{\infty} c_n e^{in\bar{\omega}t},$$

where

$$c_n = \frac{1}{T} \int_{t_1}^{t_1+T} F(t) e^{-in\bar{\omega}t} \, dt. \tag{5.19}$$

In a slightly different notation, substituting $C_n = Tc_n$ and selecting the interval from zero to T for the periodic function, we obtain

$$F(t) = \frac{1}{T} \sum_{n=-\infty}^{\infty} C_n e^{in\bar{\omega}t}, \tag{5.20}$$

$$C_n = \int_0^T F(t) e^{-in\bar{\omega}t} \, dt. \tag{5.21}$$

It should be noted that the exponential form for the Fourier series in eq. (5.20) has the advantage of simplicity when compared to the equivalent trigonometric series, eq. (5.2). The exponential form of the Fourier series can be used as before to determine the dynamic response of structural systems.

In extending the Fourier method to nonperiodic loadings, the first step is to assume that the loading is periodic with period T. This assumption introduces an approximation in the analysis. To minimize errors in the analysis of nonperiodic loads, the load period may be extended with the inclusion of a large interval of zero load in the period T; the resulting load history would thus appear similar to that shown in Fig. 5.4. The specification of the load period T also serves to define the lowest frequency that may be considered in the analysis, namely

$$\bar{\omega} = \Delta\omega = \frac{2\pi}{T}. \tag{5.22}$$

The period is then divided into N equivalent time increments $\Delta t = T/N$ and the load is defined for discrete times $t_j = j\Delta t$ where $j = 1, 2, 3 \ldots$. With the use of these relations, the exponential factor in eq. (5.20) may be written as

$$e^{in\bar{\omega}t} = e^{in\Delta\omega j\Delta t} = e^{2\pi i(nj/N)}. \tag{5.23}$$

Consequently, eq. (5.20) takes the form[1]

$$F(t_j) = \frac{\Delta\omega}{2\pi} \sum_{n=0}^{N-1} C_n e^{2\pi i(nj/N)} \tag{5.24}$$

in which the highest frequency to be considered has been arbitrarily set at $(N-1)$ $\Delta\omega$. The amplitude of the coefficient C_n in discrete form can be obtained by merely writing the sum of a finite series of discrete terms for the integral in eq. (5.21), with the following result

$$C_n = \Delta t \sum_{j=0}^{N-1} F(t_j) e^{-2\pi i(nj/N)}. \tag{5.25}$$

Equations (5.24) and (5.25) are the discrete forms of the Fourier transform. We can see that the loading in eq. (5.24) has been expressed as a summation of harmonic forces each of frequency $\omega_n = n\Delta\omega$ and amplitude

$$A_n = \frac{\Delta\omega}{2\pi} C_n, \tag{5.26}$$

where C_n is a function of $\omega_n = n\bar{\omega}$ and is given by eq. (5.21) or eq. (5.25). Having represented an arbitrary loading by a discrete exponential Fourier series, we may then also write the response to harmonic loading in exponential form. Again, only the steady-state response will be considered. The introduction of the unit exponential forcing function $E_n = e^{i\omega_n t}$ into the equation of motion, eq. (3.13), leads to

$$m\ddot{y} + c\dot{y} + ky = e^{i\omega_n t} \tag{5.27}$$

which has a steady-state solution of the form

$$y(t) = H(\omega_n) e^{i\omega_n t}. \tag{5.28}$$

When eq. (5.28) is introduced into eq. (5.27), it is found that the function $H(\omega_n)$ which will be designated as the *complex frequency response function* takes the form

$$H(\omega_n) = \frac{1}{k - m\omega_n^2 + ic\omega_n}. \tag{5.29}$$

[1]For a mathematical derivation of the discrete Fourier series, the reader may consult G. Dahlquist and A. Bjork, *Numerical Methods*, Printice-Hall, Englewood Cliffs, N.J., 1974.

Upon introducing the frequency ratio

$$r_n = \frac{\omega_n}{\omega}$$

and the damping ratio

$$\xi = \frac{c}{c_{cr}} = \frac{c}{2\sqrt{Km}} \, ,$$

eq. (5.29) becomes

$$H(\omega_n) = \frac{1}{k(1 - r_n^2 + 2ir_n\xi)} \tag{5.30}$$

or upon substituting $\omega_n = n\Delta\omega$, $r_1 = \overline{\omega}/\omega = \Delta\omega/\omega$,

$$H(n\Delta\omega) = \frac{1}{k(1 - n^2 r_1^2 + 2inr_1\xi)} \, . \tag{5.31}$$

Therefore, the response at time $t_j = j\Delta t$ to a harmonic force of amplitude $A_n = \Delta\omega C_n/2\pi$, as indicated in eq. (5.24), is given by

$$y_n(t_j) = \frac{\Delta\omega}{2\pi} \cdot \frac{C_n\, e^{2\pi i\,(nj/N)}}{k(1 - n^2 r_1^2 + 2inr_1\xi)}$$

and the total response due to the N component frequencies by

$$y(t_j) = \frac{\Delta\omega}{2\pi k} \sum_{n=0}^{N-1} \frac{C_n\, e^{2\pi i\,(nj/N)}}{1 - n^2 r_1^2 + 2inr_1\xi} \tag{5.32}$$

where C_n is expressed in discrete form by eq. (5.25). The evaluation of the sums necessary to determine the response using the discrete Fourier transform is greatly simplified by the fact that the exponential functions involved are harmonic and extend over a range of N^2. Only discrete values of j and n are introduced into these sums.

5.6   FAST FOURIER TRANSFORM

A numerical technique is available which is efficient for computer determination of the response in the frequency domain. The algorithm of this method is known as the fast Fourier transform (FFT).[2] The corresponding computer program is reproduced as a subroutine of computer Program 3 which is listed in the Appendix. The response in the frequency domain of a single degree-of-freedom system

[2]Cooley, J. W., Lewis, P. A. W., and Welch, P. D., *IEEE Trans. Education*, Vol. E-12, No. 1, pp. 27–34, March 1969.

to a general force is given by eq. (5.32) and the coefficients required are computed from eq. (5.25). It can be seen that either eq. (5.32) or eq. (5.25) may be represented, except for sign, by the exponential function

$$A(j) = \sum_{n=0}^{N-1} A^{(0)}(n) \, W_N^{jn},$$ (5.33)

where

$$W_N = e^{2\pi i/N}.$$ (5.34)

The evaluation of the sum will be most efficient if the number of time increments N into which the period T is divided is a power of 2, that is,

$$N = 2^M,$$ (5.35)

where M is an integer. In this case, the integers j and n can be expressed in binary form. For the purpose of illustration, we will consider a very simple case where the load period is divided into only eight time increments, that is, $N = 8, M = 3$. In this case, the indices will have the binary representation

$$j = j_0 + 2j_1 + 4j_2,$$

$$n = n_0 + 2n_1 + 4n_2,$$ (5.26)

and eq. (5.33) may be written as

$$A(j) = \sum_{n_2=0}^{1} \sum_{n_1=0}^{1} \sum_{n_0=0}^{1} A^{(0)}(n) \, W_8^{(j_0 + 2j_1 + 4j_2)(n_0 + 2n_1 + 4n_2)}.$$ (5.37)

The exponential factor can be written as

$$W_8^{jn} = W_8^{8(j_1 n_2 + 2j_2 n_2 + j_2 n_1)} W_8^{4n_2 j_0} W_8^{2n_1(2j_1 + j_0)} W_8^{n_0(4j_2 + 2j_1 + j_0)}.$$

We note that the first factor on the right-hand side is unity since from eq. (5.34)

$$W_8^{8I} = e^{2\pi i (8/8)I} = \cos 2\pi I + i \sin 2\pi I = 1,$$

where $I = j_1 n_2 + 2j_2 n_2 + j_2 n_1$ is an integer. Therefore, only the remaining three factors need to be considered in the summations. These summations may be performed conveniently in sequence by introducing a new notation to indicate the successive steps in the summation process. Thus the first step can be indicated by

$$A^{(1)}(j_0, n_1, n_0) = \sum_{n_2=0}^{1} A^{(0)}(n_2, n_1, n_0) \, W_8^{4n_2 j_0},$$

where $A^{(0)}(n_2, n_1, n_0) = A^{(0)}(n)$ in eq. (5.33). Similarly, the second step is

$$A^{(2)}(j_0, j_1, n_0) = \sum_{n_1=0}^{1} A^{(1)}(j_0, n_1, n_0) \, W_8^{2n_1(2n_1 + j_0)},$$

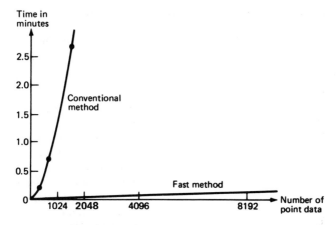

Fig. 5.8   Time required for Fourier transform using conventional and fast method (from Cooley, J. W., Lewis, P. A. W., and Welch, P. D., *IEEE Transactions and Education*, Vol. E-12, No. 1, March 1969).

and the third step (final step for $M = 3$) is

$$A^{(3)}(j_0, j_1, j_2) = \sum_{n_0 = 0}^{1} A^{(2)}(j_0, j_1, n_0) W_8^{n_0(4j_2 + 2j_1 + j_0)}.$$

The final result $A^{(3)}(j_0, j_1, j_2)$ is equal to $A(j)$ in eq. (5.33). This process, indicated for $N = 8$, can readily be extended to any integer $N = 2^M$. The method is particularly efficient because the results of one step are immediately used in the next step, thus reducing storage requirements and also because the exponential takes the value of unity in the first factor of the summation. The reduction in computational time which results from this formulation is significant when the time interval is divided into a large number of increments. The comparative times required for computing the Fourier series by a conventional program and by the fast Fourier transform algorithm are illustrated in Fig. 5.8. It is seen here how, for large values of N, one can rapidly consume so much computer time as to make the conventional method unfeasible.

5.7   PROGRAM 3—RESPONSE OF SINGLE DEGREE-OF-FREEDOM SYSTEM IN THE FREQUENCY DOMAIN (FREQRESP)

The computer program described in this section calculates the response in the frequency domain for a damped single degree-of-freedom system. The exciting force is given as a discrete function of time. When the control index INT is set equal to 1, interpolation between data points at a specified constant time incre-

TABLE 5.9   Description of Input Variables for Program 3.

Variable	Symbol in Text	Description
M	M	Exponent in $N = 2^M$
T	T	Period
AK	k	Stiffness coefficient
C	c	Damping coefficient
AM	m	Mass
INT		Interpolation index: $0 \to$ no interpolation, $1 \to$ interpolate
NEQ		Number of points defining the force function
TC(I)	t_i	Time at point i
P(I)	F_i	Force at point i

TABLE 5.10   Input Data Symbols and Format for Program 3.

Format	Variables
(I10, 4F10.2, 2I5)	M T AK C AM INT NEQ
(8F10.2)	TC(1) P(1) TC(2) P(2) $\cdots$ TC(NEQ) P(NEQ)

ment Δt will be performed. The computer output consists of two tables: (1) a table giving the first N complex Fourier coefficients of the expansion of the exciting force, and (2) a table giving the displacement history of the steady-state motion of the response. This second table also gives the forcing function as calculated from eq. (5.24), thus providing a check of the computations. The program is written in the Fortran language. A list of the input variables and symbols used in the program is given in Table 5.9. The corresponding algebraic symbols that were used in the text are also given. Input data and corresponding formats are given in Table 5.10. The main body of this program performs the task of forming the coefficients $A^{(0)}(n)$ of eq. (5.33) and of calling the subroutine FET which uses the fast Fourier transform algorithm to calculate the summation indicated in eq. (5.33). This subroutine is called three times by the main program corresponding to the calculations of the coefficients C_n in eq. (5.25), the response $y(t_j)$ in eq. (5.32), and the forcing function $F(t_j)$ in eq. (5.24).

Illustrative Example 5.4.   To present an example illustrating the use of computer Program 3, consider again the tower structure of the previous examples but now subjected to an impulsive triangular force as shown in Fig. 5.9. As indicated in this figure, the assumed period of excitation force is extended to 0.64 sec which is well beyond the time when the impulse has terminated. The input data for this example are given in Table 5.11.

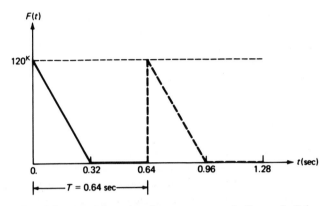

Fig. 5.9   Load function for water tower in Example 5.4.

TABLE 5.11   Input Data for Example 5.4.

Data Listing						
3	0.64	100000.	632.	100.	1	3
0.	120000.	0.32	0.0	0.64		0.0

TABLE 5.12   Output Results for Example 5.4.

FORCE FOURIER COEFFICIENTS

N	REAL	IMAG
0	24000.00	0.00
1	12994.09	-11588.22
2	4800.00	-4799.89
3	6205.89	-1988.22
4	4800.00	0.00
5	6205.88	1988.23
6	4800.00	4799.98
7	12994.08	11588.21

TIME	DISPL. REAL	DISPL. IMAG.	FORCE REAL	FORCE IMAG.
0.000	0.5652	-0.9596	119999.80	0.00
0.080	1.0465	0.6276	89999.81	-0.00
0.160	-0.0363	0.1845	59999.97	0.00
0.240	0.7095	0.1200	30000.08	0.02
0.320	0.0928	0.5015	0.09	0.00
0.400	0.1458	-0.2852	0.03	0.01
0.480	0.4208	0.0610	0.04	-0.00
0.560	0.0557	-0.2529	0.04	-0.02

The output for this example consists of the first eight complex Fourier coefficients of the excitation function and of the time history response in terms of the displacement. As a check of the calculations, the output also gives the value of the excitation force obtained by means of the inverse Fourier transform. The computer output for this example is shown in Table 5.12.

5.8   SUMMARY

In general any periodic function may be expanded into a Fourier series, eq. (5.1), whose terms are sine and cosine functions of successive multiples of the fundamental frequency. The coefficients of these functions may be calculated by integrating over a period the product of the periodic function multiplied by a sine or cosine function, eq. (5.3). The response of the dynamic system is then obtained as the superposition of the response for each term of the Fourier series expansion of the excitation function. The extension of the Fourier series to nonperiodic functions results in integrals which are known as Fourier transforms. The discrete form of these transforms, eqs. (5.24) and (5.25), permits their use in numerical applications. An extremely efficient algorithm known as the fast Fourier transform (FFT) can save as much as 99% of the computer time otherwise consumed in the evaluation of Fourier complex coefficients for the excitation function and for the response of a dynamic system.

PROBLEMS

5.1   Determine the first three terms of the Fourier series expansion for the time-varying force shown in Fig. P5.1.

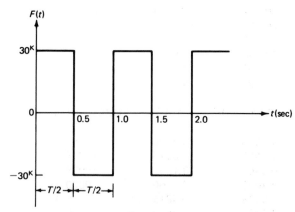

Fig. P5.1

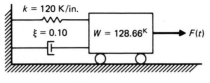

Fig. P5.2

5.2 Determine the steady-state response for the damped spring-mass system shown in Fig. P5.2 that is acted upon by the forcing function of Problem 5.1.

5.3 The spring-mass system of Fig. P5.2 is acted upon by the time-varying force shown in Fig. P5.3. Assume that the force is periodic of period $T = 1$ sec and determine the steady-state response of the system by applying Fourier series expansion of $F(t)$.

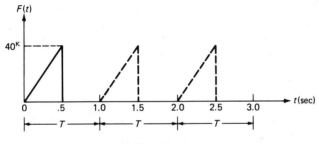

Fig. P5.3

5.4 The cantilever beam shown in Fig. P5.4(a) carries a concentrated weight at its free end and its subjected to a periodic acceleration at its support which is the rectified sine function of period $T = 0.4$ sec and amplitude $y_0 = 180$ in/sec^2. Determine: (1) the Fourier series expansion of the forcing function, and (2) the steady-state response considering only three terms of the series. Neglect damping in the system and assume the beam massless.

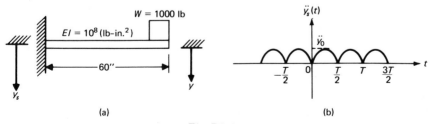

(a)   (b)

Fig. P5.4

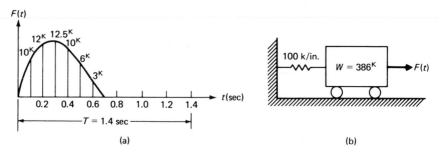

Fig. P5.5

5.5   The forcing function shown in Fig. P5.5(a) is assumed to be periodic in the extended interval $T = 1.4$ sec. Use Program 2 to determine the first ten Fourier coefficients and the steady-state response of a structure modeled by the undamped oscillator shown in Fig. P5.5(b).

5.6   Use Program 3 to determine: (1) the Fourier series expansion of the forcing function shown in Fig. P5.6(a), and (2) the steady-state response calculated in the frequency domain for the spring-mass system in Fig. P5.6(b). Assume 15% of the critical damping.

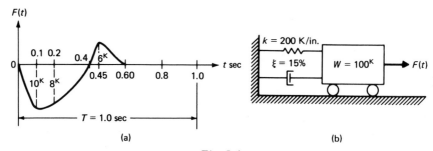

Fig. 5.6

6

Generalized coordinates for single degree-of-freedom system

In the preceding chapters we concentrated our efforts in obtaining the response to dynamic loads of structures modeled by the simple oscillator, that is, structures which may be analyzed as a damped or undamped spring-mass system. Our plan in the present chapter is to discuss the conditions under which a structural system consisting of multiple interconnected rigid bodies or having distributed mass and elasticity can still be modeled as a one-degree-of-freedom system. We begin by presenting an alternative method to the direct application of Newton's Law of Motion, the *principle of virtual work*.

6.1 PRINCIPLE OF VIRTUAL WORK

An alternative approach to the direct method employed thus far for the formulation of the equations of motion is the use of the principle of virtual work. This principle is particularly useful for relatively complex structural systems which contain many interconnected parts. The principle of virtual work was originally stated for a system in equilibrium. Nevertheless, the principle can be readily applied to dynamic systems by the simple recourse to D'Alembert's Principle which establishes dynamic equilibrium by the inclusion of the inertial forces in the system.

The principle of virtual work may be stated as follows: For a system that is in equilibrium, the work done by all the forces during an assumed displacement (virtual displacement) which is compatible with the system constraints, is equal to zero. In general, the equations of motion are obtained by introducing virtual displacements corresponding to each degree of freedom and equating the resulting work done to zero.

To illustrate the application of the principle of virtual work to obtain the equation of motion for a single degree-of-freedom system, let us consider the damped oscillator shown in Fig. 6.1(a) and its corresponding free body diagram in Fig. 6.1(b). Since the inertial force has been included among the external forces, the system is in "equilibrium" (dynamic equilibrium). Consequently, the principle of virtual work is applicable. If a virtual displacement δy is assumed to have taken place, the total work done by the forces shown in Fig. 6.1(b) is equal to zero, that is,

$$m\ddot{y}\delta y + c\dot{y}\delta y + ky\delta y - F(t)\,\delta y = 0$$

or

$$\{m\ddot{y} + c\dot{y} + ky - F(t)\}\,\delta y = 0. \qquad (6.1)$$

Since δy is arbitrarily selected as not equal to zero, the other factor in eq. (6.1) must equal zero. Hence,

$$m\ddot{y} + c\dot{y} + ky - F(t) = 0. \qquad (6.2)$$

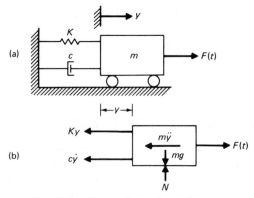

Fig. 6.1   Damped simple oscillator.

Thus we obtained in eq. (6.2) the differential equation for the motion of the damped oscillator.

6.2 GENERALIZED SINGLE DEGREE-OF-FREEDOM SYSTEM—RIGID BODY

Most frequently the configuration of a dynamic system is specified by coordinates indicating the linear or angular positions of elements of the system. However, coordinates do not necessarily have to correspond directly to displacements; they may in general be any independent quantities which are sufficient in number to specify the position of all parts of the system. These coordinates are usually called *generalized coordinates* and their number is equal to the number of degrees of freedom of the system.

The example of the rigid-body system shown in Fig. 6.2 consists of a rigid bar with distributed mass supporting a circular plate at one end. The bar is supported by springs and dampers in addition to a single frictionless support. Dynamic excitation is provided by a transverse load $F(x, t)$ varying linearly on the portion AB of the bar. Our purpose is to obtain the differential equation of motion and to identify the corresponding expressions for the parameters of the simple oscillator representing this system.

Since the bar is rigid, the system in Fig. 6.2 has only one degree of freedom, and, therefore, its dynamic response can be expressed with one equation of motion. The generalized coordinate could be selected as the vertical displacement of any point such as A, B, or C along the bar, or may be taken as the angular position of the bar. This last coordinate designated by $\theta(t)$ is selected as the generalized coordinate of the system. The corresponding free body diagram showing all the forces including the inertial forces and the inertial moments is shown in Fig. 6.3. In evaluating the displacements of the different forces, it is assumed that the displacements of the system are small and, therefore, vertical displacements are simply equal to the product of the distance to support D multiplied by the angular displacement $\theta = \theta(t)$.

The displacements resulting at the points of application of the forces in Fig.

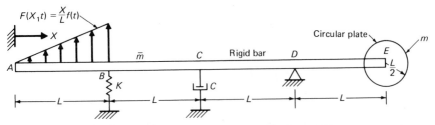

Fig. 6.2   Example of single degree-of-freedom rigid system.

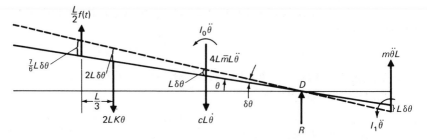

Fig. 6.3   Displacements and resultant forces for system in Fig. 6.2.

6.3 due to a virtual displacement $\delta\theta$ are indicated in this figure. By the principle of virtual work, the total work done by the forces during this virtual displacement is equal to zero. Hence

$$\delta\theta [I_0\ddot\theta + I_1\ddot\theta + 4L^3\overline{m}\ddot\theta + mL^2\ddot\theta + cL^2\dot\theta + 4kL^2\theta - \tfrac{7}{6}L^2 f(t)] = 0$$

or, since $\delta\theta$ is arbitrarily set not equal to zero, it follows that

$$(I_0 + I_1 + 4L^3\overline{m} + mL^2)\ddot\theta + cL^2\dot\theta + 4kL^2\theta - \tfrac{7}{6}L^2 f(t) = 0, \qquad (6.3)$$

where

$$I_0 = \tfrac{1}{12}(4\overline{m}L)(4L)^2 = \text{mass moment of inertia of the rod,}$$

$$I_1 = \tfrac{1}{2}m\left(\frac{L}{2}\right)^2 \qquad = \text{mass moment of inertia of the circular plate.}$$

The differential equation, eq. (6.3), governing the motion of this system may conveniently be written as

$$I^*\ddot\theta + C^*\dot\theta + K^*\theta = F^*(t), \qquad (6.4)$$

where I^*, C^*, K^*, and $F^*(t)$ are, respectively, the generalized inertia, generalized damping, generalized stiffness, and generalized load for this system. These quantities are given in eq. (6.3) by the factors corresponding to the acceleration, velocity, displacement, and force terms, namely,

$$I^* = I_0 + I_1 + 4\overline{m}L^3 + mL^2,$$

$$C^* = cL^2,$$

$$K^* = 4kL^2,$$

$$F^*(t) = \tfrac{7}{6}L^2.$$

Illustrative Example 6.1.   For the system shown in Fig. 6.4, determine the generalized physical properties M^*, C^*, K^* and generalized loading $F^*(t)$. Let $Y(t)$ at the point A_2 in Fig. 6.4 be the generalized coordinate of the system.

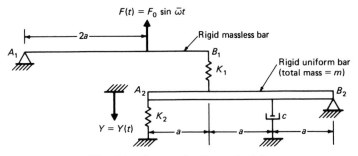

Fig. 6.4 System for Example 6.1.

The free body diagram for the system is depicted in Fig. 6.5 which shows all the forces on the two bars of the system including the inertial force and the inertial moment. The generalized coordinate is $Y(t)$ and the displacement of any point in the system should be expressed in terms of this coordinate; nevertheless, for convenience, we select also the auxiliary coordinate $Y_1(t)$ as indicated in Fig. 6.5.

The summation of the moments about point A_1 of all the forces acting on bar $A_1 - B_1$ and the summation of moments about B_2 of the forces on bar $A_2 - B_2$, give the following equations:

$$k_1(\tfrac{2}{3} Y - Y_1)3a = 2aF_0 \sin \bar{\omega} t \tag{6.5}$$

$$\frac{I_0}{3a} \ddot{Y} + \frac{3}{4} ma\ddot{Y} + a/3c\dot{Y} + k_1 \left(\frac{2}{3} Y - Y_1\right) 2a + 3ak_2 Y = 0. \tag{6.6}$$

Substituting Y_1 from eq. (6.5) into eq. (6.6), we obtain the differential equation for the motion of the system in terms of the generalized coordinate $Y(t)$, namely

$$M^*\ddot{Y}(t) + C^*\dot{Y}(t) + K^*Y(t) = F^*(t),$$

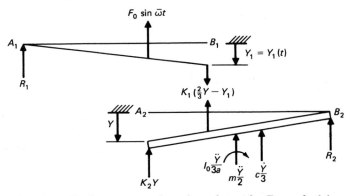

Fig. 6.5 Displacements and resultant forces for Example 6.1.

where the generalized quantities are given by

$$M^* = \frac{I_0}{3a^2} + \frac{3m}{4},$$

$$C^* = \frac{c}{3},$$

$$K^* = 3k_2,$$

and

$$F^*(t) = \frac{4}{3} F_0 \sin \bar{\omega} t.$$

6.3   GENERALIZED SINGLE DEGREE-OF-FREEDOM SYSTEM—DISTRIBUTED ELASTICITY

The example presented in the preceding section had only one degree of freedom in spite of the complexity of the various parts of the system because the two bars were interconnected through a spring and one of the bars was massless so that only one coordinate sufficed to completely specify the motion. If the bars were not rigid, but could deform in flexure, the system would have an infinite number of degrees of freedom. However, a single degree-of-freedom anlaysis could still be made, provided that only a single shape could be developed during motion, that is, provided that the knowledge of the displacement of a single point in the system determines the displacement of the entire system.

As an illustration of this method for approximating the analysis of a system with an infinite number of degrees of freedom with a single degree of freedom, consider the cantilever beam shown in Fig. 6.6. In this illustration, the physical properties of the beam are the flexural stiffness $EI(x)$ and its mass per unit of length $m(x)$. It is assumed that the beam is subjected to an arbitrary distributed forcing function $p(x, t)$ and to an axial compressive force N.

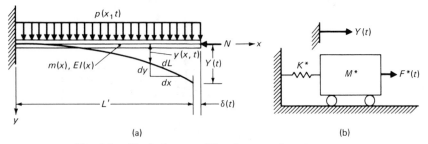

(a)   (b)

Fig. 6.6   Single degree-of-freedom continuous system.

In order to approximate the motion of this system with a single coordinate, it is necessary to assume that the beam deflects during its motion in a prescribed shape. Let $\phi(x)$ be the function describing this shape and, as a generalized coordinate, $Y(t)$ describing the displacement of the motion corresponding to the free end of the beam. Therefore, the displacement at any point x along the beam is

$$y(x, t) = \phi(x)\, Y(t), \tag{6.7}$$

where $\phi(L) = 1$.

The equivalent one-degree-of-freedom system [Fig. 6.6(b)] may be defined simply as the system for which the kinetic energy, potential energy (strain energy), and work done by the external forces have at all times the same values in the two systems.

The kinetic energy T of the beam in Fig. 6.6 vibrating in the pattern indicated by eq. (6.7) is

$$T = \int_0^L \tfrac{1}{2}\, m(x)\, \{\phi(x)\, \dot{Y}(t)\}^2 \, dx. \tag{6.8}$$

Equating this expression for the kinetic energy of the continuous system to the kinetic energy of the equivalent single degree-of-freedom system $\frac{1}{2} M^* \dot{Y}(t)^2$ and solving the resulting equation for the generalized mass, we obtain

$$M^* = \int_0^L m(x)\,\phi^2(x)\, dx. \tag{6.9}$$

The flexural strain energy V of a prismatic beam may be determined as the work done by the bending moment $M(x)$ undergoing an angular displacement $d\theta$. This angular displacement is obtained from the well-known formula for the flexural curvature of a beam, namely

$$\frac{d^2 y}{dx^2} = \frac{d\theta}{dx} = \frac{M(x)}{EI} \tag{6.10}$$

or

$$d\theta = \frac{M(x)}{EI}\, dx \tag{6.11}$$

since $dy/dx = \theta$, where θ, being assumed small, is taken as the slope of the elastic curve. Consequently, the strain energy is given by

$$V = \int_0^L \tfrac{1}{2}\, M(x)\, d\theta. \tag{6.12}$$

The factor $\frac{1}{2}$ is required for the correct evaluation of the work done by the flexural moment increasing from zero to its final value $M(x)$ (average value $M(x)/2$). Now, utilizing eqs. (6.10) and (6.11) in eq. (6.12), we obtain

$$V = \int_0^L \frac{1}{2} EI(x) \left(\frac{d^2 y}{dx^2}\right)^2 dx. \tag{6.13}$$

Finally, equating the potential energy, eq. (6.13), for the continuous system to the potential energy of the equivalent system and using eq. (6.7) results in

$$\tfrac{1}{2} K^* Y(t)^2 = \int_0^L \tfrac{1}{2} EI(x) \{\phi''(x) Y(t)\}^2 dx$$

or

$$K^* = \int_0^L EI(x) \{\phi''(x)\}^2 dx \tag{6.14}$$

where

$$\phi''(x) = \frac{d^2 y}{dx^2}.$$

The generalized force $F^*(t)$ may be found from the virtual displacement $\delta Y(t)$ of the generalized coordinate $Y(t)$ upon equating the work performed by the external forces in the structure to the work done by the generalized force in the equivalent single degree-of-freedom system. The work of the distributed external force $p(x, t)$ due to this virtual displacement is given by

$$W = \int_0^L p(x, t) \, \delta y \, dx.$$

Substituting $\delta y = \phi(x) \delta Y$ from eq. (6.7) gives

$$W = \int_0^L p(x, t) \, \phi(x) \, \delta Y \, dx. \tag{6.15}$$

The work of the generalized force $F^*(t)$ in the equivalent system corresponding to the virtual displacement δY of the generalized coordinate is

$$W^* = F^*(t) \, \delta Y. \tag{6.16}$$

Equating eq. (6.15) with eq. (6.16) and cancelling the factor δY, which is taken to be different from zero, we obtain the generalized force as

$$F^*(t) = \int_0^L p(x, t) \, \phi(x) \, dx. \tag{6.17}$$

Similarly, to determine the generalized damping coefficient, assume a virtual displacement and equate the work of the damping forces in the physical system with the work of the damping force in the equivalent single degree-of-freedom system. Hence

$$C^* \dot{Y} \delta Y = \int_0^L c(x) \dot{y} \, \delta y \, dx$$

where $c(x)$ is the distributed damping coefficient per unit length along the beam. Substituting $\delta y = \phi(x) \, \delta Y$ and $\dot{y} = \phi(x) \, \dot{Y}$ from eq. (6.7) and cancelling the common factors, we obtain

$$C^* = \int_0^L c(x)[\phi(x)]^2 \, dx \qquad (6.18)$$

which is the expression for the generalized damping coefficient.

To calculate the potential energy of the axial force N which is unchanged during the vibration of the beam and consequently is a conservative force, it is necessary to evaluate the horizontal component of the motion $\delta(t)$ of the free end of the beam. For this purpose, we consider a differential element of length dL along the beam as shown in Fig. 6.6(a). This element may be expressed as

$$dL = (dx^2 + dy^2)^{1/2}$$

or

$$dL = (1 + (dy/dx)^2)^{1/2} \, dx. \qquad (6.19)$$

Now, integrating over the horizontal projection of the length of beam (L') and expanding in series the binomial expression, we obtain

$$L = \int_0^{L'} \left(1 + \left(\frac{dy}{dx}\right)^2\right)^{1/2} dx = \int_0^{L'} \left\{1 + \frac{1}{2}\left(\frac{dy}{dx}\right)^2 - \frac{1}{8}\left(\frac{dy}{dx}\right)^4 + \dots\right\} dx.$$

Retaining only the first two terms of the series results in

$$L = L' + \int_0^L \frac{1}{2}\left(\frac{dy}{dx}\right)^2 dx \qquad (6.20)$$

or

$$\delta(t) = L - L' = \int_0^L \frac{1}{2}\left(\frac{dy}{dx}\right)^2 dx. \qquad (6.21)$$

The reader should realize that eqs. (6.20) and (6.21) involve approximations since the series was truncated and the upper limit of the integral in the final expression was conveniently set equal to the initial length of the beam L instead of to its horizontal component L'.

Now we define a new stiffness coefficient to be called the generalized geometric stiffness, K_G^*, as the stiffness of the equivalent system required to store the same potential energy as the potential energy stored by the normal force N, that is,

$$\tfrac{1}{2} K_G^* Y(t)^2 = N\delta(t).$$

Substituting $\delta(t)$ from eq. (6.21) and the derivative dy/dx from eq. (6.7), we have

$$\frac{1}{2} K_G^* Y(t)^2 = \frac{1}{2} N \int_0^L \left\{ Y(t) \frac{d\phi}{dx} \right\}^2 dx$$

or

$$K_G^* = N \int_0^L \left(\frac{d\phi}{dx} \right)^2 dx. \tag{6.22}$$

Equations (6.9), (6.14), (6.17), (6.18), and (6.22) give, respectively, the generalized expression for the mass, stiffness, force, damping, and geometric stiffness for a beam with distributed properties and load, modeling it as a simple oscillator.

For the case of an axial compressive force the potential energy in the beam decreases with a loss of stiffness in the beam. The opposite is true for a tensile axial force which results in an increase of the flexural stiffness of the beam. Customarily, the geometric stiffness is determined for a compressive axial force. Consequently, the combined generalized stiffness is then given by

$$K_c^* = K^* - K_G^*. \tag{6.23}$$

Finally, the differential equation for the equivalent system may be written as

$$M^* \ddot{Y}(t) + C^* \dot{Y}(t) + K_c^* Y(t) = F^*(t). \tag{6.24}$$

The critical buckling load N_{cr} is defined as the axial compressive load that reduces the combined stiffness to zero, that is,

$$K_c^* = K^* - K_G^* = 0.$$

The substitution of K^* and K_G^* from eqs. (6.14) and (6.22) gives

$$\int_0^L EI \left(\frac{d^2\phi}{dx^2} \right)^2 dx - N_{cr} \int \left(\frac{d\phi}{dx} \right)^2 dx = 0$$

and solving for the critical buckling load, we obtain

$$N_{cr} = \frac{\displaystyle\int_0^L EI(d^2\phi/dx^2)^2 \, dx}{\displaystyle\int_0^L (d\phi/dx)^2 \, dx} . \qquad (6.25)$$

To provide an example of the determination of the equivalent one degree of freedom for a system with distributed mass and stiffness, consider the water tower in Fig. 6.7 to have uniformly distributed mass $\overline{m}$ and stiffness EI along its length with a concentrated mass $M = \overline{m}L$ at the top. The tower is subjected to an earthquake ground motion excitation of acceleration $a_g(t)$ and to an axial compressive load due to the weight of its distributed mass and concentrated mass at the top. Neglect damping in the system. Assume that during the motion the shape of the tower is given by

$$\phi(x) = 1 - \cos\frac{\pi x}{2L} . \qquad (6.26)$$

Selecting the lateral displacement $Y(t)$ at the top of the tower as the generalized coordinate as shown in Fig. 6.7, we obtain for the displacement at any point

$$y(x, t) = Y(t)\,\phi(x) = Y(t)\left(1 - \cos\frac{\pi x}{2L}\right). \qquad (6.27)$$

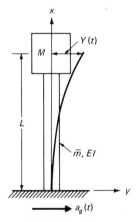

Fig. 6.7   Water tower with distributed properties for Example 6.2.

The generalized mass and the generalized stiffness of the tower are computed, respectively, from eqs. (6.9) and (6.14) as

$$M^* = \overline{m}L + \overline{m} \int_0^L \left(1 - \cos\frac{\pi x}{2L}\right)^2 dx,$$

$$M^* = \frac{\overline{m}L}{2\pi}(5\pi - 8), \tag{6.28}$$

and

$$K^* = \int_0^L EI\left(\frac{\pi}{2L}\right)^4 \cos^2\frac{\pi x}{2L}\, dx,$$

$$K^* = \frac{\pi^4 EI}{32L^3}. \tag{6.29}$$

The axial force is due to the weight of the tower above a particular section, including the concentrated weight at the top, and may be expressed as

$$N(x) = \overline{m}Lg\left(2 - \frac{x}{L}\right), \tag{6.30}$$

where g is the gravitational acceleration. Since the normal force in this case is a function of x, it is necessary in using eq. (6.22) to include $N(x)$ under the integral sign. The geometric stiffness coefficient K_G^* is then given by

$$K_G^* = \int_0^L \overline{m}Lg\left(2 - \frac{x}{L}\right)\left(\frac{\pi}{2L}\right)^2 \sin^2\frac{\pi x}{2L}\, dx$$

which, upon integration yields

$$K_G^* = \frac{\overline{m}g}{16}(3\pi^2 - 4). \tag{6.31}$$

Consequently, the combined stiffness from eqs. (6.29) and (6.31) is

$$K_c^* = K^* - K_G^* = \frac{\pi^4 EI}{32L^3} - \frac{\overline{m}g}{16}(3\pi^2 - 4). \tag{6.32}$$

By setting $K_c^* = 0$, we obtain

$$\frac{\pi^4 EI}{32L^3} - \frac{\overline{m}g}{16}(3\pi^2 - 4) = 0$$

which gives the critical load

$$(\overline{m}g)_{cr} = \frac{\pi^4 EI}{2(3\pi^2 - 4) L^3}.$$ (6.33)

The equation of motion in terms of the relative motion $u = y(t) - y_g(t)$ is given by eq. (3.38) for the undamped system as

$$M^* \ddot{u} + K_c^* u = F_{eff}^*(t),$$ (6.34)

where M^* is given by eq. (6.28), K_c^* by eq. (6.32), and the effective force by eq. (6.17) for the distributed mass and by $-\overline{m}La_g(t)$ for the concentrated mass at the top of the tower. Hence

$$F_{eff}^*(t) = \int_0^L p_{eff}(x, t) \, \phi(x) \, dx - \overline{m}La_g(t)$$

where $p_{eff}(x, t) = -\overline{m}a_g(t)$ is the effective distributed force. Then

$$F_{eff}^* = \int_0^L - \overline{m}a_g(t) \, \phi(x) \, dx - \overline{m}La_g(t).$$

Substitution of $\phi(x)$ from eq. (6.26) into the last equation yields upon integration

$$F_{eff}^* = -\frac{2\overline{m}_g(t) L}{\pi} (\pi - 1).$$ (6.35)

Illustrative Example 6.2.   As a numerical example of calculating the response of a system with distributed properties, consider the water tower shown in Fig. 6.7 excited by a sinusoidal ground acceleration $a_g(t) = 20 \sin 6.36t$ (in/sec^2). The numerical values for this example are

$$\overline{m} = 0.1 \text{ k} \cdot \text{sec}^2/\text{in per unit of length,}$$

$$EI = 1.2 \; 10^{13} \text{ k} \cdot \text{in}^2,$$

$$L = 100 \text{ ft} = 1200 \text{ in,}$$

$$\overline{\omega} = 6.36 \text{ rad/sec.}$$

From eq. (6.28) the generalized mass is

$$M^* = \frac{0.1 \times 1200}{2\pi} (5\pi - 8) = 147.21 \; \frac{\text{k} \cdot \text{sec}^2}{\text{in}},$$

and from eq. (6.32) the generalized stiffness is

$$K_c^* = \frac{\pi^4 1.2 \; 10^{13}}{32 \times (1200)^3} - \frac{0.1 \times 386}{16} (3\pi^2 - 4) = 21,077 \text{ k/in.}$$

The natural frequency is

$$\omega = \sqrt{K_c^*/M^*} = 11.96 \text{ rad/sec}$$

and the frequency ratio

$$r = \frac{\overline{\omega}}{\omega} = \frac{6.36}{11.96} = 0.532.$$

From eq. (6.35) the effective force is

$$F_{eff}^* = -\frac{2(0.1)(1200)(\pi - 1)}{\pi} 20.0 \sin 6.36t$$

or

$$F_{eff}^* = -3272 \sin 6.36t \text{ (kip)}.$$

Hence the response (neglecting damping) in terms of relative motion is

$$u = \frac{F_{eff}^*/k^*}{1 - r^2} \sin \overline{\omega}t$$

$$= \frac{3272/21,077}{1 - (0.532)^2} \sin 6.36t$$

$$= 0.217 \sin 6.36t \text{ in.} \qquad \text{(Ans.)}$$

6.4 SUMMARY

The concept of generalized coordinate presented in this chapter permits the analysis of multiple interconnected rigid or elastic bodies with distributed properties and load as single degree-of-freedom systems. The analysis as one-degree-of-freedom systems can be made provided that by the specification of a single coordinate (the generalized coordinate) the configuration of the whole system is determined. Such a system may then be modeled as the simple oscillator with its various parameters of mass, stiffness, damping, and load, calculated to be dynamically equivalent to the actual system to be analyzed. The solution of this model provides the response in terms of the generalized coordinate.

The principle of virtual work which is applicable to systems in static or dynamic equilibrium is a powerful method for obtaining the equations of motion as an alternative to the direct application of Newton's Law. The principle of virtual work states that for a system in equilibrium the summation of the work done by all its forces during any arbitrary displacement compatible with the constraints of the system is equal to zero.

PROBLEMS

6.1 For the system shown in Fig. P6.1 determine the generalized mass M^*, damping C^*, stiffness K^*, and the generalized load $F^*(t)$. Select $Y(t)$ as the generalized coordinate.

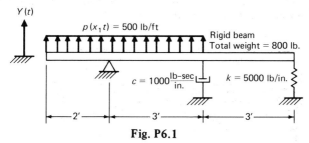

Fig. P6.1

6.2 Determine the generalized quantities M^*, C^*, K^*, and $F^*(t)$ for the structure shown in Fig. P6.2. Select $Y(t)$ as the generalized coordinate.

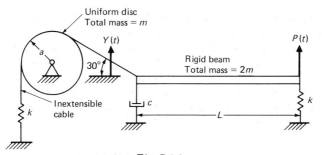

Fig. P6.2

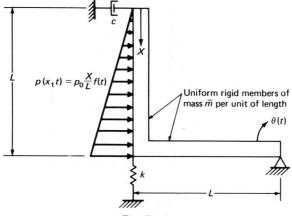

Fig. P6.3

6.3   Determine the generalized quantities M^*, C^*, K^*, and $F^*(t)$ for the structure shown in Fig. P6.3. Select $\theta(t)$ as the generalized coordinate.

6.4   For the elastic cantilever beam shown in Fig. P6.4, determine the generalized quantities M^*, K^*, and $F^*(t)$. Neglect damping. Assume that the deflected shape is given by $\phi(x) = 1 - \cos(\pi x/2L)$ and select $Y(t)$ as the generalized coordinate as shown in Fig. P6.4.

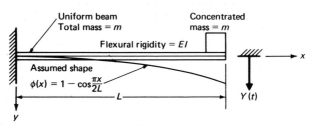

Fig. P6.4

6.5   Determine the generalized geometric stiffness K^*_G for the system in Fig. P6.4 if an axial tensile force N is applied at the free end of the beam along the x direction. What is the combined generalized stiffness K^*_c?

6.6   A concrete conical post of diameter d at the base and height L is shown in Fig. P6.6. It is assumed that the wind produces a dynamic pressure $p_0(t)$ per unit of projected area along a vertical plane. Determine the generalized quantities M^*, K^*, K^*_G, and $F^*(t)$. (Modulus of elasticity $E_c = 3 \times 10^6$ psi; concrete specific weight $\gamma = 150$ lb/ft^3.)

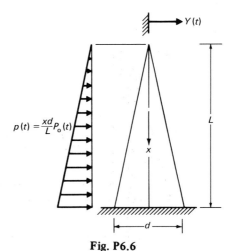

Fig. P6.6

7

Nonlinear structural response

In discussing the dynamic behavior of single degree-of-freedom systems, we assumed that in the model representing the structure, the restoring force was proportional to the displacement. We also assumed the dissipation of energy through a viscous damping mechanism in which the damping force was proportional to the velocity. In addition, the mass in the model was always considered to be unchanging with time. As a consequence of these assumptions, the equation of motion for such a system resulted in a linear, second-order ordinary differential equation with constant coefficients, namely,

$$m\ddot{y} + c\dot{y} + ky = F(t). \tag{7.1}$$

It was illustrated that for particular forcing functions such as harmonic function, it was relatively simple to solve this equation and that a general solution always existed in terms of Duhamel's integral. Equation (7.1) thus represents the dynamic behavior of many structures modeled as a single degree-of-freedom system. There are, however, physical situations for which this linear model does not adequately represent the dynamic characteristics of the structure. The analysis in such cases requires the introduction of a model in which the spring force or the damping force may not remain proportional, respectively, to the displacement or to the velocity. Consequently, the resulting equation of motion will no longer be linear and its mathematical solution, in general, will have a much greater complexity, often requiring a numerical procedure for its integration.

7.1   NONLINEAR SINGLE DEGREE-OF-FREEDOM MODEL

Consider in Fig. 7.1(a), the model for a single degree-of-freedom system and in Fig. 7.1(b), the corresponding free body diagram. The dynamic equilibrium in the system is established by equating to zero the sum of the inertial force $F_I(t)$, the damping force $F_D(t)$, the spring force $F_s(t)$, and the external force $F(t)$. Hence at time t_i the equilibrium of these forces is expressed as

$$F_I(t_i) + F_D(t_i) + F_s(t_i) = F(t_i) \qquad (7.2)$$

and at short time Δt later as

$$F_I(t_i + \Delta t) + F_D(t_i + \Delta t) + F_s(t_i + \Delta t) = F(t_i + \Delta t). \qquad (7.3)$$

Subtracting eq. (7.2) from eq. (7.3) results in the differential equation of motion in terms of increments, namely

$$\Delta F_I + \Delta F_D + \Delta F_s = \Delta F_i \qquad (7.4)$$

where the incremental forces in this equation are defined as follows:

$$\Delta F_I = F_I(t_i + \Delta t) - F_I(t_i)$$
$$\Delta F_D = F_D(t_i + \Delta t) - F_D(t_i)$$
$$\Delta F_s = F_s(t_i + \Delta t) - F_s(t_i)$$
$$\Delta F_i = F(t_i + \Delta t) - F(t_i). \qquad (7.5)$$

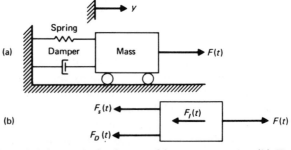

Fig. 7.1   (a) Model for a single degree-of-freedom system. (b) Free body diagram showing the inertial force, the damping force, the spring force, and the external force.

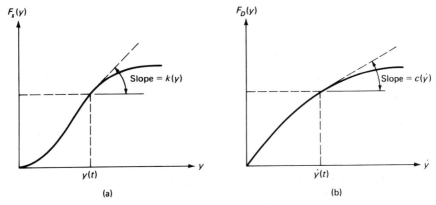

Fig. 7.2   (a) Nonlinear stiffness. (b) Nonlinear damping.

If we assume that the damping force is a function of the velocity and the spring force a function of displacement as shown graphically in Fig. 7.2, while the inertial force remains proportional to the acceleration, we may then express the incremental forces in eqs. (7.5) as

$$\Delta F_I = m\Delta\ddot{y}_i$$

$$\Delta F_D = c_i\Delta\dot{y}_i$$

$$\Delta F_s = k_i\Delta y_i \qquad (7.6)$$

where the incremental displacement Δy_i, incremental velocity $\Delta\dot{y}_i$, and incremental acceleration $\Delta\ddot{y}$ are given by

$$\Delta y_i = y(t_i + \Delta t) - y(t_i), \qquad (7.7)$$

$$\Delta\dot{y}_i = \dot{y}(t_i + \Delta t) - \dot{y}(t_i), \qquad (7.8)$$

$$\Delta\ddot{y}_i = \ddot{y}(t_i + \Delta t) - \ddot{y}(t_i). \qquad (7.9)$$

The coefficient k_i in eqs. (7.6) is defined as the current evaluation for the derivative of the spring force with respect to the displacement, namely,

$$k_i = \left(\frac{dF_s}{dy}\right)_{y=y_i}. \qquad (7.10)$$

Similarly, the coefficient c_i is defined as the current value of the derivative of the damping force with respect to the velocity, that is,

$$c_i = \left(\frac{dF_D}{d\dot{y}}\right)_{\dot{y}=\dot{y}_i}. \qquad (7.11)$$

These two coefficients k_i and c_i are graphically depicted as the slopes of the corresponding curves shown in Fig. 7.2.

The substitution of eqs. (7.6) into eq. (7.4) results in a convenient form for the incremental equation, namely

$$m\Delta\ddot{y}_i + c_i\Delta\dot{y}_i + k_i\Delta y_i = \Delta F_i \qquad (7.12)$$

where the coefficients c_i and k_i are calculated for values of velocity and displacement corresponding to time t_i and assumed to remain constant during the increment of time Δt. Since, in general, these two coefficients do not remain constant for that time increment, eq. (7.12) is an approximate equation.

7.2   INTEGRATION OF THE NONLINEAR EQUATION OF MOTION

Among the many methods available for the solution of the nonlinear equation of motion, probably one of the most effective is the step-by-step integration method. In this method, the response is evaluated at successive increments Δt of time, usually taken of equal length of time for computational convenience. At the beginning of each interval, the condition of dynamic equilibrium is established. Then, the response for a time increment Δt is evaluated approximately on the basis that the coefficients $k(y)$ and $c(\dot{y})$ remain constant during the interval Δt. The nonlinear characteristics of these coefficients are considered in the analysis by reevaluating these coefficients at the beginning of each time increment. The response is then obtained using the displacement and velocity calculated at the end of the time interval as the initial conditions for the next time step.

As we have said for each time interval, the stiffness coefficient $k(y)$ and the damping coefficient $c(\dot{y})$ are evaluated at the initiation of the interval but are assumed to remain constant until the next step; thus the nonlinear behavior of the system is approximated by a sequence of successively changing linear systems. It should also be obvious that the assumption of constant mass is unnecessary; it could just as well also be represented by a variable coefficient.

There are many procedures available for performing the step-by-step integration of eq. (7.8). Two of the most popular methods are the *constant acceleration method* and the *linear acceleration method.* As the names of these methods imply, in the first method the acceleration is assumed to remain constant during the time interval Δt, while in the second method, the acceleration is assumed to vary linearly during the interval. As may be expected, the constant acceleration method is simpler, but less accurate when compared with the linear acceleration method for the same value of the time increment. We shall present here in detail only the linear acceleration method. This method has been found to yield excellent results with relatively little computational effort.

7.3   LINEAR ACCELERATION STEP-BY-STEP METHOD

In the linear acceleration method, it is assumed that the acceleration may be expressed by a linear function of time during the time interval Δt. Let t_i and $t_{i+1} = t_i + \Delta t$ be, respectively, the designation for the time at the beginning and at the end of the time interval Δt. In this type of analysis, the material properties of the system c_i and k_i may include any form of nonlinearity. Thus it is not necessary for the spring force to be only a function of displacement nor for the damping force to be specified only as a function of velocity. The only restriction in the analysis is that we evaluate these coefficients at an instant of time t_i and then assume that they remain constant during the increment of time Δt. When the acceleration is assumed to be a linear function of time for the interval of time t_i to $t_{i+1} = t_i + \Delta t$ as depicted in Fig. 7.3, we may express the acceleration as

$$\ddot{y}(t) = \ddot{y}_i + \frac{\Delta \ddot{y}_i}{\Delta t}(t - t_i), \tag{7.13}$$

where $\Delta \ddot{y}_i$ is given by eq. (7.9). Integrating eq. (7.13) twice with respect to time between the limits t_i and t yields

$$\dot{y}(t) = \dot{y}_i + \ddot{y}_i(t - t_i) + \frac{1}{2}\frac{\Delta \ddot{y}_i}{\Delta t}(t - t_i)^2 \tag{7.14}$$

and

$$y(t) = y_i + \dot{y}_i(t - t_i) + \frac{1}{2}\ddot{y}_i(t - t_i)^2 + \frac{1}{6}\frac{\Delta \ddot{y}_i}{\Delta t}(t - t_i)^3 \tag{7.15}$$

The evaluation of eqs. (7.14) and (7.15) at time $t = t_i + \Delta t$ gives

$$\Delta \dot{y}_i = \ddot{y}_i \Delta t + \tfrac{1}{2}\Delta \ddot{y}_i \Delta t \tag{7.16}$$

and

$$\Delta y_i = \dot{y}_i \Delta t + \tfrac{1}{2}\ddot{y}_i \Delta t^2 + \tfrac{1}{6}\Delta \ddot{y}_i \Delta t^2 , \tag{7.17}$$

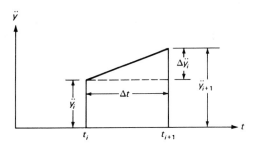

Fig.. 7.3   Assumed linear variation of acceleration during time interval.

where Δy_i and $\Delta \dot{y}_i$ are defined in eqs. (7.7) and (7.8), respectively. Now to use the incremental displacement Δy as the basic variable in the analysis, eq. (7.17) is solved for the incremental acceleration $\Delta \ddot{y}_i$ and then substituted in to eq. (7.16) to obtain

$$\Delta \ddot{y}_i = \frac{6}{\Delta t^2} \Delta y_i - \frac{6}{\Delta t} \dot{y}_i - 3\ddot{y}_i \tag{7.18}$$

and

$$\Delta \dot{y}_i = \frac{3}{\Delta t} \Delta y_i - 3\dot{y}_i - \frac{\Delta t}{2} \ddot{y}_i. \tag{7.19}$$

The substitution of eqs. (7.18) and (7.19) into eq. (7.12) leads to the following form of the equation of motion:

$$m \left\{ \frac{6}{\Delta t^2} \Delta y_i - \frac{6}{\Delta t} \dot{y}_i - 3\ddot{y}_i \right\} + c_i \left\{ \frac{3}{\Delta t} \Delta y_i - 3\dot{y}_i - \frac{\Delta t}{2} \ddot{y}_i \right\} + k_i \Delta y_i = \Delta F_i. \tag{7.20}$$

Finally, transferring in eq. (7.20) all the terms containing the unknown incremental displacement Δy_i to the left-hand side gives,

$$\bar{k}_i \Delta y_i = \Delta \bar{F}_i \tag{7.21}$$

in which

$$\bar{k}_i = k_i + \frac{6m}{\Delta t^2} + \frac{3c_i}{\Delta t}, \tag{7.22}$$

and

$$\Delta \bar{F}_i = \Delta F_i + m \left\{ \frac{6}{\Delta t} \dot{y}_i + 3\ddot{y}_i \right\} + c_i \left\{ 3\dot{y}_i + \frac{\Delta t}{2} \ddot{y}_i \right\}. \tag{7.23}$$

It should be noted that eq. (7.21) is equivalent to the static incremental-equilibrium equation, and may be solved for the incremental displacement by simply dividing the equivalent incremental load $\Delta \bar{F}_i$ by the equivalent spring constant $\bar{k}_i$, that is,

$$\Delta y_i = \frac{\Delta \bar{F}_i}{\bar{k}_i}. \tag{7.24}$$

To obtain the displacement $y_{i+1} = F(t_i + \Delta t)$ at time $t_{i+1} = t_i + \Delta t$, this value of Δy_i is substituted into eq. (7.7) yielding

$$y_{i+1} = y_i + \Delta y_i. \tag{7.25}$$

Then the incremental velocity $\Delta \dot{y}_i$ is obtained from eq. (7.19) and the velocity at time $t_{i+1} = t_i + \Delta t$ from eq. (7.8) as

$$\dot{y}_{i+1} = \dot{y}_i + \Delta \dot{y}_i. \tag{7.26}$$

Finally the acceleration $\ddot{y}_{i+1}$ at the end of the time step is obtained directly from the differential equation of motion, eq. (7.1), where the equation is written for time $t_{i+1} = t_i + \Delta t$. Hence from eq. (7.1), it follows that

$$\ddot{y}_{i+1} = \frac{1}{m} \{F(t_{i+1}) - c_{i+1}\dot{y}_{i+1} - k_{i+1}y_{i+1}\}, \tag{7.27}$$

where the coefficients c_{i+1} and k_{i+1} are now evaluated at time t_{i+1}.

After the displacement, velocity and acceleration have been determined at time $t_{i+1} = t_i + \Delta t$, the outlined procedure is repeated to calculate these quantities at the following time step $t_{i+2} = t_{i+1} + \Delta t$, and the process is continued to any desired final value of time. The reader should, however, realize that this numerical procedure involves two significant approximations: (1) the acceleration is assumed to vary linearly during the time increment Δt; and (2) the damping and stiffness properties of the system are evaluated at the initiation of each time increment and assumed to remain constant during the time interval. In general, these two assumptions introduce errors which are small if the time step is short. However, these errors generally might tend to accumulate from step to step. This accumulation of errors should be avoided by imposing a total dynamic equilibrium condition at each step in the analysis. This is accomplished by expressing the acceleration at each step using the differential equation of motion in which the displacement and velocity as well as the stiffness and damping coefficients are evaluated at that time step.

There still remains the problem of the selection of the proper time increment Δt. As in any numerical method, the accuracy of the step-by-step integration method depends upon the magnitude of the time increment selected. The following factors should be considered in the selection of Δt: (1) the natural period of the structure; (2) the rate of variation of the loading function; and (3) the complexity of the stiffness and damping functions.

In general, it has been found that sufficiently accurate results can be obtained if the time interval is taken to be no longer than one-tenth of the natural period of the structure. The second consideration is that the interval should be small enough to represent properly the variation of the load with respect to time. The third point that should be considered is any abrupt variation in the rate of change of the stiffness or damping function. For example, in the usual assumption of elastoplastic materials, the stiffness suddenly changes from linear elastic to a yielding plastic phase. In this case, to obtain the best accuracy, it would be desirable to select smaller time steps in the neighborhood of such drastic changes.

7.4   ELASTOPLASTIC BEHAVIOR

If any structure modeled as a single degree-of-freedom system (spring-mass system) is allowed to yield plastically, then the restoring force exerted is likely to be of the form shown in Fig. 7.4(a). There is a portion of the curve in which linear elastic behavior occurs, whereupon for any further deformation, plastic yielding takes place. When the structure is unloaded, the behavior is again elastic until further reverse loading produces compressive plastic yielding. The structure may be subjected to cyclic loading and unloading in this manner. Energy is dissipated during each cycle by an amount which is proportional to the area under the curve (hysteresis loop) as indicated in Fig. 7.4(a). This behavior is often simplified by assuming a definite yield point beyond which additional displacement takes place at a constant value for the restoring force without any further increase in the load. Such behavior is known as *elastoplastic* behavior; the corresponding force–displacement curve is shown in Fig. 7.4(b).

For the structure modeled as a spring-mass system, expressions of the restoring force for a system with elastoplastic behavior are easily written. These expressions depend on the magnitude of the restoring force as well as upon whether the motion is such that the displacement is increasing ($\dot{y} > 0$) or decreasing ($\dot{y} < 0$). Referring to Fig. 7.4(b) in which a general elastoplastic cycle is represented, we assume that the initial conditions are zero ($y_0 = 0$, $\dot{y}_0 = 0$) for the unloaded structure. Hence initially as the load is applied, the system behaves elastically along curve E_0. The displacement y_t, at which plastic behavior in tension may be initiated and the displacement y_c, at which plastic behavior in compression may be initiated are calculated, respectively, from

$$y_t = R_t/k \qquad (7.28)$$

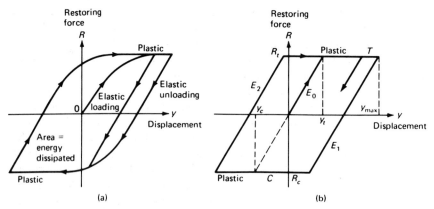

Fig. 7.4   Elastic-plastic structural models. (a) General plastic behavior. (b) Elastoplastic behavior.

and

$$y_c = R_c/k,$$

where R_t and R_c are the respective values of the forces which produce yielding in tension and compression and k, the elastic stiffness of the structure. The system will remain on curve E_0 as long as the displacement y satisfies

$$y_c < y < y_t. \tag{7.29}$$

If the displacement y increases to y_t, the system begins to behave plastically in tension along curve T on Fig. 7.4(b); it remains on curve T as long as the velocity $\dot{y} > 0$. When $\dot{y} < 0$, the system reverses to elastic behavior on a curve such as E_1 with new yielding points given by

$$y_t = y_{max},$$
$$y_c = y_{max} - (R_t - R_c)/k, \tag{7.30}$$

in which y_{max} is the maximum displacement along curve T, which occurs when $\dot{y} = 0$.

Conversely, if y decreases to y_c, the system begins a plastic behavior in compression along curve C and it remains on this curve as long as $\dot{y} < 0$. The system returns to an elastic behavior when the velocity again changes direction and $\dot{y} > 0$. In this case, the new yielding limits are given by

$$y_c = y_{min},$$
$$y_t = y_{min} + (R_t - R_c)/k \tag{7.31}$$

in which y_{min} is the minimum displacement along curve C, which occurs when $\dot{y} = 0$. The same condition given by eq. (7.29) is valid for the system to remain operating along any elastic segment such as $E_0, E_1, E_2, \ldots$ as shown in Fig. 7.4(b).

We are now interested in calculating the restoring force at each of the possible segments of the elastoplastic cycle. The restoring force on an elastic phase of the cycle $(E_0, E_1, E_2, \ldots)$ may be calculated as

$$R = R_t - (y_t - y)k; \tag{7.32}$$

on a plastic phase in tension as

$$R = R_t; \tag{7.33}$$

and on the plastic compressive phase as

$$R = R_c. \tag{7.34}$$

The algorithm for the step-by-step linear integration method of a single degree-of-freedom system assuming an elastoplastic behavior is outlined in the following.

7.5 ALGORITHM FOR STEP-BY-STEP SOLUTION FOR ELASTOPLASTIC SINGLE DEGREE-OF-FREEDOM SYSTEM

Initialize and input data:

(1) Read input values for k, m, c, R_t, R_c, and a list giving the time t_i and magnitude of the excitation F_i.

(2) Set $y_0 = 0$ and $\dot{y}_0 = 0$.

(3) Calculate initial acceleration:

$$\ddot{y}_0 = \frac{F(t = 0)}{m}. \tag{7.35}$$

(4) Select time step Δt and calculate constants:

$$a_1 = 3/\Delta t, \quad a_2 = 6/\Delta t, \quad a_3 = \Delta t/2, \quad a_4 = 6/\Delta t^2.$$

(5) Calculate initial yield points:

$$y_t = R_t/k,$$
$$y_c = R_c/k. \tag{7.36}$$

For each time step:

(1) Check if the system remains in the previous state of elastic or plastic behavior using the following code:

KEY = 0 (elastic behavior);

KEY = -1 (plastic behavior in compression);

KEY = 1 (plastic behavior in tension). $\tag{7.37}$

(2) Calculate the displacement y and velocity $\dot{y}$ at the end of the time step and set the value of KEY according to the following conditions:

(a) When the system is behaving elastically at the beginning of the time step and

$$y_c < y < y_t \qquad \text{KEY} = 0,$$
$$y > y_t \qquad \text{KEY} = 1,$$
$$y < y_c \qquad \text{KEY} = -1.$$

(b) When the system is behaving plastically in tension at the beginning of the time step and

$$\dot{y} > 0 \qquad \text{KEY} = 1,$$
$$\dot{y} < 0 \qquad \text{KEY} = 0.$$

(c) When the system is behaving plastically in compression at the beginning of the time step and

$$\dot{y} < 0 \qquad\qquad \text{KEY} = -1,$$
$$\dot{y} > 0 \qquad\qquad \text{KEY} = 0.$$

(3) Calculate the effective stiffness:

$$\bar{k}_i = k_p + a_4 m + a_1 c_i \qquad\qquad (7.38)$$

where

$$k_p = k \text{ for elastic behavior (KEY = 0)}$$
$$k_p = 0 \text{ for plastic behavior (KEY = 1 or -1).} \qquad (7.39)$$

(4) Calculate the incremental effective force:

$$\overline{\Delta F}_i = \Delta F_i + (a_2 m + 3 c_i) \dot{y}_i + (3m + a_3 c_i) \ddot{y}_i. \qquad (7.40)$$

(5) Solve for the incremental displacement:

$$\Delta y_i = \overline{\Delta F}_i / \bar{k}_i. \qquad\qquad (7.41)$$

(6) Calculate the incremental velocity:

$$\Delta \dot{y}_i = \frac{3}{\Delta t} \Delta y_i - 3\dot{y}_i - \frac{\Delta t}{2} \ddot{y}_i. \qquad (7.42)$$

(7) Calculate displacement and velocity at the end of time interval:

$$y_{i+1} = y_i + \Delta y_i. \qquad\qquad (7.43)$$
$$\dot{y}_{i+1} = \dot{y}_i + \Delta \dot{y}_i. \qquad\qquad (7.44)$$

(8) Calculate acceleration $\ddot{y}_{i+1}$ at the end of time interval using the dynamic equation of equilibrium:

$$\ddot{y}_{i+1} = \frac{1}{m} [F(t_{i+1}) - c_{i+1} \dot{y}_{i+1} - R] \qquad (7.45)$$

at which

$$R = R_t - (y_t - y_{i+1})k \qquad \text{if KEY = 0,}$$
$$R = R_t \qquad\qquad\qquad \text{if KEY = 1,} \qquad (7.46)$$

or

$$R = R_c \qquad\qquad\qquad \text{if KEY = -1.}$$

Illustrative Example 7.1.   To illustrate the hand calculations in applying the step-by-step integration method described above, consider the single degree-of-freedom system in Fig. 7.5 with elastoplastic behavior subjected to the loading history as shown. For this example, we assume that the damping coefficient remains constant ($\xi = 0.087$). Hence the only nonlinearities in the system arise from the changes in stiffness as yielding occurs.

The stiffness of the system during elastic behavior is

$$k = \frac{12EI}{L^3} = \frac{12 \times 30 \times 10^3 \times 2 \times 100}{(15 \times 12)^3} = 12.35 \text{ k/in,}$$

and the damping coefficient

$$c = \xi c_{cr} = (0.087)(2)\sqrt{0.2 \times 12.35} = 0.27 \text{ k} \cdot \text{sec/in.}$$

Initial displacement and initial velocity are $y_0 = \dot{y}_0 = 0$. Initial acceleration is

$$\ddot{y}_0 = \frac{F(0)}{k} = 0.$$

Yield displacements are

$$y_t = \frac{R_t}{k} = \frac{15}{12.35} = 1.215 \text{ in,}$$

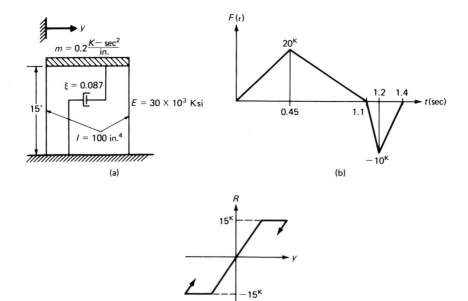

Fig. 7.5   Frame with elastoplastic behavior subjected to dynamic loading. (a) Frame. (b) Loading. (c) Elastoplastic behavior.

and

$$y_c = -1.215 \text{ in.}$$

The natural period is $T = 2\pi\sqrt{m/k} = 0.8$ sec (for the elastic system). For numerical convenience, we select $\Delta t = 0.1$ sec. The effective stiffness from eq. (7.38) is

$$\bar{k} = k_p + \frac{6}{0.1^2} 0.2 + \frac{3}{0.1} 0.27,$$

or

$$\bar{k} = k_p + 128.22, \tag{7.47}$$

where

$$k_p = k = 12.35 \text{ (elastic behavior)},$$

$$k_p = 0 \text{ (plastic behavior)}. \tag{7.48}$$

The effective incremental loading from eq. (7.40) is

$$\overline{\Delta F} = \Delta F + \left(\frac{6}{\Delta t} m + 3c\right)\dot{y} + \left(3m + \frac{\Delta t}{2} c\right)\ddot{y},$$

$$\overline{\Delta F} = \Delta F + 12.822\dot{y} + 0.6137\ddot{y}. \tag{7.49}$$

The velocity increment given by eq. (7.42) becomes

$$\Delta\dot{y} = 30\Delta y - 3\dot{y} - 0.05\ddot{y}.$$

The necessary calculations may be conveniently arranged as illustrated in Table 7.1. In this example with elastoplastic behavior, the response changes abruptly as the yielding starts and stops. To obtain better accuracy, it would be desirable to subdivide the time step in the neighborhood of the change of state; however, an iterative procedure would be required to establish the length of the subintervals. This refinement has not been used in the present analysis nor in the computer program described in the next section. The stiffness computed at the initiation of the time step has been assumed to remain constant during the entire time increment. The reader is again cautioned that a significant error may arise during phase transitions unless the time step is selected relatively small.

7.6   PROGRAM 4—ELASTOPLASTIC SINGLE DEGREE-OF-FREEDOM SYSTEM (STEPS)

A computer program for the analysis of a damped single degree-of-freedom system with elastoplastic behavior subjected to an arbitrary excitation function (force applied at the mass or acceleration at the support) is presented in this

TABLE 7.1 Nonlinear Response—Linear Acceleration Step-by-Step Method for Example 7.1.

t (sec)	F (kip)	y (in)	KEY	$\dot{y}$ (in/sec)	R (kip)	$\ddot{y}$ (in/sec²)	k_p (kip)	$\bar{k}$ (kip)	ΔF (kip)	$\Delta \bar{F}$ (kip)	Δy (in)	$\Delta \dot{y}$ (in/sec)
0	0	0	0	0	0	0	12.35	140.57	4.444	4.444	0.0316	0.9485
0.1	4.444	0.0316	0	0.9485	0.390	18.972	12.35	140.57	4.444	28.249	0.2010	2.2359
0.2	8.888	0.2326	0	3.1844	2.871	25.723	12.35	140.57	4.444	61.050	0.4343	2.193
0.3	13.333	0.6669	0	5.3760	8.233	18.134	12.35	140.57	4.444	84.510	0.6012	1.000
0.4	17.777	1.2681	1	6.3768	15.00	5.152	0	128.22	0.685	85.609	0.6677	0.6422
0.5	18.462	1.9358	1	7.0190	15.00	7.691	0	128.22	-3.077	91.641	0.7147	-0.0001
0.6	15.358	2.6505	1	7.0189	15.00	-7.693	0	128.22	-3.077	82.199	0.6409	-1.440
0.7	12.308	3.2916	1	5.5791	15.00	-21.105	0	128.22	-3.077	55.506	0.4329	-2.695
0.8	9.231	3.7244	1	2.8840	15.00	-32.797	0	128.22	-3.077	13.773	0.1074	-3.789
0.9	6.154	3.8319	0	-0.9054	15.00	-42.990	12.35	140.57	-3.077	-41.069	-0.2922	-3.899
1.0	3.077	3.5397	0	-4.8048	11.39	-34.998	12.35	140.57	-3.077	-86.162	-0.6130	-2.225
1.1	0	2.9268	0	-7.0295	3.825	-9.497	12.35	140.57	-10	-105.96	-0.7538	-1.051
1.2	-10	2.1729	0	-8.0806	-5.481	-11.525	12.35	140.57	5	-105.68	-0.7518	2.263
1.3	-5	1.4211	0	-5.8177	-14.76	56.784	12.35	140.57	5	-34.746	-0.2472	7.198
1.4	0	1.1739	-1	1.3860	-15.00	73.109	0	128.22	0	62.568	0.4880	6.842
1.5	0	1.6619	0	8.2227	-15.00	63.735	12.35	140.57	0	144.55	1.0283	2.995

TABLE 7.2   Description of Input Variables for Program 4.

Variable	Symbol in Text	Description
NYTPE		Excitation index: 0 → force at mass, 1 → acceleration at support
NEQ		Number of points defining the excitation function
SK	k	Elastic stiffness
SM	m	Mass
SC	c	Damping coefficient
DT	ΔT	Time increment
RT	R_t	Maximum restoring force in tension
RC	R_c	Maximum restoring force in compression
TC(I)	t_i	Time at point i
X(I)	F_i	Force (or acceleration) at time t_i

section. The step-by-step linear acceleration method is used to integrate the equation of motion according to the algorithm described in the previous section. Table 7.2 gives a list of the input variables and symbols used in the text. Table 7.3 describes the input data cards and corresponding formats.

The listing of the program is given in the Appendix. The program begins by reading and printing the data and by setting the initial values to the various constants and variables in the equations. Then by a process of linear interpolation, values of the forcing function are computed at time increments equal to the selected time step for the integration process. In the main body of the program, the displacement, velocity, and acceleration are computed at each time step. The nonlinear behavior of the restoring force is appropriately considered in the calculation by the variable "KEY" which is tested through a series of conditional statements in order to determine the correct expressions for the yield points and the magnitude of the restoring force in the system. The output consists of a table giving the displacement, velocity, and acceleration at time increments Δt. The last column of the table shows the value of the index KEY which provides the information about the state of the elastoplastic system. As indicated before, KEY = 0 for elastic behavior and KEY = 1 or KEY = -1 for plastic behavior, respectively, in tension or in compression.

TABLE 7.3   Input Data Symbols and Format for Program 4.

Format	Variables
(2I5, 6F10.3)	NTYPE NEQ SK SM SC DT RT RC
(8F10.2)	TC(I), X(I), (I = 1, NEQ) (as many cards as needed)

TABLE 7.4   Input Data for Example 7.2.

Symbol	Quantity	Units	Description
NTYPE	0		Force at mass
NEQ	6		Number of points defining the force
SK	12.35	k/in	Stiffness coefficient
SM	0.2	k · sec²/in	Mass
SC	0.27	k · sec/in	Damping coefficient
DT	0.1	sec	Time step
RT	15.0	kip	Maximum restoring force tension
RC	-15.0	kip	Maximum restoring force compression

TABLE 7.5   Listing of Input Data for Example 7.2.

0	6	12.35	0.2	0.27	0.10	15.0	-15.0	
	0.00	0.00	0.45	20.00	1.10	0.00	1.20	-10.00
	1.40	0.00	2.00	0.00				

TABLE 7.6   Response for the System in Fig. 7.5 with Elastoplastic Behavior.

TIME	DISPL. (in)	VELOC. (in/sec)	ACC. (in/sec²)	KEY
0.100	0.0316	0.9485	18.9709	0
0.200	0.2326	3.1834	25.7261	0
0.300	0.6669	5.3764	18.1344	0
0.400	1.2681	6.3768	5.1526	1
0.500	1.9358	7.0190	7.6916	1
0.600	2.6505	7.0189	-7.6929	1
0.700	3.2916	5.5791	-21.1048	1
0.800	3.7245	2.8840	-32.7972	1
0.900	3.8319	-0.9054	-42.9903	0
1.000	3.5397	-4.8048	-34.9975	0
1.100	2.9268	-7.0295	-9.4969	0
1.200	2.1729	-8.0806	-11.5254	0
1.300	1.4211	-5.8177	56.7840	0
1.400	1.1739	1.3806	73.1088	-1
1.500	1.6619	8.2227	63.7348	0
1.600	2.6902	11.2173	-3.8437	0
1.700	3.6966	7.9504	-61.4929	0
1.800	4.1534	0.8761	-76.2002	1

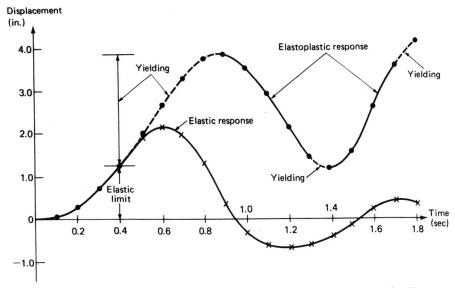

Fig. 7.6 Comparison of elastoplastic behavior with elastic response for Example 7.2.

Illustrative Example 7.2. Using the computer Program 4, find the response of the structure in Example 7.1. Then repeat the calculation assuming elastic behavior. Plot and compare results for the elastoplastic behavior with the elastic response. The basic input data required are given in Table 7.4. The appearance of the actual data required by the computer program is shown in Table 7.5 as an aid to the reader.

The same computer program is used to obtain the response for a completely elastic behavior. It is only necessary to assign a large value to the maximum restoring forces in tension and in compression. These assigned values should be large enough in order for the structure to remain in the elastic range. For the present example, $R_t = 100$ kip and $R_c = -100$ kip were deemed adequate in this case. The computer result for the elastoplastic behavior is given in Table 7.6. In order to visualize and facilitate a comparison between the elastic and inelastic responses for this example, the displacements are plotted in Fig. 7.6, with the response during yielding shown as dashed lines.

7.7 SUMMARY

Structures are usually designed on the assumption that the structure is linearly elastic and that it remains linearly elastic when subjected to any expected dynamic excitation. However, there are situations in which the structure has to be designed for an eventual excitation of large magnitude such as strong motion earthquake or the effects of nuclear explosion. In these cases, it is not realistic to assume that the structure will remain linearly elastic and it is then necessary to design the structure to withstand deformation beyond the elastic limit.

The simplest and most accepted assumption for the design beyond the elastic limit is to assume an elastoplastic behavior. In this type of behavior, the structure is elastic until the restoring force reaches a maximum value (tension or compression) at which it remains constant until the motion reverses its direction and returns to an elastic behavior.

There are many methods to solve numerically the differential equation of this type motion. The step-by-step linear acceleration presented in this chapter provides satisfactory results with relatively simple calculations. However, these calculations are tedious and time consuming when performed by hand. The use of the computer and the availability of a computer program, such as the one described in this chapter, reduce the effort to a simple routine of data preparation.

PROBLEMS

7.1   The single degree-of-freedom of Fig. P7.1(a) is subjected to the foundation acceleration history in Fig. P7.1(b). Determine the maximum relative displacement of the columns. Assume elastoplastic behavior of Fig. P7.1(c).

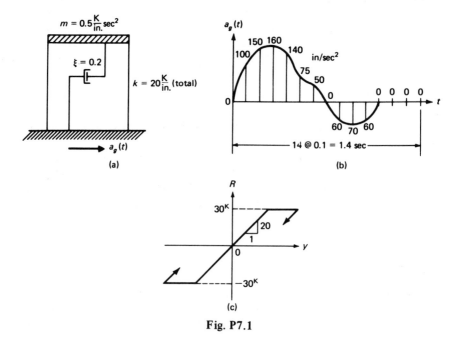

Fig. P7.1

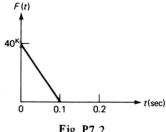

Fig. P7.2

7.2   Determine the displacement history for the structure in Fig. P7.1 when subjected to the impulse loading of Fig. P7.2 applied horizontally at the mass.

7.3   Repeat Problem 7.2 for the impulse loading shown in Fig. P7.3 applied horizontally at the mass.

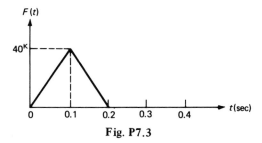

Fig. P7.3

7.4   Repeat Problem 7.2 for the acceleration history shown in Fig. P7.4 applied horizontally to the foundation.

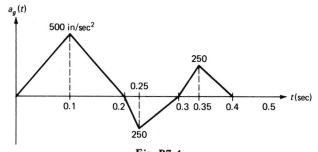

Fig. P7.4

7.5   Solve Problem 7.1 assuming elastic behavior of the structure. (Hint: Use computer Program 4 with $R_t = 100$ kip and $R_c = -100$ kip.)

7.6   Solve Problem 7.2 for elastic behavior of the structure. Plot the time-displacement response and compare with results from Problem 7.2.

7.7   Determine the ductility ratio from the results of Problem 7.2. (Ductility ratio is defined as the ratio of the maximum displacement to the displacement at the yield point.)

8

Response spectra

In this chapter, we shall introduce the concept of *response spectrum*, which in recent years has gained wide acceptance in structural dynamic practice, particularly in earthquake engineering design. Stated briefly, the response spectrum is a plot of the maximum response (maximum displacement, velocity, acceleration, or any other quantity of interest) to a specified load function for all possible single degree-of-freedom systems. The abscissa of the spectrum is the natural frequency (or period) of the system and the ordinate, the maximum response. A plot of this type is shown in Fig. 8.1, in which a one story building is subjected to a ground displacement indicated by the function $y_s(t)$. The response spectral curve shown in Fig. 8.1(a) gives, for any single degree-of-freedom system, the maximum displacement of the mass m relative to the displacement at the support. Thus to determine the response from an available spectral chart, for a specified excitation, we need only to know the natural frequency of the system.

Max relative displ.
$|y - y_s|_{max}$

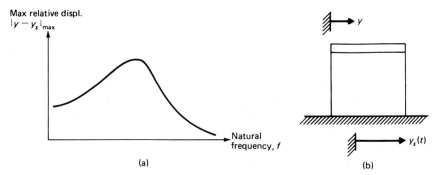

Natural
frequency, f

(a)

(b)

Fig. 8.1   (a) Typical response spectrum. (b) Single degree-of-freedom system subjected to ground excitation.

8.1   CONSTRUCTION OF RESPONSE SPECTRUM

To illustrate the construction of a response spectral chart, consider in Fig. 8.2(a) the undamped oscillator subjected to one-half period of the sinusoidal exciting force shown in Fig. 8.2(b). The system is assumed to be initially at rest. The duration of the sinusoidal impulse is denoted by t_d. The differential equation of motion is obtained by equating to zero the sum of the forces in the corresponding free body diagram shown in Fig. 8.2(c), that is,

$$m\ddot{y} + ky = F(t) \tag{8.1}$$

in which

$$F(t) = \begin{cases} F_0 \sin \bar{\omega}t & \text{for} \quad 0 \leqslant t \leqslant t_d \\ 0 & \text{for} \quad t > t_d \end{cases} \tag{8.2}$$

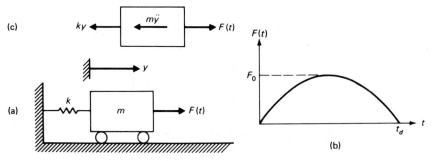

Fig. 8.2   (a) Undamped simple oscillator subjected to load $F(t)$. (b) Loading function $F(t) = F_0 \sin \bar{\omega}t \ (0 \leqslant t \leqslant t_d)$. (c) Free body diagram.

and

$$\bar{\omega} = \frac{\pi}{t_d}. \tag{8.3}$$

The solution of eq. (8.1) may be found by any of the methods studied in the preceding chapters such as the use of Duhamel's integral (Chapter 4) or the step-by-step linear acceleration method (Chapter 7). However, in this example due to the simplicity of the exciting force, we can obtain the solution of eq. (8.1) by the standard method of integration of a linear differential equation, that is, the superposition of the complementary solution y_c and the particular solution y_p

$$y = y_c + y_p. \tag{8.4}$$

The complementary solution of eq. (8.1) (right-hand side equals zero) is given by

$$y_c = A \cos \omega t + B \sin \omega t \tag{8.5}$$

in which $\omega = \sqrt{k/m}$ is the natural frequency. The particular solution for the time interval $0 \leqslant t \leqslant t_d$ is suggested by the right-hand side of eq. (8.1) to be of the form

$$y_p = C \sin \bar{\omega} t. \tag{8.6}$$

The substitution of eq. (8.6) into eq. (8.1) and solution of the resulting identity gives

$$C = \frac{F_0}{k - m\bar{\omega}^2}. \tag{8.7}$$

Combining eqs. (8.4) through (8.7), we obtain the response for $0 \leqslant t \leqslant t_d$ as

$$y = A \cos \omega t + B \sin \omega t + \frac{F_0 \sin \bar{\omega} t}{k - m\bar{\omega}^2}. \tag{8.8}$$

Introducing the initial conditions $y(0) = 0$ and $\dot{y}(0) = 0$ into eq. (8.8) and calculating the constants of integration A and B, we obtain

$$y = \frac{F_0/k}{1 - \left(\frac{\bar{\omega}}{\omega}\right)^2} \left[\sin \bar{\omega} t - \frac{\bar{\omega}}{\omega} \sin \omega t \right]. \tag{8.9}$$

It is convenient to introduce the following notation:

$$y_{st} = \frac{F_0}{k}, \quad \bar{\omega} = \frac{\pi}{t_d}, \quad \omega = \frac{2\pi}{T}.$$

Then eq. (8.9) becomes

$$\frac{y}{y_{st}} = \frac{1}{1 - \left(\dfrac{T}{2t_d}\right)^2} \left[\sin \pi \frac{t}{t_d} - \frac{T}{2t_d} \sin 2\pi \frac{t}{T} \right] \qquad \text{for} \quad 0 \leqslant t \leqslant t_d.$$

(8.10a)

After a time t_d, the external force becomes zero and the system is then in free vibration. Therefore, the response for $t > t_d$ is of the form given by eq. (8.5) with the constants of integration determined from known values of displacement and velocity calculated from eq. (8.10a) at time $t = t_d$. The final expression for the response is then given by

$$\frac{y}{y_{st}} = \frac{T/t_d}{\left(\dfrac{T}{2t_d}\right)^2 - 1} \cos \pi \frac{t_d}{T} \sin 2\pi \left(\frac{t}{T} - \frac{t_d}{2T} \right) \qquad \text{for} \quad t \geqslant t_d.$$

(8.10b)

It may be seen from eq. (8.10) that the response in terms of y/y_{st} is a function of the ratio of the pulse duration to the natural period of the system (t_d/T) and of time expressed as t/T. Hence for any fixed value of the parameter t_d/T, we can obtain the maximum response from eq. (8.10). The plot in Fig. 8.3 of these maximum values as a function of t_d/T is the response spectrum for the half-sinusoidal force duration considered in this case. It can be seen from the response

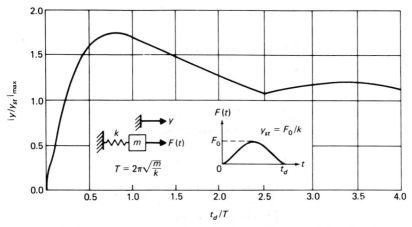

Fig. 8.3 Response spectrum for half-sinusoidal force of duration t_d.

spectrum in Fig. 8.3 that the maximum value of the response (amplification factor) $y/y_{st} = 1.76$ occurs for this particular pulse when $t_d/T = 0.8$.

Due to the simplicity of the input force, it was possible in this case to obtain a closed solution and to plot the response spectrum in terms of dimensionless ratios, thus making this plot valid for any impulsive force described by one-half of the sine cycle. However, in general, for an arbitrary input load, we cannot expect to obtain such a general plot of the response spectrum and we normally have to be satisfied with the response spectrum plotted for a completely specified input excitation.

8.2   RESPONSE SPECTRUM FOR SUPPORT DISTURBANCE

An important problem in structural dynamics is the analysis of a system subjected to excitation applied to the base or support of the structure. An example of such input excitation of the base acting on a damped oscillator which serves to model certain structures is shown in Fig. 8.4. The excitation in this case is given as an acceleration function which is represented in Fig. 8.5. The equation of motion which is obtained by equating to zero the sum of the forces in the corresponding free body diagram in Fig. 8.4(b) is

$$m\ddot{y} + c(\dot{y} - \dot{y}_s) + k(y - y_s) = 0 \tag{8.11}$$

or, with the usual substitution $\omega = \sqrt{k/m}$ and $\xi = c/c_{cr}$ $(c_{cr} = 2\sqrt{km})$,

$$\ddot{y} + 2\xi\omega\dot{y} + \omega^2 y = \omega^2 y_s(t) + 2\xi\omega\dot{y}_s(t). \tag{8.12}$$

Equation (8.12) is the differential equation of motion for the damped oscillator in terms of its absolute motion. A more useful formulation of this problem is to express eq. (8.12) in terms of the relative motion of the mass with respect to the

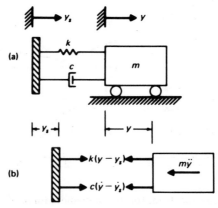

Fig. 8.4   (a) Damped simple oscillator subjected to support excitation. (b) Free body diagram.

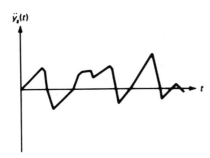
$\ddot{y}_s(t)$

Fig. 8.5   Acceleration function exciting the support of the oscillator in Fig. 8.4.

motion of the support, that is, in terms of the spring deformation. The relative displacement u is defined as

$$u = y - y_s. \tag{8.13}$$

Substitution into eq. (8.12) yields

$$\ddot{u} + 2\xi\omega\dot{u} + \omega^2 u = -\ddot{y}_s(t). \tag{8.14}$$

The formulation of the equation of motion in eq. (8.14) as a function of the relative motion between the mass and the support is particularly important since in design it is the deformation or stress in the "spring element" which is required. Besides, the input motion at the base is usually specified by means of an acceleration function (e.g., earthquake accelerograph record) thus eq. (8.14) containing in the right-hand side the acceleration of the excitation is a more convenient form than eq. (8.12) which in the right-hand side has the support displacement and the velocity.

The solution of the differential equation, eq. (8.14), may be obtained by any of the methods presented in previous chapters for the solution of one-degree-of-freedom systems. In particular, the solution is readily expressed using Duhamel's integral as

$$u(t) = -\frac{1}{\omega}\int_0^t \ddot{y}_s(\tau)\, e^{-\xi\omega(t-\tau)} \sin \omega(t-\tau)\, d\tau. \tag{8.15}$$

8.3   TRIPARTITE RESPONSE SPECTRA

It is possible to plot in a single chart using logarithmic scales the maximum response in terms of the acceleration, the relative displacement, and a third quantity known as the relative pseudovelocity. The pseudovelocity is not exactly the same as the actual velocity, but it is closely related and provides for a convenient substitute for the true velocity. These three quantities, the maximum absolute acceleration, the maximum relative displacement, and the maximum relative pseudovelocity, are sometimes known, respectively, as the spectral acceleration, spectral displacement, and spectral velocity.

It is significant that the spectral displacement S_D, that is, the maximum relative displacement is proportional to the spectral acceleration S_a, the maximum absolute acceleration. To demonstrate this fact, consider the equation of motion, eq. (8.11), which, after using eq. (8.13), becomes for the damped system

$$m\ddot{y} + c\dot{u} + ku = 0 \qquad (8.16)$$

and for the undamped system

$$m\ddot{y} + ku = 0. \qquad (8.17)$$

We observe from eq. (8.17) that the absolute acceleration is at all times proportional to the relative displacement. In particular, at maximum values, the spectral acceleration is proportional to the spectral displacement, that is, from eq. (8.17)

$$S_a = -\omega^2 S_D, \qquad (8.18)$$

where $\omega = \sqrt{k/m}$ is the natural frequency of the system, $S_a = \ddot{y}_{max}$, and $S_D = u_{max}$.

When damping is considered in the system, it may be rationalized that the maximum relative displacement occurs when the relative velocity is zero ($\dot{u} = 0$). Hence we again obtain eq. (8.18) relating spectral acceleration and spectral displacement. This relationship is by mere coincidence the same as for simple harmonic motion. The fictitious velocity associated with the apparent harmonic motion is the pseudovelocity and, for convenience, its maximum value is defined as the spectral velocity, that is

$$S_v = \omega S_D = \frac{S_a}{\omega}. \qquad (8.19)$$

Dynamic response spectra for single degree-of-freedom elastic systems have been computed for a number of input motions. A typical example of displacement response spectrum for a single degree-of-freedom system subjected to support motion is shown in Fig. 8.6. This plot is the response for the input motion given by the recorded ground acceleration of the 1940 El Centro earthquake. The acceleration record of this earthquake has been used extensively in earthquake engineering investigations. A plot of the acceleration record for this earthquake is shown in Fig. 8.7. Until the time of the San Fernando, California, earthquake of 1971, the El Centro record was one of the few records available for long and strong earthquake motions. In Fig. 8.8, the same type of data which was used to obtain the displacement response spectrum in Fig. 8.6, is plotted in terms of the spectral velocity, for several values of the damping coefficient, with the difference that the abscissa as well as the ordinate are in these cases plotted

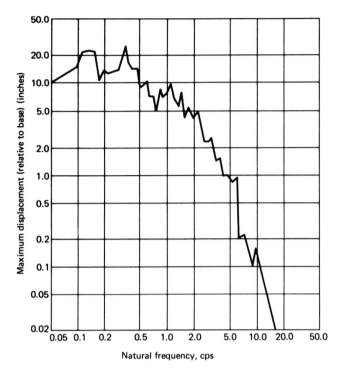

Fig. 8.6 Displacement response spectrum for elastic system subjected to the ground motion of 1940 El Centro earthquake (from *Design of Multistory Reinforced Building for Earthquake Motions* by J. A. Blum, N. M. Newmark, and L. H. Corning, Portland Cement Association, 1961).

on a logarithmic scale. In this type of plot, because of eqs. (8.18) and (8.19), it is possible to draw diagonal scales for the displacement sloping 135° with the abscissa and for the acceleration, 45° so that we can read from a single plot values of spectral acceleration, spectral velocity, and spectral displacements.

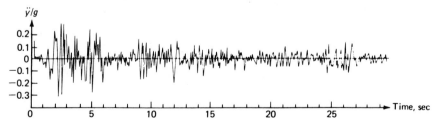

Fig. 8.7 Ground acceleration record for El Centro, California, earthquake of May 18, 1940 north–south component.

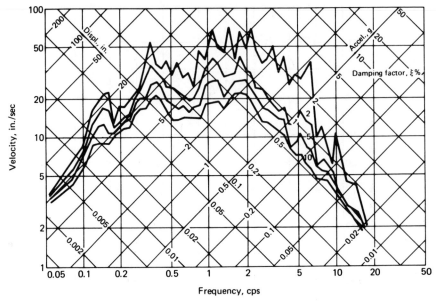

Fig. 8.8   Response spectra for elastic system for the 1940 El Centro earthquake (from *Design of Multistory Reinforced Building for Earthquake Motions* by J. A. Blum, N. M. Newmark, and L. H. Corning, Portland Cement Association, 1961).

To demonstrate the construction of a tripartite diagram such as the one of Fig. 8.8, we write eq. (8.19) in terms of the natural frequency f in cycles per second (cps) and take the logarithm of the terms, so that

$$S_v = \omega S_D = 2\pi f S_D,$$

$$\log S_v = \log f + \log (2\pi S_D). \tag{8.20}$$

For constant values of S_D, eq. (8.20) is the equation of a straight line of $\log S_v$ versus $\log f$ with a slope of $45°$. Analogously, from eq. (8.19)

$$S_v = \frac{S_a}{\omega} = \frac{S_a}{2\pi f},$$

$$\log S_v = -\log f + \log \frac{S_a}{2\pi}. \tag{8.21}$$

For a constant value of S_a eq. (8.21) is the equation of a straight line of $\log S_v$ versus $\log f$ with a slope of $135°$.

8.4   RESPONSE SPECTRA FOR ELASTIC DESIGN

In general, response spectra are prepared by calculating the response to a specified excitation of single degree-of-freedom systems with various amounts of damping.

Numerical integration with short time intervals are applied to calculate the response of the system. The step-by-step process is continued until the total earthquake record has been completed. The largest value of the function of interest is recorded and becomes the response of the system to that excitation. Changing the parameters of the system to change the natural frequency, we repeat the process and record a new maximum response. This process is repeated until all frequencies of interest have been covered and the results plotted. Since no two earthquakes are alike, this process must be repeated for all earthquakes of interest.

Until recently, there were few recorded strong earthquake motions because there were few accelerometers emplaced to measure them; the El Centro, California, earthquake of 1940 was the most severe earthquake recorded and was used as the basis for much analytical work. Recently, however, other strong earthquakes have been recorded. Maximum values of ground acceleration of about 0.32 g for the El Centro earthquake to values of more than 0.5 g for other earthquakes have been recorded. It can be expected that even larger values will be recorded as more instruments are placed closer to the epicenters of earthquakes.

Earthquakes consist of a series of essentially random ground motions. Usually the north–south, east–west, and vertical components of the ground acceleration are measured. Currently, no accurate method is available to predict the particular motion that a site can be expected to experience in future earthquakes. Thus it is reasonable to use a *consolidated response spectrum* which incorporates the spectra for several earthquakes and which represents a kind of "average" response spectrum for design. Such a consolidated response spectrum is shown in Fig. 8.9 normalized for a maximum ground acceleration of 1.0 g. This figure shows the consolidate maximum ground motion and a series of response spectral plots corresponding to various values of the damping ratio in the system.

Details for the construction of the basic spectrum for design purposes are given by Newmark and Hall[1] who have shown that smooth response spectra of idealized ground motion may be obtained by amplifying the ground motion by factors depending on the damping in the system. In general, for any given site, estimates might be made of the maximum ground acceleration, maximum ground acceleration, maximum ground velocity, and maximum ground displacement. The lines representing these maximum values are drawn on a tripartite logarithmic paper of which Fig. 8.10 is an example. The lines in this figure are shown for a maximum ground acceleration of 1.0 g, a velocity of 48 in/sec, and a displacement of 36 in. These values correspond to motions which are more intense than those generally expected in seismic design. They are, however, of proportional magnitudes which are generally correct for most practical applications. These maximum values normalized for a ground acceleration of 1.0 g are simply scaled down for other than 1.0 g acceleration of the ground. Recommended amplification factors to obtain the response spectra from maximum values of the ground

[1] Newmark, N. M., and Hall, W. J., "Procedures and criteria for earthquake resistant design," *Building Practices for Disaster Mitigation*, Dept. of Commerce, Feb. 1973.

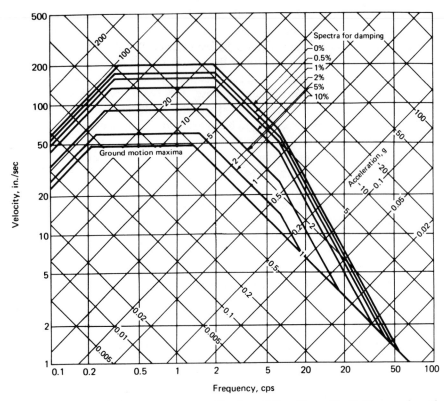

Fig. 8.9 Basic design spectra normalized to 1.0 *g* (from N. M. Newmark and W. J. Hall, "Procedures and criteria for earthquake resistant design," *Building Practices for Disaster Mitigation*, Dept. of Commerce, Feb. 1973).

motion are given in Table 8.1. For each value of the damping coefficient, the amplified displacement lines are drawn at the left, the amplified velocities at the top, and the amplified acceleration at the right of the chart. At a frequency of approximately 6 cps (Fig. 8.9), the amplified acceleration region line intersects a line sloping down toward the maximum ground acceleration value at a frequency of about 30 cps for a system with 2% damping. The lines corresponding to other values of damping are drawn parallel to the 2% damping line as shown in Fig. 8.9. The spectra so determined can be used as design spectra for elastic responses. The amplification factors of Table 8.1 were used in construction of the response spectra in Fig. 8.9.

Illustrative Example 8.1. A structure modeled as a single degree-of-freedom system has a natural period, $T = 1$ sec. Use the response spectral method to determine the maximum absolute acceleration, the maximum relative displacement, and the maximum pseudorelative velocity for: (a) a foundation motion equal to

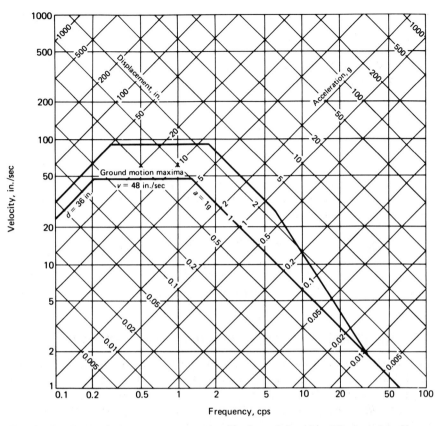

Fig. 8.10   Basic design spectrum normalized to 1.0 *g* for 5% damping (from N. M. Newmark and W. J. Hall, "Procedures and criteria for earthquake resistant design," *Building Practices for Disaster Mitigation*, Dept. of Commerce, Feb. 1973).

TABLE 8.1   Relative Values Spectrum Amplification Factors.[1]

Percent Damping	Amplification Factors		
	Displacement	Velocity	Acceleration
0	2.5	4.0	6.4
0.5	2.2	3.6	5.8
1	2.0	3.2	5.2
2	1.8	2.8	4.3
5	1.4	1.9	2.6
7	1.2	1.5	1.9
10	1.1	1.3	1.5
20	1.0	1.1	1.2

[1] See Ref., p. 143.

the El Centro earthquake of 1940, and (b) the consolidated earthquake with a maximum ground acceleration equal to 0.32 g. Assume 10% of the critical damping.

From the response spectra in Fig. 8.8 with $f = 1/T = 1.0$ cps, corresponding to the curve labeled $\xi = 0.10$, we read on the three scales the following values:

$$S_D = 3.3 \text{ in,}$$

$$S_v = 18.5 \text{ in/sec,}$$

$$S_a = 0.30 \, g.$$

From the basic design spectra in Fig. 8.9 with frequency $f = 1$ cps and 10% critical damping, we obtain after correcting for 0.32 g maximum ground acceleration in the following results:

$$S_D = 9.5 \times 0.32 = 3.04 \text{ in,}$$

$$S_v = 60 \times 0.32 = 19.2 \text{ in/sec,}$$

$$S_a = 0.95 \times 0.32 \, g = 0.304 \, g.$$

8.5 RESPONSE SPECTRA FOR INELASTIC SYSTEMS

For certain types of extreme events such as nuclear blast explosions or strong motion earthquakes, it is sometimes necessary to design structures to withstand strains beyond the elastic limit. For example, in seismic design for an earthquake of moderate intensity, it is reasonable to assume elastic behavior for a well-designed and constructed structure. However, for very strong motions, this is not a realistic assumption even for a well-designed structure. While structures can be designed to resist severe earthquakes, it is not feasible economically to design buildings to elastically withstand earthquakes of the greatest foreseeable intensity. In order to design structures for strain levels beyond the linear range, the response spectrum has been extended to include the inelastic range.[1] Generally, the elastoplastic relation between force and displacement, which was discussed in detail in Chapter 7, is used in structural dynamics. Such a force-displacement relationship is shown in Fig. 8.11. Because of the assumption of elastoplastic behavior, if the force is removed prior to the occurrence of yielding, the material will return along its loading line to the origin. However, when yielding occurs at a displacement y_t, the restoring force remains constant at a magnitude R_t. If the displacement is not reversed, the displacement may reach a maximum value y_{max}. If, however, the displacement is reversed, the elastic recovery follows along a line parallel to the initial line and the recovery proceeds elastically until a negative yield value R_c is reached in the opposite direction.

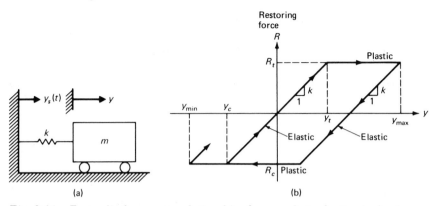

Fig. 8.11 Force-displacement relationship for an elastoplastic single degree-of-freedom system.

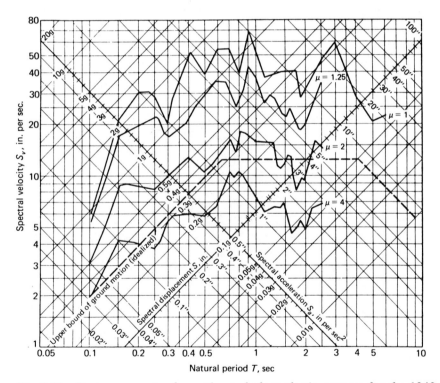

Fig. 8.12 Response spectra for undamped elastoplastic systems for the 1940 El Centro earthquake (from *Design of Multistory Reinforced Building for Earthquake Motions* by J. A. Blum, N. M. Newmark, and L. H. Corning, Portland Cement Association, 1961).

The preparation of response spectra for such an inelastic system is more diffi-cult than that for elastic systems. However, response spectra have been prepared for several kinds of input disturbances. These spectra are usually plotted as a series of curves corresponding to definite values of the ductility ratio μ. The ductility ratio μ is defined as the ratio of the maximum displacement of the structure in the inelastic range to the displacement corresponding to the yield point y_y, that is,

$$\mu = \frac{y_{max}}{y_y}. \tag{8.22}$$

The response spectra for an undamped single degree-of-freedom system subjected to a support motion equal to the El Centro 1940 earthquake is shown in Fig.

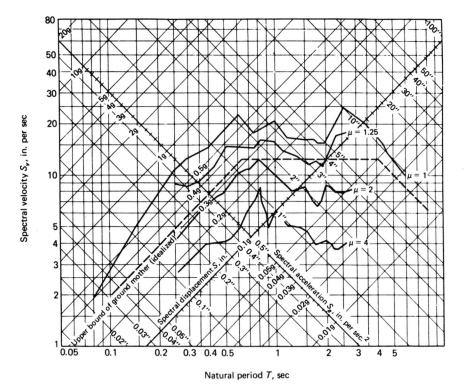

Natural period T, sec

Fig. 8.13   Response spectra for elastoplastic systems with 10% critical damp-ing for the 1940 El Centro earthquake (from *Design of Multistory Reinforced Building for Earthquake Motions* by J. A. Blum, N. M. Newmark, and L. H. Corning, Portland Cement Association, 1961).

8.12 for several values of the ductility ratio. The tripartite logarithmic scales used to plot these spectra give simultaneously for any single degree-of-freedom system of natural period T and specified ductility ratio μ, the spectral values of displacement, velocity, and acceleration. Similarly, in Fig. 8.13, is shown the response spectra for an elastoplastic system with 10% of critical damping. The spectral velocity and the spectral acceleration are read directly from the plots in Figs. 8.12 and 8.13, while the values obtained for the spectral displacement must be multiplied by the ductility ratio in order to obtain the correct value for the spectral displacement.

8.6 RESPONSE SPECTRA FOR INELASTIC DESIGN

In a previous section, the seismic design spectra for an elastic system was shown in Fig. 8.9 for several values of the damping ratio. The same procedure of constructing a basic response spectrum which consolidates the "average" effect of several earthquakes records may also be applied to design in the inelastic range. The spectra for elastoplastic systems have the same appearance as the spectra for elastic systems, but the curves are displaced downward by an amount which is related to the ductility factor μ. Figure 8.14 shows the construction of a typical design spectrum currently recommended[1] for use when inelastic action is anticipated. In this figure, the line D-V-A-A_0 is a response spectrum for elastic systems obtained from Fig. 8.9. On the basis of the elastic spectrum, two different lines are drawn to represent the spectra for inelastic action. One line labeled displacement is used to obtain maximum displacements while the other line labeled ac-

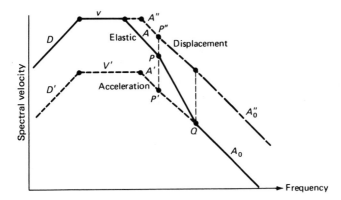

Fig. 8.14 Typical design spectra for plastic system (from N. M. Newmark and W. J. Hall, "Procedures and criteria for earthquake resistant design," *Building Practices for Disaster Mitigation*, Dept. of Commerce, Feb. 1973).

celeration may be used to obtain the maximum accelerations. For very low frequencies (or very long periods) the maximum spectral displacement approaches the maximum ground displacement, that is, for very flexible structures the displacements in the region of low frequencies are conserved at the other end of the spectrum; for very high frequencies (or very short periods) the spectral acceleration values approach magnitudes equal to the maximum ground acceleration. That is to say, for stiff structures the ground acceleration is transmitted directly to the structure. Hence, in this region, the accelerations are conserved. In the low frequency region where the displacements are conserved, the acceleration of the structure is reduced since the force for an elastoplastic structure does not increase when yielding occurs. The acceleration also reaches a maximum value when yielding occurs (Fig. 8.11). Thus the acceleration is reduced by the factor $1/\mu$ where μ again is the ductility ratio. At the other end of the spectrum, where the accelerations are conserved, the maximum accelerations of the system equal the maximum ground accelerations, while the deflections are greater than the elastic deflections. Between these extremes, the energy in the system must be conserved. The lines A and A' in Fig. 8.14 differ by a factor $\sqrt{2\mu - 1}$ which is derived from the conservation of energy methods.

The rules for construction of a response spectrum for elastoplastic behavior are given by Newmark and Hall.[1] Briefly, these rules in reference to Fig. 8.14 are:

(1) Draw on the tripartite logarithmic paper the elastic design spectra corresponding to the amount of damping specified (line D-V-A-A_0 in Fig. 8.14).

(2) Draw lines D' and V' parallel to lines D and V by dividing the ordinates of D and V by the specified ductility ratio μ.

(3) Divide the ordinate of point P on the elastic spectrum by $\sqrt{2\mu - 1}$ to locate point P'.

(4) Draw from the newly located point P', line A' at $45°$ until it intersects line V'.

(5) Join points P' and Q to complete the spectrum for the accelerations.

(6) Draw segments A'' and A_0'' obtained from the ordinates of the corresponding segments A' and A_0 multiplied by the ductility ratio μ.

In summary, line D'-V'-A'-A_0 then represents the plot for maximum accelerations for inelastic action and the line D-V-A''-A_0'' represents the plot for maximum displacements with inelastic action. These curves will correspond to the amount of damping of the basic elastic spectrum curve D-V-A-A_0 and to the specified ductility ratio used in their construction.

Illustrative Example 8.2. Calculate the response of the single degree-of-freedom system of Example 8.1, assuming that the structure is designed to with-

stand seismic motions with an elastoplastic behavior having a ductility ratio
$\mu = 4.0$.

(a) Using the response spectrum of the El Centro earthquake of 1940: From
the response spectrum corresponding to 10% of the critical damping (Fig.
8.13), we read for $T = 1$ sec and the curve labeled $\mu = 4.0$

$$S_D = 1.0 \times 4.0 = 4.0 \text{ in},$$

$$S_v = 6.2 \text{ in/sec},$$

$$S_a = 0.1 \, g.$$

The factor 4.0 is required in the calculation of S_D since as previously
noted the spectra plotted in Fig. 8.13 are correct for acceleration and for
pseudovelocity, but for displacements it is necessary to amplify the values
read from the chart by the ductility ratio.

(b) Using the consolidated design spectrum: Begin by constructing in Fig.
8.15 the required response spectrum which is based on the elastic spectrum curve with 10% of initial damping (Fig. 8.9). The construction is
accomplished following the rules given above in this section. From Fig.
8.15, corresponding to $f = 1$ cps, we obtain the following maximum values

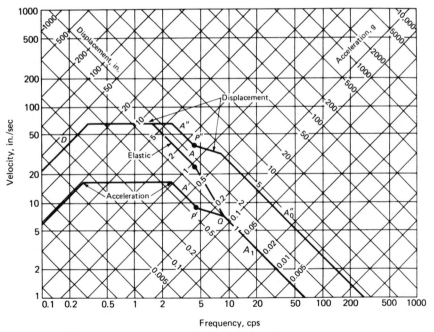

Fig. 8.15 Design spectrum for elastoplastic system with $\mu = 4.0$ and $\xi = 10\%$.

for the response

$$S_D = 10 \times 0.32 = 3.2 \text{ in},$$

$$S_v = 15.5 \times 0.32 = 4.96 \text{ in/sec},$$

$$S_a = 0.23 \times 0.32 \, g = 0.0736 \, g.$$

As can be seen, these spectral values based on the consolidated design spectrum are somewhat different than those obtained from the response spectrum of the El Centro earthquake of 1940. Also, if we compare these results for the elastoplastic behavior with the results in Example 8.1 for the elastic structure, we observe that the maximum relative displacement has essentially the same magnitude while the acceleration and the relative pseudovelocity are appreciably less. This observation is in general true for any structure when inelastic response is compared with the response based on elastic behavior.

8.7   SUMMARY

The response spectra are plots which give the maximum response for a single degree-of-freedom system subjected to a specified excitation. The construction of these plots requires the solution of single degree-of-freedom systems for a sequence of values of the natural frequency and of the damping ratio in the range of interest. Every solution provides only one point (the maximum value) of the response spectrum. In solving the single degree-of-freedom systems, use is made of Duhamel's integral (Chapter 4) for the elastic system and of the step-by-step linear acceleration method for inelastic behavior (Chapter 7). Since a large number of systems must be analyzed in order to fully plot each response spectrum, the task is lengthy and time consuming even with the use of the computer. However, once these curves are constructed and are available for the excitation of interest, the analysis for the design of structures subjected to dynamic loading is reduced to a simple calculation of the natural frequency of the system and the use of the response spectrum.

In the following chapters dealing with structures which are modeled as systems with many degrees of freedom, it will be shown that the dynamic analysis of a system with n degrees of freedom can be transformed to the problem of solving n systems in which each one is a single degree-of-freedom system. Consequently, this transformation extends the usefulness of response spectra for single degree-of-freedom systems to the solution of systems of any number of degrees of freedom.

The reader should thus realize the full importance of a thorough understanding and mastery of the concepts and methods of solutions for single degree-of-freedom systems, since these same methods are also applicable to systems of many degrees of freedom after the problem has been transformed to independent single degree-of-freedom systems.

PROBLEMS

8.1   The steel frame shown in Fig. P8.1 is subjected to horizontal force at the girder level of $(1000 \sin 10t \text{ lb})$ for a time duration of half a cycle of the forcing sine function. Use the appropriate response spectral chart to obtain the maximum displacement. Neglect damping.

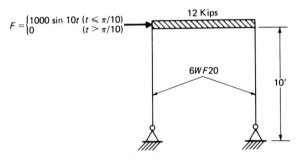

$$F = \begin{cases} 1000 \sin 10t & (t \le \pi/10) \\ 0 & (t > \pi/10) \end{cases}$$

12 Kips

6W F20

10'

Fig. P8.1

8.2   Determine the maximum stresses in the columns of the frame of Problem 8.1.

8.3   Consider the frame shown in Fig. P8.3(a) subjected to a foundation excitation produced by a half cycle of the function $a_g = 200 \sin 10t \text{ in/sec}^2$ as shown in Fig. P8.3(b). Determine the maximum horizontal displacement of the girder relative to the motion of the foundation. Neglect damping.

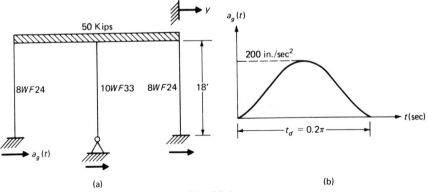

50 Kips

$a_g(t)$

200 in./sec²

8W F24   10W F33   8W F24   18'

$t_d = 0.2\pi$

$t(\text{sec})$

$a_g(t)$

(a)   (b)

Fig. P8.3

8.4 Determine the maximum stress in the columns of the frame of Problem 8.3.

8.5 The frame shown in Fig. P8.1 is subjected to the excitation produced by the El Centro earthquake of 1940. Assume 10% damping and from the appropriate chart determine the spectral values for displacement, velocity, and acceleration. Assume elastic behavior.

8.6 Repeat Problem 8.5 using the basic design spectra given in Fig. 8.9 to determine the spectral values for acceleration, velocity, and displacement.

8.7 A structure modeled as the spring-mass system shown in Fig. P8.7 is assumed to be subjected to a support motion produced by the El Centro earthquake of 1940. Assuming elastic behavior and using the appropriate response spectral chart find the maximum relative displacement between the mass and the support. Also compute the maximum force acting on the spring. Neglect damping.

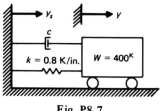

Fig. P8.7

8.8 Repeat Problem 8.7 assuming that the system has 10% of the critical damping.

8.9 Determine the force transmitted to the foundation for the system analyzed in Problem 8.8.

8.10 Consider the spring-mass system of Problem 8.7 and assume that the spring element follows and elastoplastic behavior with a maximum value for the restoring force in tension or in compression is equal to half the value of the elastic maximum force in the spring calculated in Problem 8.7. Determine the spectral values for displacement, velocity, and acceleration. Neglect damping. (Hint: Start by assuming $\mu = 2$, find spectral values, calculate μ, and find new spectral values, etc.)

8.11 Repeat Problem 8.10 for 10% damping.

8.12 A structure modeled as a single degree-of-freedom system has a natural period $T = 0.5$ sec. Use the response spectral method to determine in the elastic range the maximum absolute acceleration, the maximum relative displacement, the the maximum pseudorelative velocity for: (a) a foundation motion equal to the El Centro earthquake of 1940, and (b) the consolidated earthquake with a maximum ground acceleration equal to 0.3 g. Neglect damping.

8.13 Solve Problem 8.12 assuming elastoplastic behavior of the system with ductility ratio $\mu = 5$.

Part II

Structures modeled as shear buildings

9

The multistory shear building

In Part I we analyzed and obtained the dynamic response for structures modeled as a single degree-of-freedom system. Only if the structure can assume a unique shape during its motion will single degree model provide the exact dynamic response. Otherwise, when the structure takes more than one possible shape during motion, the solution obtained from a single degree model will be an approximation to the true dynamic behavior.

Structures cannot always be described by a single degree model and, in general, have to be represented by multiple degree models. In fact, structures are continuous systems and as such possess an infinite number of degrees of freedom. There are analytical methods to describe the dynamic behavior of continuous structures which have uniform material properties and regular geometry. These methods of analysis, though interesting in revealing information for the discrete modeling of structures, are rather complex and are applicable only to relatively simple actual structures. They require considerable mathematical analysis, including the solution of partial differential equations which will be presented in Chapter 20. For the present, we shall consider one of the most instructive and practical types of structure which involve many degrees of freedom, the multistory shear building.

9.1   STIFFNESS EQUATIONS FOR THE SHEAR BUILDING

A shear building may be defined as a structure in which there is no rotation of a horizontal section at the level of the floors. In this respect, the deflected building will have many of the features of a cantilever beam that is deflected by shear forces only; hence, the name shear building. To accomplish such deflection in a building, we must assume that: (1) the total mass of the structure is concentrated at the levels of the floors; (2) the girders on the floors are infinitely rigid as compared to the columns; and (3) the deformation of the structure is independent of the axial forces present in the columns. The first assumption transforms the problem from a structure with an infinite number of degrees of freedom (due to the distributed mass) to a structure which has only as many degrees as it has lumped masses at the floor levels. A three-story structure modeled as a shear building [Fig. 9.1(a)] will have three degrees of freedom, that is, the three horizontal displacements at the floor levels. The second assumption introduces the requirement that the joints between girders and columns are fixed against rotation. The third assumption leads to the condition that the rigid girders will remain horizontal during motion.

It should be noted that the building may have any number of bays and that it is only as a matter of convenience that we represent the shear building solely in terms of a single bay. Actually, we can further idealize the shear building as a

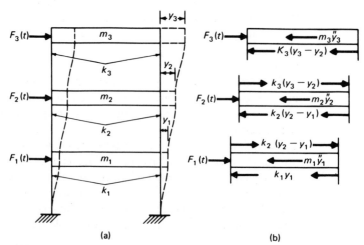

(a)   (b)

Fig. 9.1   Single bay model representation of a shear building.

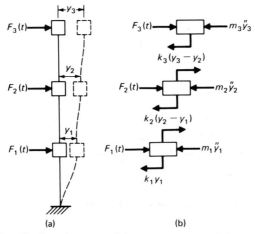

Fig. 9.2   Single column model representation of shear building.

single column [Fig. 9.2(a)], having concentrated masses at the floor levels with the understanding that only horizontal displacements of these masses are possible. Another alternative is to adopt a multimass spring system shown in Fig. 9.3(a) to represent the shear building. In any of the three representations depicted in these figures, the stiffness coefficient or spring constant k_i shown between any two consecutive masses is the force required to produce a relative unit displacement of the two adjacent floor levels.

For a uniform column with the two ends fixed against rotation, the spring constant is given by

$$k = \frac{12EI}{L^3},$$ (9.1a)

and for a column with one end fixed and the other pinned by

$$k = \frac{3EI}{L^3},$$ (9.1b)

where E is the material modulus of elasticity, I the cross-sectional moment of inertia, and L the height of the story.

It should be clear that all three representations shown in Figs. 9.1 to 9.3 for the shear building are equivalent. Consequently, the following equations of motion for the three-story shear building are obtained from any of the corresponding free body diagrams shown in these figures by equating to zero the sum of the forces acting on each mass. Hence

$$m_1\ddot{y}_1 + k_1 y_1 - k_2(y_2 - y_1) - F_1(t) = 0,$$
$$m_2\ddot{y}_2 + k_2(y_2 - y_1) - k_3(y_3 - y_2) - F_2(t) = 0,$$
$$m_3\ddot{y}_3 + k_3(y_3 - y_2) - F_3(t) = 0.$$ (9.2)

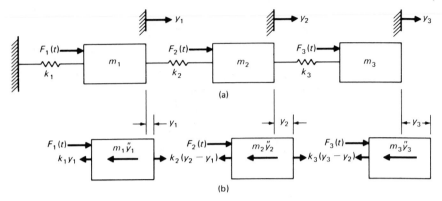

Fig. 9.3   Multimass spring model representation of a shear building.

This system of equations constitutes the *stiffness* formulation of the equations of motion for a three-story shear building. It may conveniently be written in matrix notation as

$$[M]\{\ddot{y}\} + [K]\{y\} = [F],\qquad(9.3)$$

where $[M]$ and $[K]$ are, respectively, the mass and stiffness matrices given, respectively, by

$$[M] = \begin{bmatrix} m_1 & 0 & 0 \\ 0 & m_2 & 0 \\ 0 & 0 & m_3 \end{bmatrix}\qquad(9.4)$$

$$[K] = \begin{bmatrix} k_1+k_2 & -k_2 & 0 \\ -k_2 & k_2+k_3 & -k_3 \\ 0 & -k_3 & k_3 \end{bmatrix}\qquad(9.5)$$

and $\{y\}$, $\{\ddot{y}\}$, and $\{F\}$ are, respectively, the displacement, acceleration, and force vectors given by

$$\{y\} = \begin{bmatrix} y_1 \\ y_2 \\ y_3 \end{bmatrix}, \qquad \{\ddot{y}\} = \begin{bmatrix} \ddot{y}_1 \\ \ddot{y}_2 \\ \ddot{y}_3 \end{bmatrix}, \qquad \{F\} = \begin{bmatrix} F_1(t) \\ F_2(t) \\ F_3(t) \end{bmatrix}.\qquad(9.6)$$

It should be noted that the mass matrix, eq. (9.4), corresponding to the shear building is a diagonal matrix (the nonzero elements are only in the main diagonal). The elements of the stiffness matrix, eq. (9.5), are designated *stiffness coefficients*. In general, the stiffness coefficient k_{ij} is defined as the force at coordinate i when a unit displacement is given at j, all other coordinates being fixed.

For example, the coefficient in the second row and second column of eq. (9.5) $k_{22} = k_2 + k_3$ is the force required at the second floor when a unit displacement is given to this floor.

9.2 FLEXIBILITY EQUATIONS FOR THE SHEAR BUILDING

An alternative approach in developing the equation of motion of a structure is the *flexibility* formulation. In this approach, the elastic properties of the structure are described by *flexibility coefficients* which are defined as deflections produced by a unit load applied at one of the coordinates. Specifically, the flexibility coefficient f_{ij} is defined as the displacement at coordinate i when a unit static force is applied at coordinate j. Figure 9.4 depicts the flexibility coefficients corresponding to unit force applied at one of the story levels of a shear building. Using these coefficients and applying superposition, we may state that the displacement at any coordinate is equal to the sum of the products of flexibility coefficients at that coordinate multiplied by the corresponding forces. The forces acting on the three-story shear building (including the inertial forces) are shown in Fig. 9.5. Therefore, the displacements for the three-story building may be expressed in terms of the flexibility coefficients as

$$y_1 = (F_1(t) - m_1\ddot{y}_1)f_{11} + (F_2(t) - m_2\ddot{y}_2)f_{12} + (F_3(t) - m_3\ddot{y}_3)f_{13},$$

$$y_2 = (F_1(t) - m_1\ddot{y}_1)f_{21} + (F_2(t) - m_2\ddot{y}_2)f_{22} + (F_3(t) - m_3\ddot{y}_3)f_{23},$$

$$y_3 = (F_1(t) - m_1\ddot{y}_1)f_{31} + (F_2(t) - m_2\ddot{y}_2)f_{32} + (F_3(t) - m_3\ddot{y}_3)f_{33}.$$

Rearraging the terms in these equations and using matrix notation, we obtain:

$$\{y\} = [f]\{F\} - [f][M]\{\ddot{y}\}, \tag{9.7}$$

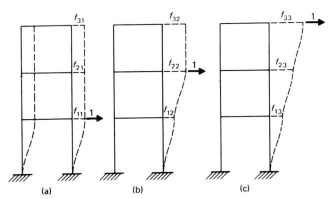

Fig. 9.4 Flexibility coefficients for a three-story shear building.

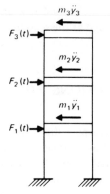

Fig. 9.5   Forces acting on a three-story shear building.

where $[M]$ is the mass matrix, eq. (9.4), $\{f\}$ is the flexibility matrix given by:

$$[f] = \begin{bmatrix} f_{11} & f_{12} & f_{13} \\ f_{21} & f_{22} & f_{23} \\ f_{31} & f_{32} & f_{33} \end{bmatrix} \tag{9.8}$$

and $\{y\}$, $\{\ddot{y}\}$, and $\{F\}$ are, respectively, the displacement, acceleration, and force vectors given by eq. (9.6).

9.3   RELATIONSHIP BETWEEN STIFFNESS AND FLEXIBILITY MATRICES

The definitions given for either stiffness or flexibility coefficients are based on static considerations in which the displacements are produced by static forces. The relation between static forces and displacements may be obtained by equating to zero the acceleration vector $\{\ddot{y}\}$ in either eq. (9.3) or eq. (9.7). Hence

$$[K]\{y\} = \{F\} \tag{9.9}$$

$$[f]\{F\} = \{y\}. \tag{9.10}$$

From these relations it follows that the stiffness matrix $[K]$ and the flexibility matrix $[f]$ are inverse matrices, that is

$$[K] = [f]^{-1}$$

or

$$[f] = [K]^{-1} \tag{9.11}$$

Consequently, the flexibility matrix $[f]$ may be obtained either by calculating the inverse of the stiffness matrix or directly from the definition of the flexibility coefficients. Taking this last approach, we have for the three-story shear building shown in Fig. 9.4(a), that

$$k_1 f_{11} = 1$$

and

$$f_{11} = f_{21} = f_{31} = \frac{1}{k_1}.$$

Analogously from Figs. 9.4(b) and 9.4(c) we obtain

$$f_{22} = f_{32} = \frac{1}{k_1} + \frac{1}{k_2}, \qquad f_{12} = \frac{1}{k_1},$$

and

$$f_{33} = \frac{1}{k_1} + \frac{1}{k_2} + \frac{1}{k_3}, \qquad f_{23} = \frac{1}{k_1} + \frac{1}{k_2}, \qquad \text{and} \qquad f_{13} = \frac{1}{k_1}$$

since the flexibility coefficients for springs in series are given by the summation of the reciprocal values of the spring constants.

Inserting these expressions for the flexibility coefficients into the flexibility matrix, eq. (9.8), results in

$$[f] = \begin{bmatrix} \dfrac{1}{k_1} & \dfrac{1}{k_1} & \dfrac{1}{k_1} \\[2ex] \dfrac{1}{k_1} & \dfrac{1}{k_1} + \dfrac{1}{k_2} & \dfrac{1}{k_1} + \dfrac{1}{k_2} \\[2ex] \dfrac{1}{k_1} & \dfrac{1}{k_1} + \dfrac{1}{k_2} & \dfrac{1}{k_1} + \dfrac{1}{k_2} + \dfrac{1}{k_3} \end{bmatrix} \qquad (9.12)$$

The extension of the flexibility matrix for a three-story shear building to any number of stories is obvious from the pattern of eq. (9.12).

9.4 SUMMARY

The shear building idealization of structures provides a simple and useful mathematical model for the analysis of dynamic systems. This model permits the representation of the structure by lumped rigid masses interconnected by elastic springs. In obtaining the equations of motion, two different formulations are possible: (1) the stiffness method in which the equations of equilibrium are expressed in terms of stiffness coefficients; and (2) the flexibility method in which the equations of compatability are written in terms of flexibility coefficients. The stiffness matrix and the flexibility matrix of a system, in reference to the same coordinates, are inverse matrices.

PROBLEMS

9.1 For the three-story shear building verify that the stiffness matrix, eq. (9.5), and the flexibility matrix, eq. (9.12), are inverse matrices.

9.2 For the two-story shear building shown in Fig. P9.2 determine the stiffness and flexibility matrices and then verify that these matrices are inverse matrices.

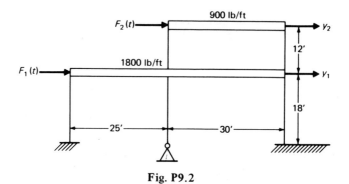

Fig. P9.2

9.3 For the three-story shear building shown in Fig. P9.3 obtain the stiffness and flexibility matrices and show that these matrices are inverse matrices.

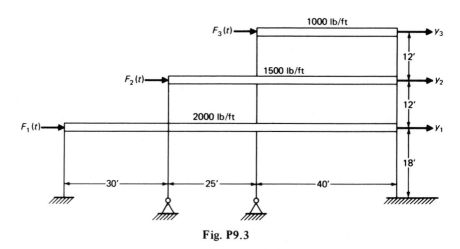

Fig. P9.3

9.4 Write the differential equation of motion using the stiffness formulation for the shear building in Fig. P9.2. Model the structure by a multimass-spring system.

9.5 Write the differential equation for the motion of the shear building in Fig. P9.3. Model the structure as a shear column with lumped masses as the floor levels.

9.6 The three-story shear building in Fig. P9.6 is subjected to a foundation motion which is given as an acceleration function $\ddot{y}_s(t)$. Obtain the stiffness differential equation of motion. Express the displacement of the floors relative to the foundation displacement (i.e., $u_i = y_i - y_s$).

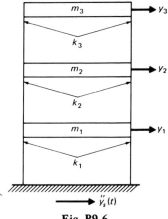

Fig. P9.6

9.7 Generalize the results of Problem 9.6 and obtain the equations of motion for a shear building of n stories.

10

Free vibration of a shear building

When free vibration is under consideration, the structure is not subjected to any external excitation (force or support motion) and its motion is governed only by the initial conditions. There are occasionally circumstances for which it is necessary to determine the motion of the structure under conditions of free vibration, but this is seldom the case. Nevertheless, the analysis of the structure in free motion provides the most important dynamic properties of the structure which are the natural frequencies and the corresponding modal shapes. We begin by considering both formulations for the equations of motion, namely, the stiffness and the flexibility equations.

10.1 NATURAL FREQUENCIES AND NORMAL MODES

The problem of free vibration requires that the force vector $\{F\}$ be equal to zero in either the stiffness, eq. (9.3), or flexibility, eq. (9.7), formulations of the equations of motion. For the stiffness equation with $\{F\} = \{0\}$, we have

$$[M] \{\ddot{y}\} + [K] \{y\} = \{0\}. \tag{10.1}$$

For free vibrations of the undamped structure, we seek solutions of eq. (10.1) in the form

$$y_i = a_i \sin (\omega t - \alpha), \qquad i = 1, 2, \ldots, n$$

or in vector notation

$$\{y\} = \{a\} \sin (\omega t - \alpha), \tag{10.2}$$

where a_i is the amplitude of motion of the ith coordinate and n is the number of degrees of freedom. The substitution of eq. (10.2) into eq. (10.1) gives

$$-\omega^2 [M] \{a\} \sin (\omega t - \alpha) + [K] \{a\} \sin (\omega t - \alpha) = \{0\},$$

or rearranging terms

$$[[K] - \omega^2 [M]] \{a\} = \{0\}, \tag{10.3}$$

which for the general case is set of n homogeneous (right-hand side equal to zero) algebraic system of linear equations with n unknown displacements a_i and an unknown parameter ω^2. The formulation of eq. (10.3) is an important mathematical problem known as an *eigenproblem*. Its nontrivial solution, that is, the solution for which not all $a_i = 0$, requires that the determinant of the matrix factor of $\{a\}$ be equal to zero; in this case,

$$|[K] - \omega^2 [M]| = 0. \tag{10.4}$$

In general, eq. (10.4) results in a polynomial equation of degree n in ω^2 which should be satisfied for n values of ω^2. This polynomial is known as the *characteristic equation* of the system. For each of these values of ω^2 satisfying the characteristic eq. (10.4), we can solve eq. (10.3) for $a_1, a_2, \ldots, a_n$ in terms of an arbitrary constant.

Analogously, for the flexibility formulation, we have for free vibration from eq. (9.7) with $\{F\} = 0$,

$$\{y\} + [f] [M] \{\ddot{y}\} = \{0\}. \tag{10.5}$$

We again assume harmonic motion as given by eq. (10.2) and substitute eq. (10.2) into eq. (10.5) to obtain

$$\{a\} = \omega^2 [f] [M] \{a\} \tag{10.6}$$

or

$$1/\omega^2 \{a\} = [D] \{a\}, \tag{10.7}$$

where $[D]$ is known as the *dynamic matrix* and is defined as

$$[D] = [f] [M]. \qquad (10.8)$$

Equation (10.7) may also be written as

$$[[D] - 1/\omega^2 [I]] \ \{a\} = 0, \qquad (10.9)$$

where $[I]$ is the unit matrix with ones in the main diagonal and zeros every-where else. For a nontrivial solution of eq. (10.9), it is required that the de-terminant of the coefficient matrix of $\{a\}$ be equal to zero, that is,

$$|[D] - 1/\omega^2 [I]| = 0. \qquad (10.10)$$

Equation (10.10) is a polynomial of degree n in $1/\omega^2$. This polynomial is the characteristic equation of the system for the flexibility formulation. For each one of the n solutions for $(1/\omega^2)$ of eq. (10.10), we can obtain from eq. (10.9) corresponding solutions for the amplitudes a_i in terms of an arbitrary constant. The necessary calculations are better explained with the use of a numerical example.

Illustrative Example 10.1.   The building to be analyzed is the simple steel rigid frame shown in Fig. 10.1. The weights of the floors and walls are indicated in the figure and are assumed to include the structural weight as well. The build-ing consists of a series of frames spaced 15 ft apart. It is further assumed that the structural properties are uniform along the length of the building and, therefore, the analysis to be made of an interior frame yields the response of the entire building.

The building is modeled as a shear building and under the assumptions stated, the entire building may be represented by the spring-mass system shown in Fig. 10.2. The concentrated weights which are each taken as the total floor

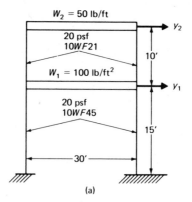

(a)

Fig. 10.1   Two-story shear building for Example 10.1.

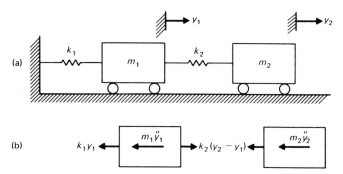

Fig. 10.2   Multimass-spring model for a two-story shear building. (a) Model. (b) Free body diagram.

weight plus that of the tributary walls are computed as follows:

$$W_1 = 100 \times 30 \times 15 + 20 \times 12.5 \times 15 \times 2 = 52{,}500 \text{ lb},$$

$$m_1 = 136 \text{ lb} \cdot \sec^2/\text{in},$$

$$W_2 = 50 \times 30 \times 15 + 20 \times 5 \times 15 \times 2 = 25{,}500 \text{ lb},$$

$$m_2 = 66 \text{ lb} \cdot \sec^2/\text{in}.$$

Since the girders are assumed to be rigid, the stiffness (spring constant) of each story is given by

$$k = \frac{12\,E(2I)}{L^3}$$

and the individual values for the steel column sections indicated are thus

$$k_1 = \frac{12 \times 30 \times 10^6 \times 248.6 \times 2}{(15 \times 12)^3} = 30{,}700 \text{ lb/in},$$

$$k_2 = \frac{12 \times 30 \times 10^6 \times 106.3 \times 2}{(10 \times 12)^3} = 44{,}300 \text{ lb/in}.$$

The equations of motion for the system which are obtained by considering in Fig. 10.2(b) the dynamic equilibrium of each mass in free vibration are

$$m_1\ddot{y}_1 + k_1 y_1 - k_2(y_2 - y_1) = 0,$$

$$m_2\ddot{y}_2 + k_2(y_2 - y_1) = 0.$$

In the usual manner, these equations of motion are solved for free vibration by substituting

$$y_1 = a_1 \sin(\omega t - \alpha)$$

$$y_2 = a_2 \sin(\omega t - \alpha) \qquad\qquad (10.11)$$

for the displacements and

$$\ddot{y}_1 = -a_1 \omega^2 \sin(\omega t - \alpha)$$
$$\ddot{y}_2 = -a_2 \omega^2 \sin(\omega t - \alpha)$$

for the accelerations. In matrix notation, we obtain

$$\begin{bmatrix} k_1 + k_2 - m_1 \omega^2 & -k_2 \\ -k_2 & k_2 - m_2 \omega^2 \end{bmatrix} \begin{bmatrix} a_1 \\ a_2 \end{bmatrix} = \begin{bmatrix} 0 \\ 0 \end{bmatrix}. \qquad (10.12)$$

For a nontrivial solution, we require that the determinant of the coefficient matrix be equal to zero, that is,

$$\begin{vmatrix} k_1 + k_2 - m_1 \omega^2 & -k_2 \\ -k_2 & k_2 - m_2 \omega^2 \end{vmatrix} = 0. \qquad (10.13)$$

The expansion of this determinant gives a quadratic equation in ω^2, namely,

$$m_1 m_2 \omega^4 - ((k_1 + k_2) m_2 + m_1 k_2) \omega^2 + k_1 k_2 = 0, \qquad (10.14)$$

or by introducing the numerical values for this example, we obtain

$$8976\omega^4 - 10,974,800\omega^2 + 1.36 \times 10^9 = 0. \qquad (10.15)$$

The roots of this quadratic are

$$\omega_1^2 = 140,$$
$$\omega_2^2 = 1082.$$

Therefore, the natural frequencies of the structure are

$$\omega_1 = 11.8 \text{ rad/sec},$$
$$\omega_2 = 32.9 \text{ rad/sec}$$

or in cycles per seconds

$$f_1 = \omega_1/2\pi = 1.88 \text{ cps},$$
$$f_2 = \omega_2/2\pi = 5.24 \text{ cps}$$

and the corresponding natural periods

$$T_1 = \frac{1}{f_1} = 0.532 \text{ sec},$$

$$T_2 = \frac{1}{f_2} = 0.191 \text{ sec}.$$

To solve eq. (10.12) for the amplitudes a_1 and a_2, we note that by equating the determinant to zero in eq. (10.13), the number of independent equations is one less. Thus in the present case, the system of two equations is reduced to one independent equation. Considering the first equation in eq. (10.12) and substituting the first natural frequency, $\omega_1 = 11.8$ rad/sec, we obtain

$$55{,}960\,a_{11} - 44{,}300\,a_{21} = 0. \tag{10.16}$$

We have introduced a second subindex in a_1 and a_2 to indicate that the value ω_1 has been used in this equation. Since in the present case there are two unknowns and only one equation, we can solve eq. (10.16) only for the relative value of a_{21} to a_{11}. This relative value is known as the normal mode or modal shape corresponding to the first frequency. For this example eq. (10.16) gives

$$\frac{a_{21}}{a_{11}} = 1.263.$$

It is customary to describe the normal modes by assigning a unit value to one of the amplitudes; thus, for the first mode we set a_{11} equal to unity so that

$$a_{11} = 1.000,$$
$$a_{21} = 1.263. \tag{10.17}$$

Similarly, substituting the second natural frequency, $\omega_2 = 32.9$ rad/sec into eq. (10.12), we obtain the second normal mode as

$$a_{12} = 1.000,$$
$$a_{22} = -1.629. \tag{10.18}$$

It should be noted that while we obtained only ratios, the amplitudes of motion could, of course, be found from initial conditions.

We have now arrived at two possible simple harmonic motions of the structure which can take place in such a way that all the masses move in phase at the same frequency, either ω_1 or ω_2. Such a motion of an undamped system is called a *normal or natural mode of vibration*. The shapes (for this example a_{21}/a_{11} and a_{22}/a_{12}) are called normal mode shapes or simply modal shapes for the corresponding natural frequencies ω_1 and ω_2. The two modes which have been obtained for this example are depicted in Fig. 10.3.

We often use the phrase *first mode* or *fundamental mode* to refer to the mode associated with the lowest frequency. The other modes are sometimes called *harmonics* or *higher harmonics*. It is evident that the modes of vibration, each having its own frequency, behave essentially as single degree-of-freedom systems. The total motion of the system, that is, the total solution of the equations of motion, eq. (10.1), is given by the superposition of the modal harmonic vibrations which in terms of arbitrary constants of integration may be written as

$$y_1(t) = C_1' a_{11} \sin(\omega_1 t - \alpha_1) + C_2' a_{12} \sin(\omega_2 t - \alpha_2),$$
$$y_2(t) = C_1' a_{21} \sin(\omega_1 t - \alpha_1) + C_2' a_{22} \sin(\omega_2 t - \alpha_2). \tag{10.19}$$

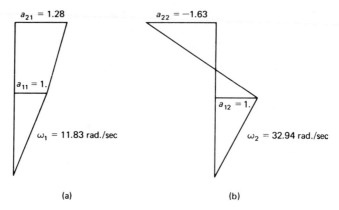

Fig. 10.3   Normal modes for Example 10.1. (a) First mode. (b) Second mode.

Here C_1', C_2' as well as α_1 and α_2 are four constants of integration to be determined from four initial conditions which are the initial displacement and velocity for each mass in the system. For a two-degree-of-freedom system, these initial conditions are

$$y_1(0) = y_{01}, \qquad \dot{y}_1(0) = \dot{y}_{01},$$
$$y_2(0) = y_{02}, \qquad \dot{y}_2(0) = \dot{y}_{02}. \qquad (10.20)$$

For computational purposes, it is convenient to eliminate the phase angles α_1 and α_2 in eq. (10.19) in favor of other constants. Expanding the trigonometric functions and renaming the constants, we obtain

$$y_1(t) = C_1 a_{11} \sin \omega_1 t + C_2 a_{11} \cos \omega_1 t + C_3 a_{12} \sin \omega_2 t + C_4 a_{12} \cos \omega_2 t,$$
$$y_2(t) = C_1 a_{21} \sin \omega_1 t + C_2 a_{21} \cos \omega_1 t + C_3 a_{22} \sin \omega_2 t + C_4 a_{22} \cos \omega_2 t.$$

$$(10.21)$$

in which C_1, C_2, C_3, and C_4 are new constants of integration. From the first two initial conditions in eq. (10.20), we have the following two equations:

$$y_{01} = C_2 a_{11} + C_4 a_{12},$$
$$y_{02} = C_2 a_{21} + C_4 a_{22}. \qquad (10.22)$$

Since the modes are independent, these equations can always be solved for C_2 and C_4. Similarly, by expressing in eq. (10.21) the velocities at time equal to zero, we find

$$\dot{y}_{01} = \omega_1 C_1 a_{11} + \omega_2 C_3 a_{12},$$
$$\dot{y}_{02} = \omega_1 C_1 a_{21} + \omega_2 C_3 a_{22}. \qquad (10.23)$$

The solution of these two sets of equations allows us to express the motion of the system in terms of the two modal vibrations, each proceeding at its own frequency, completely independent of the other, the amplitudes and phases being determined by the initial conditions.

10.2   ORTHOGONALITY PROPERTY OF THE NORMAL MODES

We shall now introduce an important property of the normal modes, the orthogonality property. This property constitutes the basis of one of the most attractive methods for solving dynamic problems of multidegree-of-freedom systems. We begin by rewriting the equations of motion in free vibration, eq. (10.3), as

$$[K] \{a\} = \omega^2 [M] \{a\}. \tag{10.24}$$

For the two-degree-of-freedom system, we obtain from eq. (10.12)

$$(k_1 + k_2) a_1 - k_2 a_2 = m_1 \omega^2 a_1,$$

$$-k_2 a_1 + k_2 a_2 = m_2 \omega^2 a_2. \tag{10.25}$$

These equations are exactly the same as eq. (10.12) but written in this form they may be given a static interpretation as the equilibrium equations for the system acted on by forces of magnitude $m_1 \omega^2 a_1$ and $m_2 \omega^2 a_2$ applied to masses m_1 and m_2, respectively. The modal shapes may then be considered as the static deflections resulting from the forces on the right-hand side of eg. (10.25) for any of the two modes. This interpretation, as a static problem, allows us to use the results of the general static theory of linear structures. In particular, we may make use of Betti's theorem, which states: For a structure acted upon by two systems of loads and corresponding displacements, the work done by the first system of loads moving through the displacements of the second system is equal to the work done by this second system of loads undergoing the displacements produced by the first load system. The two systems of loading and corresponding displacements which we shall consider are as follows:

System I:

forces $\qquad \omega_1^2 a_{11} m_1, \qquad \omega_1^2 a_{21} m_2$

and displacements $\quad a_{11}, \qquad a_{21}.$

System II:

forces $\qquad \omega_2^2 a_{12} m_1, \qquad \omega_2^2 a_{22} m_2$

and displacements $\quad a_{12}, \quad a_{22}.$

The application of Betti's theorem for these two systems yields

$$\omega_1^2 m_1 a_{11} a_{12} + \omega_1^2 m_2 a_{21} a_{22} = \omega_2^2 m_1 a_{12} a_{11} + \omega_2^2 m_2 a_{22} a_{21}$$

or

$$(\omega_1^2 - \omega_2^2)(m_1 a_{11} a_{12} + m_2 a_{21} a_{22}) = 0. \tag{10.26}$$

If the natural frequencies are different $(\omega_1 \neq \omega_2)$, it follows from eq. (10.26) that

$$m_1 a_{11} a_{12} + m_2 a_{21} a_{22} = 0$$

which is the so-called orthogonality relation between modal shapes of a two-degree-of-freedom system. For an n-degree-of-freedom system in which the mass matrix is diagonal, the orthogonality condition between any two modes i and j may be expressed as

$$\sum_{k=1}^{m} m_k a_{ki} a_{kj} = 0, \qquad \text{for} \quad i \neq j \tag{10.27}$$

and in general for any n-degree-of-freedom system as

$$\{a_i\}^T [M] \{a_j\} = 0, \qquad \text{for} \quad i \neq j \tag{10.28}$$

in which $\{a_i\}$ and $\{a_j\}$ are any two modal vectors and $[M]$ is the matrix of the system.

As mentioned before, the amplitudes of vibration in a normal mode are only relative values which may be scaled or normalized to some extent as a matter of choice. The following is an especially convenient normalization for a general system:

$$\phi_{ij} = \frac{a_{ij}}{\sqrt{\{a_j\}^T [M] \{a_j\}}} \tag{10.29}$$

which, for a system having a diagonal mass matrix may be written as

$$\phi_{ij} = \frac{a_{ij}}{\sqrt{\sum_{k=1}^{n} m_k a_{kj}^2}} \tag{10.30}$$

in which ϕ_{ij} is the normalized i component of the j modal vector.

Illustrative Example 10.2.   For the two degree shear building of illustrative Example 10.1, determine the normalized modal shapes and verify the orthogonality condition between modes.

The substitution of eqs. (10.17) and (10.18) together with the values of the masses from Example 10.1 into the normalization factor required in eq. (10.30) gives

$$\sqrt{(136)(1.00)^2 + (66)(1.263)^2} = \sqrt{241.31}$$

$$\sqrt{(136)(1.00)^2 + (66)(-1.629)^2} = \sqrt{311.08}.$$

Consequently, the normalized modes are

$$\phi_{11} = \frac{1.00}{\sqrt{241.31}} = 0.06437, \qquad \phi_{12} = \frac{1.00}{\sqrt{311.08}} = 0.0567,$$

$$\phi_{21} = \frac{1.263}{\sqrt{241.31}} = 0.0813, \qquad \phi_{22} = \frac{-1.6287}{\sqrt{311.08}} = -0.0924.$$

The normal modes may be conveniently arranged in the columns of a matrix known as the *modal matrix* of the system. For the general case of n degrees of freedom, the modal matrix is written as

$$[\Phi] = \begin{bmatrix} \phi_{11} & \phi_{12} & \cdots & \phi_{1n} \\ \phi_{21} & \phi_{22} & \cdots & \phi_{2n} \\ \phi_{n1} & \phi_{n2} & \cdots & \phi_{nn} \end{bmatrix}. \tag{10.31}$$

The orthogonality condition may then be expressed in general as

$$[\Phi]^T [M] [\Phi] = [I] \tag{10.32}$$

where $[\Phi]^T$ is the matrix transpose of $[\Phi]$ and $[M]$ the mass matrix of the system. For this example of two degrees of freedom, the modal matrix is

$$[\Phi] = \begin{bmatrix} 0.06437 & 0.0567 \\ 0.0813 & -0.0924 \end{bmatrix}. \tag{10.33}$$

To check the orthogonality condition, we simply substitute the normal modes from eq. (10.33) into eq. (10.32) and obtain

$$\begin{bmatrix} 0.06437 & 0.0813 \\ 0.0567 & -0.0924 \end{bmatrix} \begin{bmatrix} 136 & 0 \\ 0 & 66 \end{bmatrix} \begin{bmatrix} 0.06437 & 0.0567 \\ 0.0813 & -0.0924 \end{bmatrix} = \begin{bmatrix} 1 & 0 \\ 0 & 1 \end{bmatrix}.$$

10.3   SUBROUTINE JACOBI

In the preceding section on free vibration of a shear building, we have to solve an eigenproblem to determine the natural frequencies and normal modes of vibration. The direct method of solution based on the expansion of the determinant and the solution of the characteristic equation is limited in practice to systems having only a few degrees of freedom. For a system of many degrees of freedom, the algebraic and numerical work required for the solution of an eigenproblem becomes so immense as to make the direct method impossible. However, there are many numerical methods available for the calculation of eignevalues and eigenvectors of an eigenproblem. The discussion of these methods belongs in a mathematical text on numerical methods rather than in a text such as this on structural dynamics. For our purpose we have selected among the various

TABLE 10.1   Variables and Symbols for Subroutine JACOBI.

Variable	Symbol in Text	Description
A(I, I)	$[K]$	Stiffness matrix
B(I, I)	$[M]$	Mass matrix
X(I, I)	$[\Phi]$	Modal matrix
EIGV(I)	ω_i^2	Eigenvalues
D(I)		Working vector
N		Order of the matrices A and B
RTOL		Convergence tolerance (set to 10^{-12})
NSMAX		Maximum number of sweeps (set to 15)
IFPR		Index for printing during iteration $1 \rightarrow$ print, $0 \rightarrow$ do not print

methods available for a numerical solution of an eigenproblem, the *Jacobi method* which is an iterative method to calculate the eigenvalues and eigenvectors of the system. The basic Jacobi solution method has been developed for the solution of standard eigenproblems (i.e., $[M]$ being the identity matrix). The method was proposed over a century ago and has been used extensively. This method can be applied to all symmetric matrices $[K]$ with no restriction on the eigenvalues. It is possible to transform the generalized eigenproblem, $[[K] - \omega^2 [M]] \{\Phi\} = \{0\}$ into the standard form and still maintain the symmetry required for the Jacobi method. However, this transformation can be dispensed with by using a generalized Jacobi solution method which operates directly on $[K]$ and $[M]$. A generalized Jacobi computer program has been published by Bathe and Wilson[1] and its listing is reproduced as Program 5 in the Appendix. A description of the principal symbols utilized in this program is listed in Table 10.1 and a short computer program for testing subroutine JACOBI is presented in the next section.

10.4   PROGRAM 5—TESTING SUBROUTINE JACOBI

Computer Program 5 is a short program designed to test subroutine JACOBI. Input data and corresponding formats for this program are given in Table 10.2. The listing of this program, followed by subroutine JACOBI, is given in the Appendix.

The output for Program 5 is printed by subroutine JACOBI and consists of the eigenvalues (ω_i^2) and the corresponding normalized eigenvectors $\{\Phi_i\}$. If the

[1]Bathe, K. J., and Wilson, E. L., *Numerical Methods in Finite Element Analysis*, Prentice-Hall, Englewood Cliffs, N.J., 1976.

TABLE 10.2   Input Data and Formats for Program 5.

Format	Variables
(2I10)	N,IFPR
(8F10.4)	A(I,J) (read by rows)
(8F10.4)	B(I,J) (read by rows)

TABLE 10.3   Input Data and Computer Results for Example 10.3.

INPUT DATA

3	1	
18.8600	−12.0000	5.1430
−12.0000	15.0000	−12.0000
5.1430	−12.0000	18.8600
0.8169	0.1286	−0.0740
0.1286	0.8571	0.1286
−0.0740	0.1286	0.8169

EIGENVALUES

SWEEP= 1

| 0.60062D 02 | 0.21996D 01 | 0.15515D 02 |

SWEEP= 2

| 0.60803D 02 | 0.19628D 01 | 0.15397D 02 |

SWEEP= 3

| 0.60803D 02 | 0.19628D 01 | 0.15397D 02 |

SWEEP= 4

| 0.60803D 02 | 0.19628D 01 | 0.15397D 02 |

SWEEP= 5

| 0.60803D 02 | 0.19628D 01 | 0.15397D 02 |

EIGENVECTORS

0.72280D 00	0.43298D 00	−0.74915D 00
−0.77197D 00	0.79669D 00	−0.11363D−15
0.72280D 00	0.43209D 00	0.74915D 00

index IFPR is set equal to one, then current values of the eigenvalues during the iteration process are also printed.

Illustrative Example 10.3.   To present an example illustrating the use of computer Program 5, consider a structure (not a shear building) with three degrees of freedom for which the stiffness and mass matrices are the following:

$$[K] = \begin{bmatrix} 18.8600 & -12.0000 & 5.1430 \\ -12.0000 & 15.0000 & -12.0000 \\ 5.1430 & -12.0000 & 18.8600 \end{bmatrix}$$

$$[M] = \begin{bmatrix} 0.8169 & 0.1286 & -0.0740 \\ 0.1286 & 0.8571 & 0.1286 \\ -0.0740 & 0.1286 & 0.8169 \end{bmatrix}.$$

The listing of the input data for this example, followed by the computer results is given in Table 10.3.

10.5   SUMMARY

The motion of an undamped dynamic system in free vibration is governed by a homogeneous system of differential equations which in matrix notation is

$$[M] \{\ddot{y}\} + [K] \{y\} = \{0\}.$$

The process of solving this system of equations leads to a homogeneous system of linear algebraic equations of the form

$$([K] - \omega^2 [M]) \{a\} = \{0\},$$

which mathematically is known as an eigenproblem.

For a nontrivial solution of this problem, it is required that the determinant of the coefficients of the unknown $\{a\}$ be equal to zero, that is,

$$|[K] - \omega^2 [M]| = 0.$$

The roots ω_i^2 of this equation provide the natural frequencies ω_i. It is then possible to solve for the unknowns $\{a\}_i$ in terms of relative values. The vectors $\{a\}_i$ corresponding to the roots ω_i^2 are the modal shapes (eigenvectors) of the dynamic system. The arrangement in matrix format of the modal shapes constitutes the modal matrix $[\Phi]$ of the system. It is particularly convenient to normalize the eigenvectors to satisfy the following condition:

$$\{\phi\}_i^T [M] \{\phi\}_i = 1, \qquad i = 1, 2, \ldots, n,$$

where the normalized modal vectors $\{\phi\}_i$ are obtained by dividing the components of the vector $\{a\}_i$ by $\sqrt{\{a\}_i^T [M] \{a\}_i}$.

The modal vectors satisfy the important condition of orthogonality

$$\{\phi\}_i^T [M] \{\phi\}_j = 0, \qquad \text{for} \quad i \neq j.$$

The above relations are equivalent to

$$[\Phi]^T [M] [\Phi] = [I]$$

in which $[\Phi]$ is the modal matrix of the system.

For a dynamic system with only a few degrees of freedom, the natural frequencies and modal shapes may be determined by expanding the determinant and calculating the roots of the resulting characteristic equation. However, for a system with a large number of degrees of freedom, this direct method of solution becomes impractical. It is then necessary to resort to other numerical methods which usually require an iteration process. Among the various methods available for the solution of an eigenproblem, we have selected the Jacobi method. Computer Program 5 listed in the Appendix is a testing program for subroutine JACOBI. This subroutine may be called by any of the computer programs presented in this text for the dynamic analysis of structures.

PROBLEMS

10.1   Determine the natural frequencies and normal modes for the two-story shear building shown in Fig. P10.1.

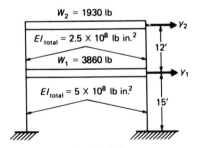

$W_2 = 1930$ lb

$EI_{total} = 2.5 \times 10^8$ lb in.^{2}

$W_1 = 3860$ lb

$EI_{total} = 5 \times 10^8$ lb in.^{2}

12'

y_2

y_1

15'

Fig. P10.1

10.2   A certain structure has been modeled as a three-degree-of-freedom system having the numerical values indicated in Fig. P10.2. Determine the natural frequencies of the structure and the corresponding nomal modes. Check your answer using computer Program 5.

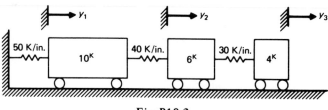

Fig. P10.2

10.3   Assume a shear building model for the frame shown in Fig. P10.3 and determine the natural frequencies and normal modes.

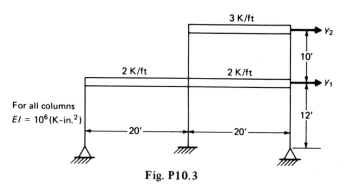

Fig. P10.3

10.4   A movable structural frame is supported on rollers as shown in Fig. P10.4. Determine natural periods and corresponding normal modes. Assume shear building model and check answers using Program 5.

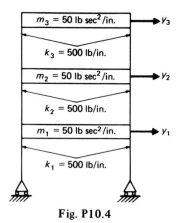

Fig. P10.4

11

Forced motion of shear buildings

In the preceding chapter, we have shown that the free motion of a dynamic system may be expressed in terms of free modal vibrations. Our present interest is to demonstrate that the forced motion of such a system may also be expressed in terms of the normal modes of vibration and that the total response may be obtained as the superposition of the solution of independent modal equations. In other words, our aim in this chapter is to show that the normal modes may be used to transform the system of *coupled* differential equations into a set of *uncoupled* differential equations in which each equation contains only one dependent variable. Thus the *modal superposition method* reduces the problem of finding the response of a multidegree-of-freedom system to the determination of the response of single degree-of-freedom systems.

11.1 MODAL SUPERPOSITION METHOD

We have shown that any free motion of a multidegree-of-freedom system may be expressed in terms of free modal vibrations. It will now be demonstrated that the forced motion of such a system may also be expressed in terms of the normal modes of vibration. We return to the equations of motion, eq. (9.3), which, for the particular case of a two-degree-of-freedom shear building are

$$m_1 \ddot{y}_1 + (k_1 + k_2) y_1 - k_2 y_2 = F_1(t),$$

$$m_2 \ddot{y}_2 - k_2 y_1 + k_2 y_2 = F_2(t). \tag{11.1}$$

We seek to transform this coupled system of equation into a system of independent or uncoupled equations in which each equation contains only one unknown function of time. It is first necessary to express the solution in terms of the normal modes multiplied by some factors determining the contribution of each mode. In the case of free motion, these factors were sinusoidal functions of time; in the present case, for forced motion, they are general functions of time which we designate as $z_i(t)$. Hence the solution of eq. (11.1) is assumed to be of the form

$$y_1(t) = a_{11} z_1(t) + a_{12} z_2(t),$$

$$y_2(t) = a_{21} z_1(t) + a_{22} z_2(t). \tag{11.2}$$

Upon substitution into eq. (11.1), we obtain

$$m_1 a_{11} \ddot{z}_1 + (k_1 + k_2) a_{11} z_1 - k_2 a_{21} z_1 + m_1 a_{12} \ddot{z}_2$$

$$+ (k_1 + k_2) a_{12} z_2 - k_2 a_{22} z_2 = F_1(t),$$

$$m_2 a_{21} \ddot{z}_1 - k_2 a_{11} z_1 + k_2 a_{21} z_1 + m_2 a_{22} \ddot{z}_2 - k_2 a_{12} z_2 + k_2 a_{22} z_2 = F_2(t). \tag{11.3}$$

To determine the appropriate factors $z_1(t)$ and $z_2(t)$ which will uncouple eq. (11.3), it is advantageous to make use of the orthogonality relations to separate the modes. The orthogonality relations are used by multiplying the first of eq. (11.3) by a_{11} and the second by a_{21}. Addition of these equations after multiplication and simplification by using eq. (10.25) and (10.27) yields

$$(m_1 a_{11}^2 + m_2 a_{21}^2) \ddot{z}_1 + \omega_1^2 (m_1 a_{11}^2 + m_2 a_{21}^2) z_1 = a_{11} F_1(t) + a_{21} F_2(t). \tag{11.4a}$$

Similarly, multiplying the first of eqs. (11.3) by a_{12} and the second by a_{22}, we obtain

$$(m_1 a_{12}^2 + m_2 a_{22}^2) \ddot{z}_2 + \omega_2^2 (m_1 a_{12}^2 + m_2 a_{22}^2) z_2 = a_{12} F_1(t) + a_{22} F_2(t). \tag{11.4b}$$

The results obtained in eqs. (11.4) permit a simple physical interpretation. The force which is effective in exciting a mode is equal to the work done by the external force displaced by the modal shape in question. From the mathematical point of view, what we have accomplished is to separate or uncouple, by a change of variables, the original system of differential equations. Consequently, each of these equations, eq. (11.4a) or eq. (11.4b), corresponds to a single degree-of-freedom system which may be written as

$$M_1 \ddot{z}_1 + K_1 z_1 = P_1(t),$$

$$M_2 \ddot{z}_2 + K_2 z_2 = P_2(t), \tag{11.5}$$

where $M_1 = m_1 a_{11}^2 + m_2 a_{21}^2$ and $M_2 = m_1 a_{12}^2 + m_2 a_{22}^2$ are the modal masses; $K_1 = \omega_1^2 M_1$ and $K_2 = \omega_2^2 M_2$, the modal spring constants; and $P_1(t) = a_{11} F_1(t) + a_{21} F_2(t)$ and $P_2(t) = a_{12} F_1(t) + a_{22} F_2(t)$, the modal forces. Alternatively, if we use the previous normalization, eq. (10.30), these equations may be written simply as

$$\ddot{z}_1 + \omega_1^2 z_1 = P_1(t),$$

$$\ddot{z}_2 + \omega_2^2 z_2 = P_2(t), \tag{11.6}$$

where P_1 and P_2 are now given by

$$P_1 = \phi_{11} F_1(t) + \phi_{21} F_2(t),$$

$$P_2 = \phi_{12} F_1(t) + \phi_{22} F_2(t). \tag{11.7}$$

The solution for the uncoupled differential equations, eqs. (11.5) or eqs. (11.6), may now be found by any of the methods presented in the previous chapters. In particular, Duhamel's integral provides a general solution for these equations regardless of the functions describing the forces acting on the structure. Also, maximum values of the response for each modal equation may readily be obtained using available response spectra. However, the superposition of modal maximum responses presents a problem. The fact is that these modal maximum values will in general not occur simultaneously as the transformation of coordinates, eq. (11.2), requires. To obviate the difficulty, it is necessary to use an approximate method. An upper limit for the maximum response may be obtained by adding the absolute values of the maximum modal contributions, that is, by substituting z_1 and z_2 in eqs. (11.2) for the maximum modal responses ($z_{1\,max}$ and $z_{2\,max}$) and adding the absolute values of the terms in these equations, so that

$$y_{1\,max} = |\phi_{11} z_{1\,max}| + |\phi_{12} z_{2\,max}|,$$

$$y_{2\,max} = |\phi_{21} z_{1\,max}| + |\phi_{22} z_{2\,max}|. \tag{11.8}$$

The results obtained by this method will overestimate the maximum response. Another method, which is widely accepted and which gives a reasonable estimate of the maximum response from these spectral values is the square root of the sum of the squares of the modal contributions. Thus the maximum displacements may be appoximated by

$$y_{1\,max} = \sqrt{(\phi_{11} z_{1\,max})^2 + (\phi_{12} z_{2\,max})^2}$$

and

$$y_{2\,max} = \sqrt{(\phi_{21}z_{1\,max})^2 + (\phi_{22}z_{2\,max})^2}. \qquad (11.9)$$

Illustrative Example 11.1. The two-story frame of Example 10.1 is acted upon at the floor levels by triangular impulsive forces shown in Fig. 11.1. For this frame, determine the maximum floor displacements and the maximum shear forces in the columns.

The results obtained in Examples 10.1 and 10.2 for the free vibration of this frame gave the following values for the natural frequencies and normalized modes:

$$\omega_1 = 11.8 \text{ rad/sec}, \qquad \omega_2 = 32.9 \text{ rad/sec},$$

$$\phi_{11} = 0.06437, \qquad \phi_{12} = 0.0567,$$

$$\phi_{21} = 0.08130, \qquad \phi_{22} = -0.0924.$$

The forces acting on the frame which are shown in Fig. 11.1(b) may be expressed by

$$F_1(t) = 10,000(1 - t/t_d)\,\text{lb},$$

$$F_2(t) = 20,000(1 - t/t_d)\,\text{lb}, \qquad \text{for} \quad t \leqslant 0.1 \text{ sec}$$

in which $t_d = 0.1$ sec and

$$F_1(t) = F_2(t) = 0, \qquad \text{for} \quad t > 0.1 \text{ sec}.$$

The substitution of these values into the uncoupled equations of motion, eqs. (11.6), gives

$$\ddot{z}_1 + 139.24 z_1 = 2270\,f(t)$$

$$\ddot{z}_2 + 1082.41 z_2 = -1281\,f(t)$$

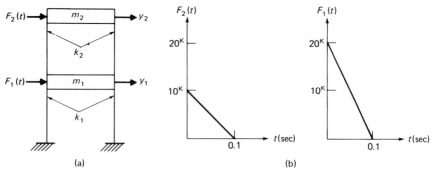

Fig. 11.1 Shear building with impulsive loadings. (a) Two-story shear building. (b) Impulsive loadings.

in which $f(t) = 1 - t/t_d$ for $\leqslant 0.1$ and $f(t) = 0$ for $t > 0.1$. The maximum values for z_1 and z_2 are then obtained from available spectrum charts such as the one shown in Fig. 4.5. For this example,

$$\frac{t_d}{T_1} = \frac{0.1}{0.532} = 0.188,$$

and

$$\frac{t_d}{T_2} = \frac{0.1}{0.191} = 0.524.$$

From Fig. 4.5, we obtain

$$(DLF)_{1\,max} = \frac{z_{1\,max}}{z_{1\,st}} = 0.590,$$

$$(DLF)_{2\,max} = \frac{z_{2\,max}}{z_{2\,st}} = 1.22.$$

where in this case the static deflections are calculated as

$$z_{1\,st} = \frac{F_{01}}{\omega_1^2} = \frac{2270}{139.24} = 16.3, \qquad z_{2\,st} = \frac{F_{02}}{\omega_2^2} = \frac{1281}{1082.41} = 1.18.$$

Then the maximum modal response is

$$z_{1\,max} = 0.590 \times 16.3 = 9.62, \qquad z_{2\,max} = 1.22 \times 1.18 = 1.44.$$

As indicated above these maximum modal values do not occur simultaneously and therefore cannot simply be superimposed to obtain the maximum response of the system. However, an upper limit for the absolute maximum displacement may be calculated with eqs. (11.8) as

$$y_{1\,max} = |0.06437 \times 9.62| + |0.0567 \times 1.44| = 0.70 \text{ in},$$

$$y_{2\,max} = |0.08130 \times 9.62| + |-0.0924 \times 1.44| = 0.92 \text{ in}.$$

A second acceptable estimate of the maximum response is obtained by taking the square root of the sum of modal contributions as indicated by eqs. (11.9). For this example, we have

$$y_{1\,max} = \sqrt{(0.06437 \times 9.62)^2 + (0.0567 \times 1.44)^2} = 0.62 \text{ in},$$

$$y_{2\,max} = \sqrt{(0.08130 \times 9.62)^2 + (-0.0924 \times 1.44)^2} = 0.79 \text{ in}. \quad (11.10)$$

The maximum shear force V_{max} in the columns is given by

$$V_{max} = \frac{12EI\,\Delta y}{L^3}$$

in which Δy is the difference between the displacements at the two ends of the column. When absolute displacements are calculated as in eqs. (11.10), then Δy should be computed as the sum of the absolute displacements at the ends of the columns. Using the displacements obtained in eqs. (11.10), we have for the columns of the first story

$$V_{1\,\text{max}} = \frac{12 \times 30 \times 10^6 \times 248.6 \times 0.62}{(15 \times 12)^3} = 9514 \text{ lb}$$

and, for the second story with $\Delta y = 0.79 + 0.62 = 1.41$,

$$V_{2\,\text{max}} = \frac{12 \times 30 \times 10^6 \times 106.3 \times 1.41}{(10 \times 12)^3} = 21{,}636 \text{ lb.}$$

11.2   RESPONSE OF A SHEAR BUILDING TO BASE MOTION

The response of a shear building to the base or foundation motion is conveniently obtained in terms of floor displacements relative to the base motion. For the two-story shear building of Fig. 11.2(a), which is modeled as shown in Fig. 11.2(b), the equations of motion obtained by equating to zero the sum of forces in the free body diagrams of Fig. 11.2(c) are the following:

$$m_1 \ddot{y}_1 + k_1(y_1 - y_s) - k_2(y_2 - y_1) = 0$$

$$m_2 \ddot{y}_2 + k_2(y_2 - y_1) = 0 \qquad (11.11)$$

where $y_s = y_s(t)$ is the displacement imposed at the foundation of the structure. Expressing the floor displacements relative to the base motion, we have

$$u_1 = y_1 - y_s,$$

$$u_2 = y_2 - y_s. \qquad (11.12)$$

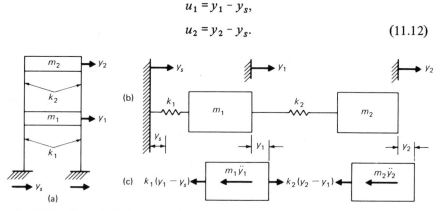

Fig. 11.2   Shear building with base motion. (a) Two-story shear building. (b) Mathematical model. (c) Free body diagram.

Then differentiation yields

$$\ddot{y}_1 = \ddot{u}_1 + \ddot{y}_s,$$

$$\ddot{y}_2 = \ddot{u}_2 + \ddot{y}_s. \tag{11.13}$$

Substitution of eqs. (11.12) and (11.13) into eqs. (11.11) results in

$$m_1\ddot{u}_1 + (k_1 + k_2)u_1 - k_2u_2 = -m_1\ddot{y}_s,$$

$$m_2\ddot{u}_2 - k_2u_1 + k_2u_2 = -m_2\ddot{y}_s. \tag{11.14}$$

We note that the right-hand sides of eqs. (11.14) are proportional to the same function of time, $\ddot{y}_s(t)$. This fact leads to a somewhat simpler solution compared to the solution of eqs. (11.6) which may contain different functions of time in each equation. For the base motion of the shear building, eqs. (11.4) may be written as

$$\ddot{z}_1 + \omega_1^2 z_1 = -\frac{m_1 a_{11} + m_2 a_{21}}{m_1 a_{11}^2 + m_2 a_{21}^2}\ddot{y}_s(t),$$

$$\ddot{z}_2 + \omega_2^2 z_2 = -\frac{m_1 a_{12} + m_2 a_{22}}{m_1 a_{12}^2 + m_2 a_{22}^2}\ddot{y}_s(t), \tag{11.15}$$

or

$$\ddot{z}_1 + \omega_1^2 z_1 = \Gamma_1 \ddot{y}_s(t),$$

$$\ddot{z}_2 + \omega_2^2 z_2 = \Gamma_2 \ddot{y}_s(t), \tag{11.16}$$

where Γ_1 and Γ_2 are called *participation factors* and are given by

$$\Gamma_1 = -\frac{m_1 a_{11} + m_2 a_{21}}{m_1 a_{11}^2 + m_2 a_{21}^2}$$

$$\Gamma_2 = -\frac{m_1 a_{12} + m_2 a_{22}}{m_1 a_{12}^2 + m_2 a_{22}^2}. \tag{11.17}$$

The relation between the modal displacements z_1, z_2 and the relative displacement u_1, u_2 is given from eqs. (11.2) as

$$u_1 = a_{11}z_1 + a_{12}z_2,$$

$$u_2 = a_{21}z_1 + a_{22}z_2. \tag{11.18}$$

In practice it is convenient to introduce a change of variables in eqs. (11.16) such that the second members of these equations equal $\ddot{y}_s(t)$. The required change of variables to accomplish this simplification is

$$z_1 = \Gamma_1 g_1,$$

$$z_2 = \Gamma_2 g_2, \tag{11.19}$$

which when introduced into eqs. (11.16) gives

$$\ddot{g}_1 + \omega_1^2 g_1 = \ddot{y}_s(t),$$

$$\ddot{g}_2 + \omega_2^2 g_2 = \ddot{y}_s(t). \tag{11.20}$$

Finally, solving for $g_1(t)$ and $g_2(t)$ in the uncoupled eqs. (11.20) and substituting the solution into eqs. (11.18) and (11.19) give the response as

$$u_1(t) = \Gamma_1 a_{11} g_1(t) + \Gamma_2 a_{12} g_2(t),$$

$$u_2(t) = \Gamma_1 a_{21} g_1(t) + \Gamma_2 a_{22} g_2(t). \tag{11.21}$$

When the maximum modal response $g_{1\,max}$ and $g_{2\,max}$ are obtained from spectral charts, we may estimate the maximum values $u_{1\,max}$ and $u_{2\,max}$ from eqs. (11.9) as

$$u_{1\,max} = \sqrt{(\Gamma_1 a_{11} g_{1\,max})^2 + (\Gamma_2 a_{12} g_{2\,max})^2},$$

$$u_{2\,max} = \sqrt{(\Gamma_1 a_{21} g_{1\,max})^2 + (\Gamma_2 a_{22} g_{2\,max})^2}. \tag{11.22}$$

Illustrative Example 11.2. Determine the response of the frame of Example 11.1 shown in Fig. 11.2 when it is subjected to a suddenly applied constant acceleration $\ddot{y}_s = 0.28\,g$ at its base.

The natural frequencies and corresponding normal modes from calculations in Examples 10.1 and 10.2 are

$$\omega_1 = 11.83 \text{ rad/sec}, \qquad \omega_2 = 32.9 \text{ rad/sec},$$

$$\phi_{11} = 0.06437, \qquad \phi_{12} = 0.0567,$$

$$\phi_{21} = 0.08130, \qquad \phi_{22} = -0.0924.$$

The acceleration acting at the base of this structure is

$$\ddot{y}_s = 0.28 \times 386 = 108.47 \text{ in/sec}^2.$$

The participation factors are calculated from eqs. (11.17) with the denominators set equal to unity since the modes are normalized. These factors are then

$$\Gamma_1 = -(136 \times 0.06437 + 66 \times 0.08130) = -14.120,$$

$$\Gamma_2 = -(136 \times 0.0567 - 66 \times 0.0924) = -1.613. \tag{a}$$

The modal equations, eqs. (11.20), are

$$\ddot{g}_1 + 140 g_1 = 108.47,$$

$$\ddot{g}_2 + 1082 g_2 = 108.47, \tag{b}$$

and their solution assuming zero initial conditions for velocity and displacement, is given by eqs. (4.5) as

$$g_1(t) = \frac{108.47}{140} (1 - \cos 11.83t),$$

$$g_2(t) = \frac{108.47}{1082} (1 - \cos 32.89t). \tag{c}$$

The response in terms of the relative motion of the stories at the floor levels with respect to the displacement of the base is given as a function of time by eqs. (11.21) as

$$u_1(t) = -14.120 \times 0.06437 \times 0.775 (1 - \cos 11.83t)$$

$$- 1.613 \times 0.0567 \times 0.100 (1 - \cos 32.89t),$$

$$u_2(t) = -14.120 \times 0.08130 \times 0.775 (1 - \cos 11.83t)$$

$$+ 1.614 \times 0.0924 \times 0.100 (1 - \cos 32.89t),$$

or, upon simplification as

$$u_1 = -0.7135 + 0.704 \cos 11.83t + 0.009 \cos 32.89t,$$

$$u_2 = -0.874 + 0.900 \cos 11.83t - 0.015 \cos 32.89t. \tag{d}$$

In this example due to the simple excitation function (a constant acceleration), it was possible to obtain a closed solution of the problem as a function of time. For a complex excitation function such as the one produced by an actual earthquake, it would be necessary either to resort to numerical integration to obtain the response or to use response spectra if available. The maximum modal response is obtained for the present example when the cosine functions in eqs. (c) are equal to minus one. In this case the maximum modal response is then

$$g_{1\,max} = 1.55,$$

$$g_{2\,max} = 0.20 \tag{e}$$

and the maximum response, calculated from the approximate formulas, eqs. (11.22), is

$$u_{1\,max} = 1.409 \text{ in},$$

$$u_{2\,max} = 1.800 \text{ in}. \tag{f}$$

The possible maximum values for the response calculated from eqs. (d) by setting the cosine functions to their maximum value, result in

$$u_{1\,max} = 1.426 \text{ in},$$

$$u_{2\,max} = 1.789 \text{ in}, \tag{g}$$

which for this particular example certainly compares very well with the approximate results obtained in eqs. (f) above.

11.3   PROGRAM 6–TESTING SUBROUTINE MODAL

Subroutine MODAL which is designed to give the response of a multidegree-of-freedom system using the modal superposition method is described in this section. This subroutine may be called by any of the computer programs which are

TABLE 11.1   Variables and Symbols for Subroutine MODAL.

Variable	Symbol in Text	Description
ND	N	Number of degrees of freedom
GR	g	Excitation index: For support excitation, g = acceleration of gravity. For forced excitation, $g = 0$
EIGEN(I)	ω_i^2	Square of natural frequencies (eigenvalues)
X(I, J)	$[\Phi]$	Modal matrix (eigenvectors)
DT		Time step of integration
TMAX		Maximum time of integration
NQ(L)		Number of points defining the excitation at coordinate L
M(I, J)		Mass matrix
T(I)	t_i	Time at point i
P(I)	$P(t_i)$	Force or acceleration (g's) at time t_i
XIS(I)	ξ_i	Damping ratios

TABLE 11.2   Input Data and Format for Program 6.

Format	Variables	Program
(I10, F10.0)	ND, GR	Testing
(8F10.4)[a]	M(I, J) (read by rows)	
(8F10.4)	EIGEN(I), (I = 1, ND)	
(8F10.4)	X(I, J) (read by rows)	
(2F10.4, 12I5)	DT, TMAX, NQ(L) (L = 1 . . . NG), where NG = ND when forces are at coordinates or NG = 1 when acceleration is at support	Subroutine MODAL
(8F10.2)	T(I), P(I) (I = 1, NQ(L)) (one card per forcing function)	
(8F10.3)	XSI(I), (I = 1, ND)	

[a]Skip this read statement if GR = 0.0.

presented in the following chapters for the dynamic analysis of structures. A short computer program labeled as Program 6 which is convenient in testing subroutine MODAL is also described in this section. The full listing of Program 6 followed by the listing of subroutine MODAL is given in the Appendix.

The principal symbols used in subroutine MODAL are described in Table 11.1 and a listing of the input data and formats necessary to run Program 6 is given in Table 11.2.

Illustrative Example 11.3. Use computer Program 6 to find the response of the frame analyzed in Example 11.2.

The natural frequencies and the modal matrix for this structure as calculated in Example 10.1 are

$$\omega_1 = 11.83 \text{ rad/sec}$$

$$\omega_2 = 32.9 \text{ rad/sec}$$

and

$$[\Phi] = \begin{bmatrix} 0.0644 & 0.0567 \\ 0.0813 & -0.0924 \end{bmatrix}.$$

The mass matrix is

$$[M] = \begin{bmatrix} 136 & 0 \\ 0 & 66 \end{bmatrix}.$$

The input data for Program 6 prepared according to Table 11.2 is listed in Table 11.3 and the computer results are then given in Table 11.4.

TABLE 11.3   Input Data for Example 11.3.

Data Listing			
2	386.0		
136.0	0.0		
0.0	66.0		
140.0	1082.0		
0.0644	0.0567		
0.0813	−0.0924		
0.05	1.0	2	
0.00	0.28	1.00	0.28
0.00	0.00		

TABLE 11.4   Computer Output for Example 11.3.

RESPONSE FOR ELASTIC SYSTEM

TIME	DISPLACEMENTS AT NODAL COORDINATES			
t	u_1		u_2	
0.000	0.0000D	00	0.0000D	00
0.050	-0.1292D	00	-0.1347D	00
0.100	-0.4550D	00	-0.5218D	00
0.150	-0.8516D	00	-0.1054D	01
0.200	-0.1204D	01	-0.1519D	01
0.250	-0.1405D	01	-0.1738D	01
0.300	-0.1364D	01	-0.1672D	01
0.350	-0.1086D	01	-0.1358D	01
0.400	-0.6894D	00	-0.8658D	00
0.450	-0.3135D	00	-0.3531D	00
0.500	-0.6267D	-01	-0.3318D	-01
0.550	-0.2012D	-01	-0.1817D	-01
0.600	-0.2245D	00	-0.2737D	00
0.650	-0.6048D	00	-0.7155D	00
0.700	-0.1008D	01	-0.1232D	01
0.750	-0.1302D	01	-0.1640D	01
0.800	-0.1410D	01	-0.1763D	01
0.850	-0.1286D	01	-0.1572D	01
0.900	-0.9521D	00	-0.1169D	01

11.4   PROGRAM 7—SEISMIC RESPONSE OF AN ELASTIC SHEAR BUILDING (SRESB)

In this section, a computer program is described to calculate the dynamic response in the linear range for a structure modeled as a shear building and subjected to excitation at its base. The modal superposition method of analysis is used to uncouple the differential equations of motion and to obtain the modal equations. Subroutine JACOBI (Chapter 10) is called to solve the eigenproblem resulting in eigenvalues (ω_i^2) and the eigenvectors which form the modal matrix [Φ]. Subroutine MODAL, which is called next, solves the resulting modal equations using Duhamel's integral method as previously described in Chapter 4. Finally at each step, the solutions of the modal equations are combined to obtain the response in terms of the original coordinates of the shear building.

A short list of the principal input variables and symbols used in the program is given in Table 11.5. The corresponding algebraic symbols that were used in the equations are also given. Formats for the input data cards are described in Table 11.6.

TABLE 11.5   Description of Input Variables for Program 7.

Variable	Symbol in Text	Description
E	E	Modulus of elasticity
GR	g	Acceleration of gravity
ND		Number of degrees of freedom
IFPR		Index for intermediate printing in Jacobi: $1 \to$ print, $0 \to$ do not print
SI	I_i	Moment inertia story i
SL	L_i	Height of story i
SM(I, I)	M_i	Mass of floor level i
DT		Time step of integration
TMAX		Maximum time of integration
NQ		Number of points defining the excitation
T(I)		Time at point i
P(I)		Acceleration at point i (g's)
XIS(I)	ξ_i	Modal damping ratios

TABLE 11.6   Input Data Symbols and Formats for Program 7.

Format	Variables
2F10.0, 2I5	E, GR, ND, IFPR
3F10.2	SI, SL, SM(I, I) (One card per story)
2F10.4, 12I5	DT, TMAX, NQ
8F10.3	T(L), P(L) (L = 1, NQ)
8F10.3	XSI(I) (I = 1, ND)

TABLE 11.7   Data Input for Example 11.2.

	Data Listing			
	30000000.	386.	2	0
497.2	180.	136.		
212.0	120.	66.		
0.01	0.19	2		
0.00	0.28	1.00	0.28	
0.	0.			

Illustrative Example 11.4.   To present an example in the use of Program 7, let us obtain the response of the two-story shear building of Example 11.2, which is shown in Fig. 11.2. Table 11.7 gives the required data input and Table 11.8 shows the computer output for this example.

TABLE 11.8   Computer Output for Example 11.2.

SEISMIC RESPONSE OF ELASTIC SHEAR BUILDING

TIME	DISPLACEMENTS	
t	u_1	u_2
0.000	0.0000	0.0000
0.010	-0.0054	-0.0054
0.020	-0.0215	-0.0217
0.030	-0.0480	-0.0488
0.040	-0.0842	-0.0867
0.050	-0.1295	-0.1352
0.060	-0.1828	-0.1943
0.070	-0.2433	-0.2633
0.080	-0.3097	-0.3419
0.090	-0.3811	-0.4291
0.100	-0.4563	-0.5239
0.110	-0.5342	-0.6250
0.120	-0.6139	-0.7309
0.130	-0.6944	-0.8397
0.140	-0.7748	-0.9496
0.150	-0.8542	-1.0586
0.160	-0.9319	-1.1648
0.170	-1.0070	-1.2662
0.180	-1.0787	-1.3610
0.190	-1.1460	-1.4478

11.5   HARMONIC FORCED EXCITATION

When the excitation, that is, the external forces or base motion is harmonic
(sine or cosine function), the analysis is quite simple and the response can
readily be found without the use of modal analysis. Let us consider the two-
story shear building as shown in Fig. 11.3 subjected to a single harmonic force

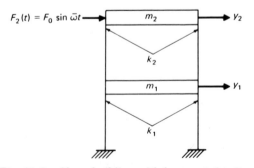

Fig. 11.3   Shear building with harmonic loading.

$F = F_0 \sin \bar{\omega}t$ which is applied at the level of the second floor. In this case eqs. (11.1) with $F_1(t) = 0$ and $F_2 = F_0 \sin \bar{\omega}t$ become

$$m_1 \ddot{y}_1 + (k_1 + k_2)y_1 - k_2 y_2 = 0,$$
$$m_2 \ddot{y}_2 - k_2 y_1 + k_2 y_2 = F_0 \sin \bar{\omega}t. \qquad (11.23)$$

For the steady-state response we seek a solution of the form

$$y_1 = Y_1 \sin \bar{\omega}t,$$
$$y_2 = Y_2 \sin \bar{\omega}t. \qquad (11.24)$$

After substitution of eqs. (11.24) into eqs. (11.23) and cancellation of the common factor $\sin \bar{\omega}t$, we then obtain

$$(k_1 + k_2 - m_1 \bar{\omega}^2) Y_1 - k_2 Y_2 = 0,$$
$$-k_2 Y_1 + (k_2 - m_2 \bar{\omega}^2) Y_2 = F_0, \qquad (11.25)$$

which is a system of two equations in two unknowns, Y_1 and Y_2. This system always has a unique solution except in the case when the determinant formed by the coefficients of the unknowns is equal to zero. The reader should remember that in this case the forced frequency $\bar{\omega}$ would equal one of the natural frequencies, since this determinant when equated to zero is precisely the condition used for determining the natural frequencies. In other words, unless the structure is forced to vibrate at one of the resonant frequencies, the algebraic system of eqs. (11.23) has a unique solution for Y_1 and Y_2.

Illustrative Example 11.5. Determine the steady-state response of the two-story shear building of Example 10.1 when a force $F_2(t) = 10,000 \sin 20t$ is applied to the second story as shown in Fig. 11.3.

The natural frequencies for this frame were determined in Example 10.1 to be

$$\omega_1 = 11.8 \text{ rad/sec},$$
$$\omega_2 = 32.9 \text{ rad/sec}$$

Since the forcing frequency is 20 rad/sec, the system is not at resonance. The steady-state response is then given by solving eqs. (11.25) for Y_1 and Y_2. Substituting numerical values in this system of equations, we have

$$(75,000 - 136 \times 20^2)Y_1 - 44,300\, Y_2 = 0$$
$$-44,300\, Y_1 + (44,300 - 66 \times 20^2)Y_2 = 10,000.$$

Solving these equations simultaneously results in

$$Y_1 = 0.28 \text{ in}, \qquad Y_2 = -0.13 \text{ in}.$$

Therefore, according to eqs. (11.24) the steady-state response is

$$y_1 = 0.28 \sin 20t \text{ in,}$$

$$y_2 = -0.13 \sin 20t \text{ in.}$$ (Ans.)

Damping may be considered in the analysis by simply including damping elements in the model as it is shown in Fig. 11.4 for a two-story shear building. The equations of motion which are obtained by equating to zero the sum of the forces in the free body diagram shown in Fig. 11.4(c) are

$$m_1\ddot{y}_1 + (c_1 + c_2)\dot{y}_1 + (k_1 + k_2)y_1 - c_2\dot{y}_2 - k_2 y_2 = 0,$$

$$m_2\ddot{y}_2 - c_2\dot{y}_1 - k_2 y_1 + c_2\dot{y}_2 + k_2 y_2 = F_0 \sin \bar{\omega}t.$$ (11.26)

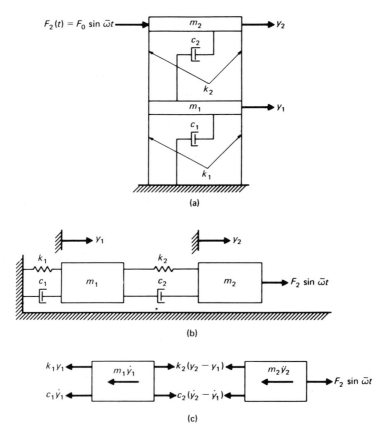

Fig. 11.4   (a) Damped shear building with harmonic load. (b) Multidegree mass-spring model. (c) Free body diagram.

As was explained in Chapter 3, for single degree-of-freedom systems, it is expedient to substitute in place of the sinusoidal force in eqs. (11.26) a complex exponential function, namely

$$F_0 e^{i\bar{\omega}t} = F_0(\cos \bar{\omega}t + i \sin \bar{\omega}t), \tag{11.27}$$

with the understanding that the real part of the solution will then be disregarded. This substitution in eqs. (11.26) yields

$$m_1\ddot{y}_1 + (c_1 + c_2)\dot{y}_1 + (k_1 + k_2)y_1 - c_2\dot{y}_2 - k_2 y_2 = 0$$

$$m_2\ddot{y}_2 - c_2\dot{y}_1 - k_2 y_1 + c_2\dot{y}_2 + k_2 y_2 = F_0 e^{i\bar{\omega}t}. \tag{11.28}$$

For the steady-state response, we seek solutions in the form of

$$y_1 = Y_1 e^{i\bar{\omega}t}$$

$$y_2 = Y_2 e^{i\bar{\omega}t}. \tag{11.29}$$

The substitution of eqs. (11.29) and the first and second derivatives of y_1 and y_2, namely

$$\dot{y}_1 = i\bar{\omega}Y_1 e^{i\bar{\omega}t}, \qquad \ddot{y}_1 = -\bar{\omega}^2 Y_1 e^{i\bar{\omega}t},$$

$$\dot{y}_2 = i\bar{\omega}Y_2 e^{i\bar{\omega}t}, \qquad \ddot{y}_2 = -\bar{\omega}^2 Y_2 e^{i\bar{\omega}t} \tag{11.30}$$

into eqs. (11.28) results in the following complex system of algebraic equations:

$$\{(k_1 + k_2 - m_1\bar{\omega}^2) + i\bar{\omega}(c_1 + c_2)\}Y_1 - (k_2 + i\bar{\omega}c_2)Y_2 = 0,$$

$$-(k_2 + i\bar{\omega}c_2)Y_1 + \{(k_2 - m_2\bar{\omega}^2) + i\bar{\omega}c_2\}Y_2 = F_0. \tag{11.31}$$

The response is then found by solving the system of eqs. (11.31) for the unknown amplitudes Y_1 and Y_2, substituting these results into eqs. (11.29), and then separating the real and imaginary part in the resulting expressions. The response in this case is given by the imaginary part since the force was expressed by a sine function. The response would be given by the real part of the solution were the external forces expressed as cosine functions. The necessary calculations are better explained through the use of a numerical example.

Illustrative Example 11.6.   Determine the steady-state response for the two-story shear building of Example 11.5 in which damping is considered in the analysis (Fig. 11.4). Assume for this example that the damping constants c_1 and c_2 are, respectively, proportional to the magnitude of spring constants k_1 and k_2 in which the factor of proportionality, $a_0 = 0.01$. Hence, we have

$$c_1 = a_0 k_1 = 307 \text{ lb} \cdot \sec/\text{in}$$

$$c_2 = a_0 k_2 = 443. \tag{a}$$

The substitution of numerical values for this example into eqs. (11.31) gives the following system of equations:

$$(20600 + 15000i)Y_1 - (44300 + 8860i)Y_2 = 0$$

$$-(44300 + 8860i)Y_1 + (17900 + 8960i)Y_2 = 10,000. \qquad \text{(b)}$$

The solution of this system of equations is

$$Y_1 = -0.2686 + 0.0007i$$

$$Y_2 = -0.1378 + 0.0631i \qquad \text{(c)}$$

or in exponential form

$$Y_1 = 0.2686 \, e^{3.139i}$$

$$Y_2 = 0.1515 \, e^{3.571i} \qquad \text{(d)}$$

Substitution of Y_1 and Y_2 from eqs. (d) into eqs. (11.29) gives

$$y_1 = 0.2686 \, e^{i(\bar{\omega}t+3.139)}$$

$$y_2 = 0.1515 \, e^{i(\bar{\omega}t+3.571)}.$$

Now, substituting $\bar{\omega} = 20$ rad/sec and recalling that for this example (force given by a sine function), we should retain only the imaginary part of the solution. We finally obtain the response as

$$y_1 = 0.2686 \sin(20t + 3.139) \text{ in,}$$

$$y_2 = 0.1515 \sin(20t + 3.571). \qquad \text{(Ans.)}$$

When these results are compared with those obtained for the undamped structure in Example 11.5, we note only a small change in the amplitude of motion. This is always the case for systems which are lightly damped and subjected to harmonic excitation of a frequency which is not close to one of the natural frequencies of the system. For this example, the forced frequency $\bar{\omega} = 20$ rad/sec is relatively far from the natural frequencies $\omega_1 = 11.83$ rad/sec or $\omega_2 = 32.94$ rad/sec which were calculated in Example 10.1.

11.6   PROGRAM 8—TESTING SUBROUTINE HARMO

Computer Program 8 is a short program for testing subroutine HARMO. This subroutine calculates the steady-state response of a system subjected to harmonic forces. The number of degree of freedom, ND, and the stiffness and mass matrices of the system are transferred from the main program to the subroutine. Damping in the system is assumed to be proportional to the stiffness and mass

TABLE 11.9   Description of Input Variables for Program 8.

Variable	Symbol in Text	Description
ND		Number of degrees of freedom
SK(I, J)	$[K]$	Stiffness matrix
SM(I, J)	$[M]$	Mass matrix
FACK	a_1	Damping stiffness factor
FACM	a_0	Damping mass factor
W	$\overline{\omega}$	Forced frequency (rad/sec)
$F_c(I), F_s(I)$		Coefficients of the force $F_c \cos \overline{\omega}t + F_s \sin \overline{\omega}t$ at coordinate i

coefficients, that is, the damping matrix is calculated from

$$[C] = a_0 [M] + a_1 [K] \tag{11.32}$$

in which a_0 and a_1 are constants specified in the input data.

Subroutine HARMO uses complex algebra to solve the resulting equations. The description of input variables for Program 8 is given in Table 11.9 and the input data symbols and formats are given in Table 11.10.

Illustrative Example 11.7.   Obtain the response of the damped two-degree-of-freedom shear building of Example 11.6 using computer Program 8.

The listing of the input data for this example followed by the output results is given in Table 11.11.

From Table 11.11 the response for this two-degree-of-freedom system is

$$y_1(t) = 0.00068 \cos 20t - 0.2686 \sin 20t$$

$$y_2(t) = -0.0631 \cos 20t - 0.1378 \sin 20t$$

TABLE 11.10   Input Data and Format for Program 8.

Format	Variables	Program
(I5)	ND	Testing
(8F10.0)	SK(I, J) (read by rows)	
(8F10.0)	SM(I, J) (read by rows)	
(8F10.2)	FACK, FACM, W	Subroutine
8F10.2	$F_c(I), F_s(I)$ (I = 1, ND)	HARMO

TABLE 11.11   Input Data and Output Results for Example 11.7.

DATA LISTING

2			
75000.	−44300.		
−44300.	44300.		
136.	0.		
0.	66.		
0.01	0.00	20.00	
0.00	0.00	0.00	100000.00

THE STEADY-STATE RESPONSE IS

COORD.	COS COMP.	SIN COMP.
1	0.6814D −03	−0.2686D 00
2	−0.6309D −01	−0.1378D 00

or

$$y_1(t) = 0.2686 \sin (20t + 3.139)$$

$$y_2(t) = 0.1516 \sin (20t + 3.571).$$

The results given by the computer, as expected, are the same as the values cal-culated in Example 11.6.

11.7 SUMMARY

For the solution of linear equations of motion, we may employ either the modal superposition method of dynamic analysis or a step-by-step numerical integra-tion procedure. The modal superposition method is restricted to the analysis of structures governed by linear systems of equations while the step-by-step methods of numerical integration are equally applicable to systems with linear or nonlinear behavior. We have deferred the presentation of the step-by-step inte-gration method to Chapter 19 on nonlinear structural response of multidegree-of-freedom systems.

In the present chapter, we have introduced the modal superposition method in obtaining the response of a shear building subjected to either force excitation or to base motion and have demonstrated that the use of the normal modes of free vibration for transforming the coordinates leads to a set of uncoupled dif-ferential equations. The solution of these equations may then be obtained by any of the methods presented in Part I for the single degree-of-freedom system. In the particular case of harmonic excitation, the response may be obtained in

closed form by simply solving a system of algebraic equations in which the unknowns are the amplitudes of the response at the various coordinates.

PROBLEMS

11.1   Determine the response as a function of time for the two-story shear
building of Problem 10.1 when a constant force of 5000 lb is suddenly
applied at the level of the second floor as shown in Fig. P11.1.

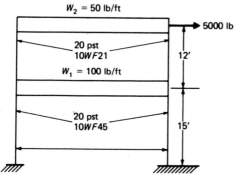

Fig. P11.1

11.2   Repeat Problem 11.1 if the excitation is applied to the base of the structure in the form of a suddenly applied acceleration of magnitude 0.5 g.

11.3   Determine the maximum displacement at the floor levels of the three-
story shear building [Fig. P11.3(a)] subjected to an impulsive triangular
load as shown in Fig. P11.3(b). The total stiffness of the columns of
each story is $k = 1500$ lb/in and the mass at each floor level is $m = 150$ lb ·
$\sec^2/$in.

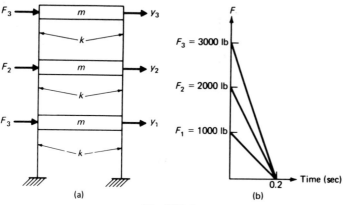

Fig. P11.3

11.4 Determine the maximum bending moments in the columns of Problem 11.3.

11.5 Use computer Program 7 to obtain the time history response of the three-story shear building in Fig. P11.3 subjected to the support acceleration plotted in Fig. P11.5. Determine the response for 1 sec after application of the excitation using a time step $\Delta t = 0.05$ sec and modal damping coefficient of 10% for all the modes.

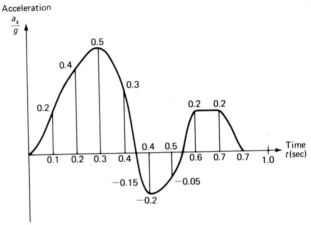

Fig. P11.5

11.6 Find the steady-state motion of the structure in Fig. P11.6 subjected to the harmonic forces indicated in the figure.

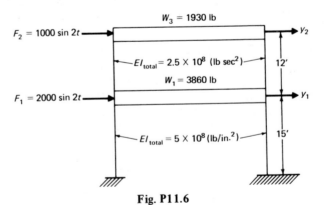

Fig. P11.6

12

Damped motion of shear buildings

In the previous chapters, we determined the natural frequencies and modal shapes for undamped structures when modeled as shear buildings. We also determined the response of these structures using the modal superposition method. In this method, as we have seen, the differential equations of motion are uncoupled by means of a transformation of coordinates which incorporates the orthogonality property of the modal shapes.

The consideration of damping in the dynamic analysis of structures complicates the problem. Not only will the differential equations of motion have additional terms due to damping forces, but the uncoupling of the equations will only be possible by imposing some restrictions or conditions on the functional expression for the damping coefficients.

The damping normally present in structures is relatively small and practically does not affect the calculation of natural frequencies and modal shapes of the system. Hence, the effect of damping is neglected in determining the natural frequencies and modal shapes of the structural systems. Therefore, in practice, the eigen-problem for the damped structure is solved by using the same methods employed for undamped structures.

12.1   EQUATIONS FOR DAMPED SHEAR BUILDING

For a viscously damped shear building, such as the three-story building shown in Fig. 12.1, the equations of motion obtained by summing forces in the corre-

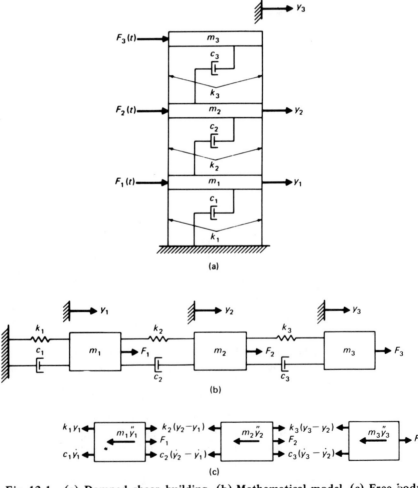

Fig. 12.1   (a) Damped shear building. (b) Mathematical model. (c) Free body diagram.

sponding free body diagrams are

$$m_1 \ddot{y}_1 + c_1 \dot{y}_1 + k_1 y_1 - c_2(\dot{y}_2 - \dot{y}_1) - k_2(y_2 - y_1) = F_1(t),$$

$$m_2 \ddot{y}_2 + c_2(\dot{y}_2 - \dot{y}_1) + k_2(y_2 - y_1) - c_3(\dot{y}_3 - \dot{y}_2) - k_3(y_3 - y_2) = F_2(t),$$

$$m_3 \ddot{y}_3 + c_3(\dot{y}_3 - \dot{y}_2) + k_3(y_3 - y_2) = F_3(t).$$

$$(12.1)$$

These equations may be conveniently written in matrix notation as

$$[M] \{\ddot{y}\} + [C] \{\dot{y}\} + [K] \{y\} = \{F(t)\}, \qquad (12.2)$$

where the matrices and vectors are as previously defined, except for the damping matrix $[C]$ which is given by

$$[C] = \begin{bmatrix} c_1 + c_2 & -c_2 & 0 \\ -c_2 & c_2 + c_3 & -c_3 \\ 0 & -c_3 & c_3 \end{bmatrix}. \qquad (12.3)$$

In the next section, we shall establish the conditions under which the damped equations of motion may be transformed to an uncoupled set of independent equations.

12.2 UNCOUPLED DAMPED EQUATIONS

To solve the differential equations of motion, eq. (12.2), we seek to uncouple these equations. We, therefore, introduce the transformation of coordinates

$$\{y\} = [\Phi] \{z\}, \qquad (12.4)$$

where $[\Phi]$ is the modal matrix obtained in the solution of the undamped free-vibration system. The substitution of eq. (12.4) and its derivatives into eq. (12.2) leads to

$$[M] [\Phi] \{\ddot{z}\} + [C] [\Phi] \{\dot{z}\} + [K] [\Phi] \{z\} = \{F(t)\}. \qquad (12.5)$$

Premultiplying eq. (12.5) by the transpose of the nth modal vector $\{\Phi\}_n^T$ yields

$$\{\Phi\}_n^T [M] [\Phi] \{\ddot{z}\} + \{\Phi\}_n^T [C] [\Phi] \{\dot{z}\} + \{\Phi\}_n^T [K] [\Phi] \{z\} = \{\Phi\}_n^T \{F(t)\}. \qquad (12.6)$$

It is noted that the orthogonality property of the modal shapes,

$$\{\Phi\}_n^T [M] \{\Phi\}_m = 0,$$

$$\{\Phi\}_n^T [K] \{\Phi\}_m = 0, \qquad m \neq n \qquad (12.7)$$

causes all components except the nth mode in the first and third terms of eq. (12.6) to vanish. A similar reduction is *assumed* to apply to the damping term in eq. (12.6), that is, if it is assumed that

$$\{\Phi\}_n^T [C] \{\Phi\}_m = 0, \qquad n \neq m, \tag{12.8}$$

then the coefficient of the damping term in eq. (12.6) will reduce to

$$\{\Phi\}_n^T [C] \{\Phi\}_n.$$

In this case eq. (12.6) may be written as

$$M_n \ddot{z}_n + C_n \dot{z}_n + K_n z_n = F_n(t)$$

or, alternatively as

$$\ddot{z}_n + 2\xi_n \omega_n \dot{z}_n + \omega_n^2 z_n = F_n(t)/M_n \tag{12.9}$$

in which case

$$M_n = \{\Phi\}_n^T [M] \{\Phi\}_n, \tag{12.10a}$$

$$K_n = \{\Phi\}_n^T [K] \{\Phi\}_n = \omega_n^2 M_n, \tag{12.10b}$$

$$C_n = \{\Phi\}_n^T [C] \{\Phi\}_n = 2\xi\omega_n M_n, \tag{12.10c}$$

$$F_n(t) = \{\Phi\}_n^T \{F(t)\}. \tag{12.10d}$$

The normalization discussed previously

$$\{\Phi\}_n^T [M] \{\Phi\}_n = 1 \tag{12.11}$$

will result in

$$M_n = 1$$

so that eq. (12.9) reduces to

$$\ddot{z}_n + 2\xi_n \omega_n \dot{z}_n + \omega_n^2 z_n = F_n(t) \tag{12.12}$$

which is a set of N uncoupled differential equations ($n = 1, 2, \ldots, N$).

12.3   CONDITIONS FOR DAMPING UNCOUPLING

In the derivation of the uncoupled damped equation, eq. (12.12), it has been assumed that the normal coordinate transformation, eq. (12.4), that serves to uncouple the inertial and elastic forces also uncouples the damping forces. It is of interest to consider the conditions under which this uncoupling will occur, that is, the form of the damping matrix $[C]$ to which eq. (12.8) applies.

When the damping matrix is of the form

$$[C] = a_0 [M] + a_1 [K] \tag{12.13}$$

in which a_0 and a_1 are arbitrary proportionality factors, the orthogonality condition will be satisfied. This may be demonstrated by applying the orthogonality condition to eq. (12.13), that is, premultiplying both sides of this equation by the transpose of the nth mode $\{\Phi\}_n^T$ and postmultiplying by the modal matrix $[\Phi]$. We obtain

$$\{\Phi\}_n^T [C] [\Phi] = a_0 \{\Phi\}_n^T [M] [\Phi] + a_1 \{\Phi\}_n^T [K] [\Phi]. \qquad (12.14)$$

The orthogonality conditions, eqs. (12.7), then reduce eq. (12.14) to

$$\{\Phi\}_n^T [C] [\Phi] = a_0 \{\Phi\}_n^T [M] \{\Phi\}_n + a_1 \{\Phi\}_n^T [K] \{\Phi\}_n$$

or, by eqs. (12.10), to

$$\{\Phi\}_n^T [C] [\Phi] = a_0 M_n + a_1 M_n \omega_n^2,$$

$$\{\Phi\}_n^T [C] [\Phi] = (a_0 + a_1 \omega_n^2) M_n$$

which shows that, when the damping matrix is of the form of eq. (12.13), the damping forces are also uncoupled with the transformation eq. (12.4). However, it can be shown that there are other matrices formed from the mass and stiffness matrices which also satisfy the orthogonality condition. In general, the damping matrix may be of the form

$$[C] = [M] \sum_i a_i ([M]^{-1} [K])^i, \qquad (12.15)$$

where i can be anywhere in the range $-\infty < i < \infty$ and the summation may include as many terms as desired. The damping matrix, eq. (12.13), can obviously be obtained as a special case of eq. (12.15). By taking two terms corresponding to $i = 0$ and $i = 1$ in eq. (12.15), we obtain the damping matrix expressed by eq. (12.13). With this form of the damping matrix it is possible to compute the damping coefficients necessary to provide uncoupling of a system having any desired damping ratios in any specified number of modes. For any mode n, the modal damping is given by eq. (12.10c), that is

$$C_n = \{\Phi\}_n^T [C] \{\Phi\}_n = 2\xi_n \omega_n M_n.$$

If $[C]$ as given by eq. (12.15) is substituted in the expression for C_n, we obtain

$$C_n = \{\Phi\}_n^T [M] \sum_i a_i ([M]^{-1} [K])^i \{\Phi\}_n. \qquad (12.16)$$

Now, using eq. (10.24) $(K \{\Phi\}_n = \omega_n^2 M \{\Phi\}_n)$ and performing several algebraic operations, we can show[1] that the damping coefficient associated with any mode n may be written as

$$C_n = \sum_i a_i \omega_n^{2i} M_n = 2\xi_n \omega_n M_n \qquad (12.17)$$

[1]Clough, R. W., and Penzien, J., *Dynamics of Structures*, McGraw-Hill, New York, 1975, p. 195.

from which

$$\xi_n = \frac{1}{2\omega_n} \sum_i a_i \omega_n^{2i}. \qquad (12.18)$$

Equation (12.18) may be used to determine the constants a_i for any desired values of modal damping ratios corresponding to any specified numbers of modes. For example, to evaluate these constants specifying the first four modal damping ratios $\xi_1, \xi_2, \xi_3, \xi_4$, we may choose $i = 1, 2, 3, 4$. In this case eq. (12.18) gives the following system of equations

$$\begin{bmatrix} \xi_1 \\ \xi_2 \\ \xi_3 \\ \xi_4 \end{bmatrix} = \frac{1}{2} \begin{bmatrix} \omega_1 & \omega_1^3 & \omega_1^5 & \omega_1^7 \\ \omega_2 & \omega_2^3 & \omega_2^5 & \omega_2^7 \\ \omega_3 & \omega_3^3 & \omega_3^5 & \omega_3^7 \\ \omega_4 & \omega_4^3 & \omega_4^5 & \omega_4^7 \end{bmatrix} \begin{bmatrix} a_1 \\ a_2 \\ a_3 \\ a_4 \end{bmatrix}. \qquad (12.19)$$

In general, eq. (12.19) may be written symbolically as

$$\{\xi\} = \tfrac{1}{2} [Q] \{a\}, \qquad (12.20)$$

where $[Q]$ is a square matrix having different powers of the natural frequencies. The solution of eq. (12.20) gives the constants $\{a\}$ as

$$\{a\} = 2[Q]^{-1} \{\xi\}. \qquad (12.21)$$

Finally the damping matrix is obtained after the substitution of eq. (12.21) into eq. (12.15).

It is interesting to observe from eq. (12.18) that in the special case when the damping matrix is proportional to the mass $\{C\} = a_0 [M]$, $(i = 0)$, the damping ratios are inversely proportional to the natural frequencies; thus the higher modes of the structures will be given very little damping. Analogously, when the damping is proportional to the stiffness matrix $([C] = a_1 [K])$, the damping ratios are directly proportional to the corresponding natural frequencies as can be seen from eq. (12.18) evaluated for $i = 1$; and in this case the higher modes of the structure will be very heavily damped.

Illustrative Example 12.1.   Determine the absolute damping coefficients for the structure presented in Example 10.1. Assume 10% of the critical damping for each mode. From Example 10.1, we have the following information.
    Natural frequencies:

$$\omega_1 = 11.83 \text{ rad/sec},$$

$$\omega_2 = 32.94 \text{ rad/sec}. \qquad (a)$$

Modal matrix:

$$[\Phi] = \begin{bmatrix} 1.00 & 1.00 \\ 1.26 & -1.63 \end{bmatrix}. \tag{b}$$

Mass matrix:

$$[M] = \begin{bmatrix} 136 & 0 \\ 0 & 66 \end{bmatrix}.$$

Stiffness matrix:

$$[K] = \begin{bmatrix} 75000 & -44300 \\ -44300 & 44300 \end{bmatrix}.$$

Using eq. (12.18) with $i = 0, 1$ to calculate the constants a_i needed in eq. (12.15), we obtain the following system of equations:

$$\begin{bmatrix} 0.1 \\ 0.1 \end{bmatrix} = \frac{1}{2} \begin{bmatrix} 11.83 & (11.83)^3 \\ 32.94 & (32.94)^3 \end{bmatrix} \begin{bmatrix} a_1 \\ a_2 \end{bmatrix}.$$

Solving this system of equations gives

$$a_1 = 0.01851,$$

$$a_2 = -0.00001146.$$

We also calculate

$$[M]^{-1} = \begin{bmatrix} 0.007353 & 0 \\ 0 & 0.01515 \end{bmatrix}$$

and

$$[M]^{-1}[K] = \begin{bmatrix} 551.475 & -325.738 \\ -671.145 & 671.145 \end{bmatrix}.$$

Then

$$\sum_{i=1}^{2} a_i([M]^{-1}[K])^i = 0.01851 \begin{bmatrix} 551.475 & -325.738 \\ -671.145 & 671.145 \end{bmatrix}$$

$$-0.00001146 \begin{bmatrix} 551.475 & -325.738 \\ -671.145 & 671.145 \end{bmatrix}^2$$

$$= \begin{bmatrix} 4.2172 & -1.4654 \\ -3.0193 & 4.7556 \end{bmatrix}.$$

Finally substituting this matrix into eq. (12.15) yields the damping matrix as

$$[C] = \begin{bmatrix} 136 & 0 \\ 0 & 66 \end{bmatrix} \begin{bmatrix} 4.2172 & -1.4654 \\ -3.0193 & 4.7556 \end{bmatrix} = \begin{bmatrix} 573.5 & -199.3 \\ -199.3 & 313.9 \end{bmatrix}.$$

There is yet a second method for evaluating the damping matrix corresponding to any set of specified modal damping ratio. The method may be explained starting with the relationship

$$[A] = [\Phi]^T [C] [\Phi] = \begin{bmatrix} 2\xi_1\omega_1 M_1 & 0 & 0 & - \\ 0 & 2\xi_2\omega_2 M_2 & 0 & - \\ 0 & 0 & 2\xi_3\omega_3 M_3 & - \\ - & - & - & - \end{bmatrix} \quad (12.22)$$

in which the modal masses $M_1, M_2, M_3 \ldots$ are equal to one if the modal matrix $[\Phi]$ has been normalized. It is evident that the damping matrix $[C]$ may be evaluated by postmultiplying and premultiplying eq. (12.22) by the inverse of the modal matrix and its inverse transpose, such that

$$[C] = [\Phi]^{-T} [A] [\Phi]^{-1}. \quad (12.23)$$

Therefore, for any specified set of modal damping ratios $\{\xi\}$, matrix $[A]$ can be evaluated from eq. (12.22) and the damping matrix $[C]$ from eq. (12.23). However, in practice, the inversion of the modal matrix is a large computational effort. Instead, taking advantage of orthogonality properties of the mode shapes, we can deduce[2] the following expression for the system damping matrix, namely

$$[C] = [M]\left(\sum_{n=1}^{N} \frac{2\xi_n\omega_n}{M_n} \{\Phi\}_n \{\Phi\}_n^T \right)[M]. \quad (12.24)$$

The damping matrix $[C]$ obtained from eq. (12.24) will satisfy the orthogonality property and, therefore, the damping term in the differential equation, eq. (12.2), will be uncoupled with the same transformation, eq. (12.4), which serves to uncouple the inertial and elastic forces.

It is of interest to note in eq. (12.24) that the contribution to the damping matrix of each mode is proportional to the modal damping ratio; thus any undamped mode will contribute nothing to the damping matrix.

We should mention at this point the circumstances under which it will be desirable to evaluate the elements of the damping matrix, as eq. (12.15) or eq. (12.24). It has been stated that absolute structural damping is a rather difficult quantity to determine or even to estimate. However, modal damping ratios may be estimated on the basis of past experience. This past experience indicates that

[2] Ibid., p. 197.

values for the modal damping ratios in structures are generally in the range of 2% to 10%, probably not exceeding 20%. In other words, on this basis and giving some consideration to the type of structure and materials utilized, we can assign numerical values to the modal damping ratios. The modal damping ratios are then used to determine the damping matrix which is needed explicitly when dynamic response is obtained by some analytical procedure other than modal analysis, e.g., step-by-step integration of a nonlinear system.

Illustrative Example 12.2.   Determine the damping matrix of Example 12.1 using the method based on eq. (12.22). To determine $[C]$ we could use either eq. (12.23) or eq. (12.24). From Example 10.2, the normalized modal matrix is

$$[\Phi] = \begin{bmatrix} 0.06437 & 0.0567 \\ 0.0813 & -0.0924 \end{bmatrix}$$

and its inverse is

$$[\Phi]^{-1} = \begin{bmatrix} 8.752 & 5.370 \\ 7.700 & -6.097 \end{bmatrix}.$$

Substituting into eq. (12.22), we obtain

$$2\xi_1 \omega_1 M_1 = (2)\,(0.1)\,(11.83)\,(1) = 2.366,$$
$$2\xi_2 \omega_2 M_2 = (2)\,(0.1)\,(32.89)\,(1) = 6.578.$$

Then by eq. (12.23)

$$[C] = \begin{bmatrix} 8.752 & 7.700 \\ 5.370 & -6.097 \end{bmatrix} \begin{bmatrix} 2.366 & 0 \\ 0 & 6.578 \end{bmatrix} \begin{bmatrix} 8.752 & 5.370 \\ 7.700 & -6.097 \end{bmatrix},$$

$$[C] = \begin{bmatrix} 572 & -198 \\ -198 & 313 \end{bmatrix}$$

(Ans.)

which checks with the damping matrix obtained in Example 12.1 for the same structure using eq. (12.15).

12.4   SUBROUTINE DAMP–CALCULATION OF THE DAMPING MATRIX

This subroutine calculates the system damping matrix $[C]$ using eq. (12.24) from specified modal damping ratios. The main program makes available to this subroutine the natural frequencies ω_n, the modal matrix $[\Phi]$, and the system mass matrix $[M]$ which are required in eq. (12.24). The only additional data needed in the subroutine DAMP are the values for the modal damping ratios which are read according to the format indicated in Table 12.1.

TABLE 12.1   Input Data for Subroutine DAMP.

Variable	Symbol in Text	Format	Description
X(I)	ξ	8F10.2	Damping ratios for modes 1 to NL

TABLE 12.2   Input Data and Formats for Program 9.

Variable	Symbol in Text	Format	Description
NL		(I10)	Number of degrees of freedom
EIGEN(I)	ω_i^2	(8F10.4)	Eigenvalues ($i = 1$, NL)
X(I,J)	$[\Phi]$	(8F10.4)	Modal matrix (read by rows)
SM(I,J)	$[M]$	(8F10.4)	Mass matrix (read by rows)
X(I)	ξ	(8F10.2)	Damping ratios (I = 1, NL)

12.5   PROGRAM 9—TESTING SUBROUTINE DAMP

A short computer program designated as Program 9 for testing subroutine DAMP is listed in the Appendix followed by the listing of this subroutine. The required input data and corresponding formats for this program are indicated in Table 12.2.

   Illustrative Example 12.3.   Use Program 9 to calculate the damping matrix for a structure with three degrees of freedom for which the squares of the natural frequencies (eigenvalues) are

$$\omega_1^2 = 1.9618, \qquad \omega_2^2 = 15.3927, \qquad \omega_3^2 = 60.7968.$$

The modal and mass matrices are

$$[\Phi] = \begin{bmatrix} 0.4330 & -0.7421 & 0.7228 \\ 0.7967 & 0.0000 & -0.7719 \\ 0.4330 & 0.7491 & 0.7228 \end{bmatrix}$$

and

$$[M] = \begin{bmatrix} 0.8169 & 0.1286 & -0.0740 \\ 0.1286 & 0.8571 & 0.1286 \\ -0.0740 & 0.1286 & 0.8169 \end{bmatrix}.$$

The listing of the input data for this example followed by computer output is given in Table 12.3.

TABLE 12.3   Input Data and Computer Output for Example 12.3.

INPUT DATA

3		
1.9618	15.3927	60.7986
0.4330	-0.7421	0.7228
0.7967	0.0000	-0.7119
0.4330	0.7421	0.7228
0.8169	0.1286	-0.0740
0.1286	0.8571	0.1286
-0.0740	0.1286	0.8169
0.10	0.10	0.10

THE DAMPING MATRIX IS

0.6921D 00	-0.2303D 00	0.6178D-02
-0.2303D 00	0.5296D 00	-0.2303D 00
0.6178D-02	-0.2303D 00	0.6921D 00

12.6   SUMMARY

The most common method of taking into account the dissipation of energy in structural dynamics is to assume in the mathematical model the presence of damping forces of magnitudes which are proportional to the relative velocity and of directions opposite to the motion. This type of damping is known as viscous damping because it is the kind of damping that will be developed by motion in an ideal viscous fluid. The inclusion of this type of damping in the equations does not alter the linearity of the differential equations of motion. Since the amount of damping commonly presented in structural systems is relatively small, its effect is neglected in the calculation of natural frequencies and mode shapes. However, to uncouple the damped differential equations of motion, it is necessary to impose some restrictions on the values of damping coefficients in the system. These restrictions are of no consequence due to the fact that in practice it is easier to determine or to estimate modal damping ratios rather than absolute damping coefficients. In addition, when solving the equations of motion by the modal superposition method, only damping ratios are required. When the solution is sought by other methods, the absolute value of the damping coefficients may be calculated from modal damping ratios by any of the various methods presented in this chapter.

PROBLEMS

12.1   The stiffness and mass matrices for a certain two-degree-of-freedom structure are

$$[K] = \begin{bmatrix} 400 & -200 \\ -200 & 200 \end{bmatrix}, \qquad [M] = \begin{bmatrix} 2 & 0 \\ 0 & 1 \end{bmatrix}.$$

Determine the damping matrix for this system corresponding to 20% of the critical damping for the first mode and 10% for the second mode. Use the method based on eqs. (12.16) and (12.17).

12.2   Repeat Problem 12.1 using the method based on eqs. (12.22) and (12.24).

12.3   The natural frequencies and corresponding normal modes (arranged in the modal matrix) for the three-story shear building shown in Fig. P12.3 are $\omega_1 = 9.31$ rad/sec, $\omega_2 = 20.94$ rad/sec, $\omega_3 = 29.00$ rad/sec, and

$$[\Phi] = \begin{bmatrix} 0.1114 & -0.1968 & -0.1245 \\ 0.2117 & -0.0277 & 0.2333 \\ 0.2703 & 0.2868 & -0.2114 \end{bmatrix}.$$

Determine the damping matrix for the system corresponding to damping ratios of 10% for all the modes.

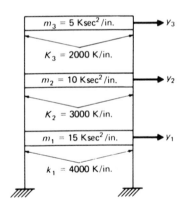

Fig. P12.3

12.4   Repeat Problem 12.3 for damping ratios of 20% for all the modes.

12.5   Repeat Problem 12.3 for the following value of modal damping ratios:

$$\xi_1 = 0.2, \qquad \xi_2 = 0.1, \qquad \xi_3 = 0.0.$$

12.6   Use Program 5 JACOBI to determine the natural frequencies and normal modes for the five-story shear building shown in Fig. P12.6; then use Program 9 DAMP to determine the damping matrix corresponding to an 8% damping ratio in all the modes.

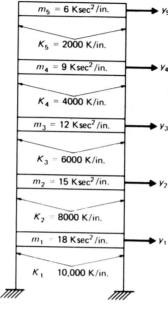

$m_5 = 6 \ K\sec^2/in.$ → y_5

$K_5 = 2000 \ K/in.$

$m_4 = 9 \ K\sec^2/in.$ → y_4

$K_4 = 4000 \ K/in.$

$m_3 = 12 \ K\sec^2/in.$ → y_3

$K_3 = 6000 \ K/in.$

$m_2 = 15 \ K\sec^2/in.$ → y_2

$K_2 = 8000 \ K/in.$

$m_1 = 18 \ K\sec^2/in.$ → y_1

$K_1 = 10,000 \ K/in.$

Fig. P12.6

12.7 Repeat Problem 12.6 for the following values of the modal damping ratios:

$\xi_1 = 0.20,$ $\qquad \xi_2 = 0.15,$ $\qquad \xi_3 = 0.10,$ $\qquad \xi_4 = 0.05,$ $\qquad \xi_5 = 0.$

13

Reduction of dynamic matrices

In the discretization process, it is sometimes necessary to divide a structure into a large number of elements because of changes in geometry, loading, or material properties. When the elements are assembled for the entire structure the number of unknown displacements, that is, the number of degrees of freedom, may indeed be very large. As a consequence, the stiffness, mass, and damping matrices are of large dimensions. In such cases the solution of the corresponding eigenproblem to determine natural frequencies and mode shapes will be difficult and, in addition, expensive. In such cases it is desirable, therefore, to reduce the size of these matrices in order to make the problem more manageable and economical.

13.1 STATIC CONDENSATION

A practical method of accomplishing the reduction of these matrices is to identify those degrees of freedom to be reduced as *dependent* coordinates and to express them in terms of the remaining independent degrees of freedom. The relation between the dependent and independent degrees of freedom is found by establishing the static relation between them, hence the name *static condensation method*. This relation provides the means to reduce the stiffness matrix. In order to reduce the mass and the damping matrices, it is assumed that the same static relation between dependent and independent degrees of freedom remains valid in the dynamic problem. Hence the same transformation based on static condensation for the reduction of the stiffness matrix is also used in reducing the mass and damping matrices. In general this method of reducing the dynamic problem is not exact and introduces errors in the results. The magnitude of these errors depends on the relative number of degrees of freedom reduced as well as on the specific selection of these degrees of freedom for a given structure. No error is introduced in reducing *massless* degrees of freedom, that is, degrees of freedom for which there is no mass allocated. The procedure of static condensation also is used in static problems to eliminate unwanted degrees of freedom such as the internal degrees of freedom of an element used with the finite element method of analysis.

Let us assume that those degrees of freedom to be reduced or condensed are the first p coordinates and proceed to carry out the Gauss-Jordan elimination for the first p unknown displacements. At this stage of the elimination process, the stiffness equation for the structure may be arranged in partition matrices as follows:

$$\begin{bmatrix} [I] & -[\overline{T}] \\ 0 & [\overline{K}] \end{bmatrix} \begin{bmatrix} \{y_p\} \\ \{y_q\} \end{bmatrix} \begin{bmatrix} \{0\} \\ \{F_q\} \end{bmatrix}, \tag{13.1}$$

where $\{y_p\}$ is the displacement vector corresponding to the p degrees of freedom to be reduced and $\{y_q\}$ the vector corresponding to the remaining q independent degrees of freedom. It should be noted that in eq. (13.1) it was assumed that the external forces were zero at the dependent degrees of freedom $\{y_p\}$. Equation (13.1) is equivalent to the two following relations:

$$\{y_p\} = [\overline{T}] \{y_q\}, \tag{13.2}$$

$$[\overline{K}] \{y_q\} = [F_q]. \tag{13.3}$$

Equation (13.2) which expresses the static relation between coordinates $\{y_p\}$ and $\{y_q\}$ may also be written as

$$\begin{bmatrix} \{y_p\} \\ \{y_q\} \end{bmatrix} = \begin{bmatrix} [\overline{T}] \\ [I] \end{bmatrix} \{y_q\} \tag{13.4}$$

or

$$\{y\} = [T] \{y_q\} \tag{13.5}$$

where

$$\{y\} = \begin{bmatrix} \{y_p\} \\ \{y_q\} \end{bmatrix}, \qquad [T] = \begin{bmatrix} [\overline{T}] \\ [I] \end{bmatrix}. \qquad (13.6)$$

Equation (13.3) which establishes the relation between coordinates $\{y_q\}$ and forces $\{F_q\}$ is the reduced stiffness equation and $[\overline{K}]$ the reduced stiffness matrix of the system. The reduced stiffness matrix may also be expressed as a transformation of the system stiffness matrix $[K]$ as

$$[\overline{K}] = [T]^T [K] [T]. \qquad (13.7)$$

Illustrative Example 13.1.   Let us consider the two-degree-of-freedom system represented by the model shown in Fig. 13.1. Use static condensation to reduce the first coordinate.

For this system the equations of equilibrium are readily obtained as:

$$\begin{bmatrix} 2k & -k \\ -k & k \end{bmatrix} \begin{bmatrix} y_1 \\ y_2 \end{bmatrix} = \begin{bmatrix} 0 \\ F_2 \end{bmatrix}. \qquad (13.8)$$

The reduction of y_1 using Gauss elimination leads to

$$\begin{bmatrix} 1 & -\tfrac{1}{2} \\ 0 & k/2 \end{bmatrix} \begin{bmatrix} y_1 \\ y_2 \end{bmatrix} = \begin{bmatrix} 0 \\ F_2 \end{bmatrix}. \qquad (13.9)$$

Comparing eq. (13.9) with eq. (13.1), we identify in this example

$$[\overline{T}] = \tfrac{1}{2},$$
$$[\overline{K}] = k/2. \qquad (13.10)$$

Consequently, from eq. (13.6) the transformation matrix is

$$[T] = \begin{bmatrix} \tfrac{1}{2} \\ 1 \end{bmatrix}. \qquad (13.11)$$

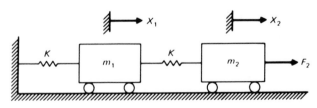

Fig. 13.1   Mathematical model for a two-degree-of-freedom system.

We can now check eq. (13.7) by simply performing the indicated multiplications, namely

$$[\bar{K}] = \begin{bmatrix} \frac{1}{2} & 1 \end{bmatrix} \begin{bmatrix} 2k & -k \\ -k & k \end{bmatrix} \begin{bmatrix} \frac{1}{2} \\ 1 \end{bmatrix} = \frac{k}{2}$$ (13.12)

which agrees with the result given in eqs. (13.10).

13.2   STATIC CONDENSATION APPLIED TO DYNAMIC PROBLEMS

We consider first the case in which the discretization of the mass has left a number of massless degrees of freedom. In this case it is only necessary to carry out the static condensation of the stiffness matrix and to delete from the mass matrix the rows and columns corresponding to the massless degrees of freedom. The condensation method in this case does not alter the original problem, thus resulting in an equivalent eigenproblem.

In the general case, that is, the case involving the condensation of degrees of freedom to which the discretization process has allocated mass, the procedure commonly used is to perform the same transformation, eq. (13.7), applied to the mass matrix. For a damped system again transformation eq. (13.7) is applied to the damping matrix of the system, that is, the reduced mass matrix is calculated as

$$[\bar{M}] = [T]^T [M] [T]$$ (13.13)

and the reduced damping matrix as

$$[\bar{C}] = [T]^T [C] [T],$$ (13.14)

where the transformation matrix $[T]$ is defined in eq. (13.6). This method of reducing the mass and damping matrices may be justified as follows. The potential elastic energy, V, and the kinetic energy, KE, of the structure may be written, respectively, as

$$V = \tfrac{1}{2} \{y\}^T [K] \{y\},$$ (13.15)

$$KE = \tfrac{1}{2} \{\dot{y}\}^T [M] \{\dot{y}\}.$$ (13.16)

Analogously, the work δW_d done by the damping forces $F_d = [C] \{\dot{y}\}$ corresponding to displacement $\{\delta y\}$ may be expressed as

$$\delta W_d = \{\delta y\}^T [C] \{\dot{y}\}.$$ (13.17)

Introduction of the transformation, eq. (13.5), in the above equations results in

$$V = \tfrac{1}{2}\{y_q\}^T [T]^T [K] [T] \{y_q\}, \tag{13.18}$$

$$KE = \tfrac{1}{2}\{\dot{y}_q\}^T [T]^T [M] [T] \{\dot{y}_q\}, \tag{13.19}$$

$$\delta W_d = \{\delta y_q\}^T [T]^T [C] [T] \{\dot{y}_q\}. \tag{13.20}$$

The respective substitution of $[\bar{K}]$, $[\bar{M}]$, and $[\bar{C}]$ from eqs. (13.7), (13.13), and (13.14) for the product of the three matrices in eqs. (13.18), (13.19), and (13.20) yields

$$V = \tfrac{1}{2}\{y_q\}^T [\bar{K}] \{y_q\}, \tag{13.21}$$

$$KE = \tfrac{1}{2}\{\dot{y}_q\}^T [\bar{M}] \{\dot{y}_q\}, \tag{13.22}$$

$$\delta W_d = \{\delta y_q\}^T [\bar{C}] \{\dot{y}_q\}. \tag{13.23}$$

These last three equations express the potential energy, the kinetic energy, and the virtual work of the damping forces in terms of the independent coordinates $\{y_q\}$. Hence the matrices $[\bar{K}]$, $[\bar{M}]$, and $[\bar{C}]$ may be interpreted, respectively, as the stiffness, mass, and damping matrices of the structure corresponding to the independent degrees of freedom $\{y_q\}$.

Illustrative Example 13.2. Find the natural frequencies and modal shapes for the three-degree-of-freedom shear building shown in Fig. 13.2. Also condense one degree of freedom and compare the resulting values obtained for natural frequencies and mode shapes. The stiffness of each story and the mass at each floor level are indicated in the figure.

The equation of motion in free vibration for this structure is given by eq. (9.3) with the force vector $\{F\} = 0$, namely,

$$[M]\{\ddot{y}\} + [K]\{y\} = [0]$$

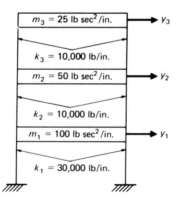

Fig. 13.2   Shear building for Example 13.2.

where the matrices $[M]$ and $[K]$ are given, respectively, by eqs. (9.4) and (9.5). Substitution of corresponding numerical values in these equations yields:

$$\begin{bmatrix} 100 & 0 & 0 \\ 0 & 50 & 0 \\ 0 & 0 & 25 \end{bmatrix} \begin{bmatrix} \ddot{y}_1 \\ \ddot{y}_2 \\ \ddot{y}_3 \end{bmatrix} + 10^3 \begin{bmatrix} 40 & -10 & 0 \\ -10 & 20 & -10 \\ 0 & -10 & 10 \end{bmatrix} \begin{bmatrix} y_1 \\ y_2 \\ y_3 \end{bmatrix} = \begin{bmatrix} 0 \\ 0 \\ 0 \end{bmatrix}.$$

After substituting $y_i = a_i \sin \omega t$ and cancelling the factor $\sin \omega t$, we obtain

$$\begin{bmatrix} 40,000 - 100\omega^2 & -10,000 & 0 \\ -10,000 & 20,000 - 50\omega^2 & 10,000 \\ 0 & 10,000 & 10,000 - 25\omega^2 \end{bmatrix} \begin{bmatrix} a_1 \\ a_2 \\ a_3 \end{bmatrix} = \begin{bmatrix} 0 \\ 0 \\ 0 \end{bmatrix} \quad (a)$$

which for a nontrivial solution requires that the determinant of the square matrix be equal to zero, that is,

$$\begin{vmatrix} 40,000 - 100\omega^2 & -10,000 & 0 \\ -10,000 & 20,000 - 50\omega^2 & 10,000 \\ 0 & 10,000 & 10,000 - 25\omega^2 \end{vmatrix} = 0.$$

The expansion of this determinant results in a third degree equation in terms of ω^2 having the following roots

$$\omega_1^2 = 84.64,$$
$$\omega_2^2 = 400,$$
$$\omega_3^2 = 536. \quad (b)$$

The natural frequencies are calculated by $f = \omega/2\pi$, so that

$$f_1 = 1.464 \text{ cps},$$
$$f_2 = 3.183 \text{ cps},$$
$$f_3 = 3.685 \text{ cps}.$$

The modal shapes are then determined by substituting each of the values for the natural frequencies into eq. (a), deleting one of the equations, and solving the remaining two equations for two of the unknowns in terms of the third unknown. As we mentioned previously, in solving for these unknowns it is expedient to set the first unknown equal to one. Performing these operations, we obtain from eqs. (b) and (a) the following values for the modal shapes

$$a_{11} = 1.00, \quad a_{12} = 1.00, \quad a_{13} = 1.00,$$
$$a_{21} = 3.18, \quad a_{22} = 0, \quad a_{23} = -2.88,$$
$$a_{31} = 4.00, \quad a_{32} = -1.00, \quad a_{33} = 4.00.$$

Condensation of coordinate y_1: The stiffness matrix for this structure is

$$\begin{bmatrix} 40,000 & -10,000 & 0 \\ -10,000 & 20,000 & -10,000 \\ 0 & -10,000 & 10,000 \end{bmatrix}.$$

Gauss elimination of the first unknown gives

$$\begin{bmatrix} 1 & -0.25 & 0 \\ 0 & 17,500 & -10,000 \\ 0 & -10,000 & 10,000 \end{bmatrix}. \tag{c}$$

A comparison of eq. (c) with eq. (13.1) indicates that

$$[\bar{T}] = [0.25 \quad 0],$$

$$[\bar{K}] = \begin{bmatrix} 17,500 & -10,000 \\ -10,000 & 10,000 \end{bmatrix}, \tag{d}$$

and from eq. (13.6)

$$[T] = \begin{bmatrix} 0.25 & 0 \\ 1 & 0 \\ 0 & 1 \end{bmatrix}. \tag{e}$$

To check we use eq. (13.7) to compute $[\bar{K}]$. Hence

$$[\bar{K}] = \begin{bmatrix} 0.25 & 1 & 0 \\ 0 & 0 & 1 \end{bmatrix} \begin{bmatrix} 40,000 & -10,000 & 0 \\ -10,000 & 20,000 & -10,000 \\ 0 & -10,000 & 10,000 \end{bmatrix} \begin{bmatrix} 0.25 & 0 \\ 1 & 0 \\ 0 & 1 \end{bmatrix},$$

$$[\bar{K}] = \begin{bmatrix} 17,500 & -10,000 \\ -10,000 & 10,000 \end{bmatrix}$$

which checks with eqs. (d). The condensed mass matrix is calculated by substituting matrix $[T]$ and its transpose from eq. (e) into eq. (13.13), so that

$$[\bar{M}] = \begin{bmatrix} 0.25 & 1 & 0 \\ 0 & 0 & 1 \end{bmatrix} \begin{bmatrix} 100 & 0 & 0 \\ 0 & 50 & 0 \\ 0 & 0 & 25 \end{bmatrix} \begin{bmatrix} 0.25 & 0 \\ 1 & 0 \\ 0 & 1 \end{bmatrix}$$

which results in

$$[\bar{M}] = \begin{bmatrix} 56.25 & 0 \\ 0 & 25 \end{bmatrix}.$$

The condensed dynamic problem is then

$$
\begin{bmatrix} 56.25 & 0 \\ 0 & 25 \end{bmatrix} \begin{bmatrix} \ddot{y}_2 \\ \ddot{y}_3 \end{bmatrix} + \begin{bmatrix} 17,500 & -10,000 \\ -10,000 & 10,000 \end{bmatrix} \begin{bmatrix} y_1 \\ y_2 \end{bmatrix} = \begin{bmatrix} 0 \\ 0 \end{bmatrix}.
$$

The natural frequencies and mode shapes are then determined from the solution of the following eigenproblem:

$$
\begin{bmatrix} 17,500 - 56.25\omega^2 & -10,000 \\ -10,000 & 10,000 - 25\omega^2 \end{bmatrix} \begin{bmatrix} a_2 \\ a_3 \end{bmatrix} = \begin{bmatrix} 0 \\ 0 \end{bmatrix}. \tag{f}
$$

Equating to zero the determinant of the square matrix in eq. (f) and solving the resulting quadratic equation in ω^2 gives

$$
\omega_1^2 = 85.20,
$$

$$
\omega_2^2 = 625.90.
$$

Then

$$
f_1 = \sqrt{85.2}/2\pi = 1.47 \text{ cps}
$$

$$
f_2 = \sqrt{625.9}/2\pi = 3.98 \text{ cps}.
$$

Corresponding mode shapes are obtained from eq. (f) after substituting the numerical values for ω_1^2 or ω_2^2 and solving the first equation for a_3 with $a_2 = 1$. We then obtain

$$
a_{21} = 1, \qquad a_{22} = 1,
$$

$$
a_{31} = 1.27, \qquad a_{32} = -1.77.
$$

For this system of only three degrees of freedom, the reduction of one coordinate gives results that compare well only for the first mode. Experience shows that the condensation process results in an eigenproblem which provides only about half of its eigenvalues and eigenvectors (natural frequencies and modal shapes) within acceptable approximate values. Also, in general, condensation of degrees of freedom which do not carry significant inertial force such as the rotational degrees of freedom, do not substantially affect the numerical values of the natural frequences and modal shapes.

13.3   PROGRAM 10—TESTING SUBROUTINE CONDE

Computer Program 10 is a short program to test subroutine CONDE. This subroutine when called by the main program reduces by static condensation the first p nodal coordinates (dependent variables) and calculates the reduced stiffness matrix $[\overline{K}]$, the reduced mass matrix $[\overline{M}]$, and the transformation matrix $[T]$. The principal variables and symbols used in the subroutine are described in

TABLE 13.1   Variables and Symbols for Subroutine CONDE.

Variable	Symbol in Text	Description
ND	N	Total number of degrees of freedom
NCR	p	Number of dependent nodal coordinates
NL	q	Number of independent nodal coordinates
SM(I,J)	$[M]$	Mass matrix
SK(I,J)	$[K]$	Stiffness matrix
T(I,J)	$[T]$	Transformation matrix

TABLE 13.2   Input Data and Formats for Program 10.

Format	Variables	
(2I10)	ND NCR	
(8F10.2)	SK(I,J)	(matrix read by rows)
(8F10.6)	SM(I,J)	(matrix read by rows)

Table 13.1. The computer program listing for Program 10 followed by subroutine CONDE is given in the Appendix. Input data and formats required in Program 10 are specified in Table 13.2.

Illustrative Example 13.3.   Repeat Example 13.2 using Program 10 to condense the first nodal coordinate.

The input data obtained from Example 13.2 are given in Table 13.3 followed by the computer results. These results consist of the reduced stiffness matrix $[\overline{K}]$, the transformation matrix $[T]$, and the reduced mass matrix $[\overline{M}]$. The results given by the computer are, as expected, identical to the hand solution for this problem presented in Example 13.2.

13.4   SUMMARY

The reduction of unwanted or dependent degrees of freedom is accomplished in practice by the static condensation method. This method consists in determining, by a partial Gauss–Jordan elimination, the reduced stiffness matrix corresponding to the independent degrees of freedom and the transformation matrix relating the dependent and independent degrees of freedom. The same transformation matrix is then used to condense by an orthogonal transformation the mass and damping matrices of the system.

TABLE 13.3   Input Data and Computer Results for Example 13.3.

INPUT DATA

3	1	
40000.	-10000.	0.
-10000.	20000.	-10000.
0.	-10000.	10000.
100.	0.	0.
0.	50.	0.
0.	0.	25.

THE REDUCED STIFFNESS MATRIX IS

0.1750D 05	-0.1000D 05
-0.1000D 05	0.1000D 05

THE TRANSFORMATION MATRIX IS

0.2500D 00	-0.0000D 00
0.1000D 01	0.0000D 00
0.0000D 00	0.1000D 01

THE REDUCED MASS MATRIX IS

0.5625D 02	0.0000D 00
0.0000D 00	0.2500D 02

PROBLEMS

13.1 Consider the problem where the stiffness and mass matrices of a certain structure are given by

$$[K] = \begin{bmatrix} 10 & -2 & -1 & 0 \\ -2 & 6 & -3 & -2 \\ -1 & -3 & 12 & -1 \\ 0 & -2 & -1 & 8 \end{bmatrix}; \qquad [M] = \begin{bmatrix} 0 & 0 & 0 & 0 \\ 0 & 0 & 0 & 0 \\ 0 & 0 & 3 & 0 \\ 0 & 0 & 0 & 2 \end{bmatrix}.$$

Use static condensation to determine the transformation and the reduced stiffness matrices corresponding to the elimination of the first two degrees of freedom (the massless degrees of freedom).

13.2 Determine the natural frequencies and corresponding normal modes for the reduced system in Problem 13.1.

13.3 Repeat Problem 13.1 for a structure having the stiffness matrix indicated in that problem, but the mass matrix given by

$$[M] = \begin{bmatrix} 1 & 0 & 0 & 0 \\ 0 & 1 & 0 & 0 \\ 0 & 0 & 3 & 0 \\ 0 & 0 & 0 & 2 \end{bmatrix}.$$

13.4 Determine the natural frequencies and corresponding normal modes for the reduced system in Problem 13.3.

13.5 For the shear building shown in Fig. P13.5, determine the transformation and the reduced stiffness and mass matrices corresponding to the condensation of the nodal coordinates y_1 and y_2 as indicated in the figure.

13.6 Determine the natural frequencies and normal modes for the reduced system in Problem 13.5.

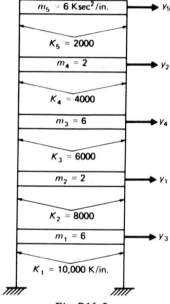

Fig. P13.5

13.7 The stiffness and matrices for a certain structure are

$$[K] = 10^6 \begin{bmatrix} 0.906 & 0.294 & 0.424 \\ 0.294 & 0.318 & 0.176 \\ 0.424 & 0.176 & 80,000 \end{bmatrix};$$

$$[M] = \begin{bmatrix} 288 & -8 & 1556 \\ -8 & 304 & 644 \\ 1556 & 644 & 80,000 \end{bmatrix}.$$

Reduce by static condensation the first two coordinates of the system and calculate the natural frequency of the resulting single degree-of-freedom system.

13.8 Determine the fundamental frequency for the three-degrees-of-freedom system of Problem 13.7 and compare this value with the frequency obtained for the reduced system of Problem 13.7.

Part III

Framed structures modeled as discrete multidegree-of-freedom systems

Part III

Framed structures modeled as discrete multidegree-of-freedom systems

14

Dynamic analysis of beams

In this chapter, we shall study the dynamic behavior of structures designated as beams, that is, structures which carry loads which are mainly transverse to the longitudinal direction, thus producing flexural stresses and lateral displacements. We begin by establishing the static characteristics for a beam segment; we then introduce the dynamic effects produced by the inertial forces. Two approximate methods are presented to take into account the inertial effect in the structure: (1) the lumped mass method in which the distributed mass is assigned to point masses, and (2) the consistent mass method in which the assignment to point masses includes rotational effects. The latter method is consistent with the static elastic deflections of the beam.

In Chapters 20 and 21, the exact theory for dynamics of beams considering the elastic and inertial distributed properties will be presented and in these chapters it will be shown mathematically that the stiffness and the consistent mass coefficients are in effect the first two terms of the exact solution expanded in a Taylor's series.

14.1   STATIC PROPERTIES FOR A BEAM SEGMENT

Consider a uniform beam segment of cross-sectional moment of inertia I, length L, and material modulus of elasticity E as shown in Fig. 14.1. We shall establish the relation between static forces and moments designated as P_1, P_2, P_3, P_4 and the corresponding linear and angular displacements $\delta_1, \delta_2, \delta_3, \delta_4$ at the ends of the beam segment as indicated in Fig. 14.1. The relation thus obtained is the stiffness matrix for a beam segment. The forces P_i and the displacements δ_i are said to be at the *nodal coordinates* defined for the beam segment.

The differential equation for small transverse displacements of a beam, which is well known from elementary studies of strength of materials, is given by

$$EI \frac{d^2 y}{dx^2} = M(x) \tag{14.1}$$

in which $M(x)$ is the bending moment at a section of the beam and y is the transverse displacement. We state first the general definition of stiffness coefficient which is designated by k_{ij}, that is, k_{ij} is the force at nodal coordinate i due to a unit displacement at nodal coordinate j while all other nodal coordinates are maintained at zero displacement.

Figure 14.2 shows the displacement curves corresponding to a unit displacement at each one of the four nodal coordinates for a beam segment indicating the corresponding stiffness coefficients. To determine the expressions for these coefficients, we begin by finding the equations of the deflected curves.

For the beam segment in Fig. 14.2(a), the bending moment at a section x is given by

$$M(x) = k_{11} x - k_{21} \tag{14.2}$$

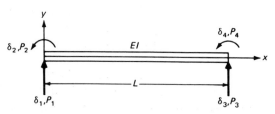

Fig. 14.1   Beam segment showing forces and displacements at the nodal co-ordinates.

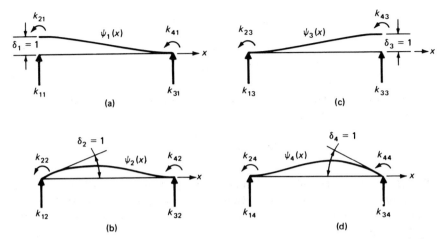

Fig. 14.2   Static deflection curves due to a unit displacement at one of the nodal coordinates.

which substituted in eq. (14.1) results in

$$EI \frac{d^2y}{dx^2} = k_{11}x - k_{21}.$$ (14.3)

Integrating eq. (14.3) twice gives

$$EI \frac{dy}{dx} = \frac{k_{11}}{2} x^2 - k_{21}x + C_1,$$ (14.4)

$$EIy = \frac{k_{11}}{6} x^3 - \frac{k_{21}}{2} x^2 + C_1 x + C_2.$$ (14.5)

There are four unknown quantities in these equations: the constants of integration C_1, C_2 and the stiffness coefficients k_{11}, k_{21}. For the evaluation of these quantities, use is made of the following boundary conditions:

$$\text{at} \quad x = 0, \quad y(0) = 1 \quad \text{and} \quad \frac{dy(0)}{dx} = 0;$$ (14.6)

$$\text{at} \quad x = L, \quad y(L) = 0 \quad \text{and} \quad \frac{dy(L)}{dx} = 0.$$ (14.7)

Introduction of these boundary conditions into eqs. (14.4) and (14.5) results in a system of four equations whose solution yields

$$C_1 = 0, \qquad k_{11} = \frac{12EI}{L^3},$$

$$C_2 = EI, \qquad k_{21} = \frac{6EI}{L^2}. \qquad (14.8)$$

The substitution of eqs. (14.8) into eq. (14.5) gives the equation of the deflected curve for the beam segment in Fig. 14.2(a) as

$$\psi_1(x) = 1 - 3\left(\frac{x}{L}\right)^2 + 2\left(\frac{x}{L}\right)^3 \qquad (14.9a)$$

in which $\psi_1(x)$ is used instead of $y(x)$ to correspond to the condition $\delta_1 = 1$ imposed on the beam segment. Proceeding in analogous fashion, we obtain for the equations of the deflected curves in the other cases depicted in Fig. 14.2 the following equations:

$$\psi_2(x) = x\left(1 - \frac{x}{L}\right)^2, \qquad (14.9b)$$

$$\psi_3(x) = 3\left(\frac{x}{L}\right)^2 - 2\left(\frac{x}{L}\right)^3, \qquad (14.9c)$$

$$\psi_4(x) = \frac{x^2}{L}\left(\frac{x}{L} - 1\right). \qquad (14.9d)$$

Since $\psi_1(x)$ is the deflection corresponding to a unit displacement $\delta_1 = 1$, the displacement resulting from an arbitrary displacement δ_1 is $\psi_1(x)\delta_1$. Analogously, the deflection resulting from nodal displacements δ_2, δ_3, and δ_4 are, respectively, $\psi_2(x)\delta_2$, $\psi_3(x)\delta_3$, and $\psi_4(x)\delta_4$. Therefore, the total deflection $y(x)$ at coordinate x due to arbitrary displacements at the nodal coordinates of the beam segment is given by superposition as

$$y(x) = \psi_1(x)\,\delta_1 + \psi_2(x)\,\delta_2 + \psi_3(x)\,\delta_3 + \psi_4(x)\,\delta_4. \qquad (14.10)$$

The deflection equations which are given by eqs. (14.9) and which correspond to unit displacements at the nodal coordinates of a beam segment may be used to determine expressions for the stiffness coefficients. For example, consider the beam in Fig. 14.2(b) which is in equilibrium with the forces producing the displacement $\delta_2 = 1.0$. For this beam in the equilibrium position, assume that a virtual displacement equal to the deflection curve shown in Fig. 14.2(a) takes place. We then apply the principle of virtual work which states that, for an elastic system in equilibrium, the work done by the external forces is equal to the work of the internal forces during the virtual displacement. In order to apply this principle, we note that the external work W_E is equal to the product of the force k_{12} displaced by $\delta_1 = 1$, that is

$$W_E = k_{12}\delta_1. \tag{14.11}$$

This work, as stated above, is equal to the work performed by the elastic forces during the virtual displacement. Considering the work performed by the bending moment, we obtain for the internal work

$$W_I = \int_0^L M(x)\, d\theta \tag{14.12}$$

in which $M(x)$ is the bending moment at section x of the beam and $d\theta$ is the relative angular displacement of this section.

For the virtual displacement under consideration, the transverse deflection of the beam is given by eq. (14.9b) which is related to the bending moment through the differential equation, eq. (14.1). Substitution of the second derivative $\psi_2''(x)$ of eq. (14.9b) into eq. (14.1) results in

$$EI\psi_2''(x) = M(x). \tag{14.13}$$

The angular deflection $d\theta$ produced during this virtual displacement is related to the resulting transverse deflection of the beam $\psi_1(x)$ by

$$\frac{d\theta}{dx} = \frac{d^2\psi_1(x)}{dx^2} = \psi_1''(x)$$

or

$$d\theta = \psi_1''(x)\, dx. \tag{14.14}$$

Equating the external virtual work W_E from eq. (14.11) with the internal virtual work W_I from eq. (14.12) after using $M(x)$ and $d\theta$ from eqs. (14.13) and (14.14), respectively, finally gives the stiffness coefficient as

$$k_{12} = \int_0^L EI\psi_1''(x)\,\psi_2''(x)\, dx. \tag{14.15}$$

In general, any stiffness coefficient associated with beam flexure, therefore, may be expressed as

$$k_{ij} = \int_0^L EI\psi_i''(x)\,\psi_j''(x)\, dx. \tag{14.16}$$

It may be seen from eq. (14.16) that $k_{ij} = k_{ji}$ since the interchange of indices requires only an interchange of the two factors $\psi_i''(x)$ and $\psi_j''(x)$ in eq. (14.16).

The equivalence $k_{ij} = k_{ji}$ is a particular case of Betti's theorem, but it is better known as *Maxwell's reciprocal theorem*.

It should be pointed out that, although the deflection functions, eqs. (14.9), were obtained for a uniform beam, in practice, they are nevertheless, also used in determining the stiffness coefficients for nonuniform beams.

Considering the case of a uniform beam segment of length L and cross-sectional moment of inertia I, we may calculate any stiffness coefficient from eq. (14.16) and the use of eqs. (14.9). In particular, the stiffness coefficient k_{12} is calculated as follows. From eq. (14.9a), we obtain

$$\psi_1''(x) = -\frac{6}{L^2} + \frac{12x}{L^3}$$

and from eq. (14.9b)

$$\psi_2''(x) = -\frac{4}{L} + \frac{6x}{L^2}.$$

Substitution in eq. (14.15) gives

$$k_{12} = EI \int_0^L \left(\frac{-6}{L^2} + \frac{12x}{L^3}\right)\left(\frac{-4}{L} + \frac{6x}{L^2}\right) dx$$

and integration gives

$$k_{12} = \frac{6EI}{L^2}.$$

Since the stiffness coefficient k_{1j} is defined as the force at the nodal coordinate 1 due to unit displacement at the coordinate j, the forces at coordinate 1 due to successive displacement $\delta_1, \delta_2, \delta_3, \delta_4$ at the four nodal coordinates of the beam segment are given, respectively, by $k_{11}\delta_1, k_{12}\delta_2, k_{13}\delta_3$, and $k_{14}\delta_4$. Therefore, the total force P_1 at coordinate 1 resulting from these nodal displacements is obtained by the superposition of the resulting forces, that is,

$$P_1 = k_{11}\delta_1 + k_{12}\delta_2 + k_{13}\delta_3 + k_{14}\delta_4.$$

Analogously the forces at the other nodal coordinates are

$$P_2 = k_{21}\delta_1 + k_{22}\delta_2 + k_{23}\delta_3 + k_{24}\delta_4,$$
$$P_3 = k_{31}\delta_1 + k_{32}\delta_2 + k_{33}\delta_3 + k_{34}\delta_4,$$
$$P_4 = k_{41}\delta_1 + k_{42}\delta_2 + k_{43}\delta_3 + k_{44}\delta_4. \tag{14.17}$$

The above equations are written conveniently in matrix notation as

$$
\begin{bmatrix} P_1 \\ P_2 \\ P_3 \\ P_4 \end{bmatrix} = \begin{bmatrix} k_{11} & k_{12} & k_{13} & k_{14} \\ k_{21} & k_{22} & k_{23} & k_{24} \\ k_{31} & k_{32} & k_{33} & k_{34} \\ k_{41} & k_{42} & k_{43} & k_{44} \end{bmatrix} \begin{bmatrix} \delta_1 \\ \delta_2 \\ \delta_3 \\ \delta_4 \end{bmatrix}
\tag{14.18}
$$

or symbolically as

$$
\{P\} = [k]\,\{\delta\}
\tag{14.19}
$$

in which $\{P\}$ and $\{\delta\}$ are, respectively, the force and the displacement vectors at the nodal coordinates of the beam element and $[k]$ is the beam element stiffness matrix.

The use of eq. (14.16) in the manner shown above to determine the coefficient k_{12} will result in the evaluation of all the coefficients of the stiffness matrix. This result for a uniform beam segment is

$$
\begin{bmatrix} P_1 \\ P_2 \\ P_3 \\ P_4 \end{bmatrix} = \frac{2EI}{L^3} \begin{bmatrix} 6 & 3L & -6 & 3L \\ 3L & 2L^2 & -3L & L^2 \\ -6 & -3L & 6 & -3L \\ 3L & L^2 & -3L & 2L^2 \end{bmatrix} \begin{bmatrix} \delta_1 \\ \delta_2 \\ \delta_3 \\ \delta_4 \end{bmatrix}
\tag{14.20}
$$

or in condensed notation

$$
\{P\} = [k]\,\{\delta\}.
\tag{14.21}
$$

14.2   SYSTEM STIFFNESS MATRIX

Thus far we have established in eq. (14.20) the stiffness equation for a uniform beam segment, that is, we have obtained the relation between nodal displacements (linear and angular) and nodal forces (forces and moments). Our next objective is to obtain the same type of relation between the nodal displacements and the nodal forces, but now for the entire structure (system stiffness equation). Furthermore, our aim is to obtain the system stiffness matrix from the stiffness matrix of each element of the system. The procedure is perhaps better explained through a specific example such as the cantilever beam shown in Fig. 14.3.

The first step in obtaining the system stiffness matrix is to divide the structure into elements. The beam in Fig. 14.3 has been divided into three elements which are numbered sequentially for identification. The second step is to identify the nodes or joints between elements and to number consecutively those nodal coordinates which are not constrained. The constrained or fixed nodal coordinates

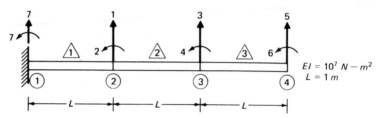

Fig. 14.3   Cantilever beam divided into three beam segments with numbered system nodal coordinates.

are the last to be labeled. All the fixed coordinates may be given the same label as shown in Fig. 14.3. In the present case, we consider only two possible displacements at each node, a vertical deflection and an angular displacement. The cantilever beam in Fig. 14.3 with its three elements results in a total of six free nodal coordinates and two fixed nodal coordinates, the latter being labeled with the number seven. The third step is to obtain systematically the stiffness matrix for each element in the system and to add the element stiffness coefficients appropriately to obtain the system stiffness matrix. This method of assembling the system stiffness matrix is called the *direct method*. In effect, any stiffness coefficient k_{ij} of the system may be obtained by adding together the corresponding stiffness coefficients associated with those nodal coordinates. Thus, for example, to obtain the system stiffness coefficient k_{33}, it is necessary to add the stiffness coefficients of beam segments △ and △ corresponding to node three. These coefficients are designated as $k_{33}^{(2)}$ and $k_{11}^{(3)}$, respectively. The upper indices serve to identify the beam segment and the lower indices to locate the appropriate stiffness coefficients in the corresponding element stiffness matrices.

Proceeding with the example in Fig. 14.3 and using eq. (14.20), we obtain the following expression for the stiffness matrix of beam segment △, namely

$$[k^{(2)}] = 10^7 \begin{array}{c} \begin{array}{cccc} 1 & 2 & 3 & 4 \end{array} \\ \begin{bmatrix} 12 & 6 & -12 & 6 \\ 6 & 4 & -6 & 2 \\ -12 & -6 & 12 & -6 \\ 6 & 2 & -6 & 4 \end{bmatrix} \begin{array}{c} 1 \\ 2 \\ 3 \\ 4 \end{array} \end{array}. \qquad (14.22)$$

For the beam segment △, the element nodal coordinates numbered one to four coincide with the assignment of system nodal coordinates also numbered one to four as may be seen in Fig. 14.3. However, for the beam segments △ and △ of this beam, the assignment of element nodal coordinates numbered one to four does not coincide with the assigned system coordinates. For example, for element △ the assigned system coordinates as seen in Fig. 14.3 are 7, 7, 1, 2; for element △, 3, 4, 5, 6. In the process of assembling the system stiffness, coefficients for element △ will be correctly allocated to coordinates 1, 2, 3, 4; for element △ to coordinates 7, 7, 1, 2; and for element △ to coordinates 3, 4, 5, 6. A simple way to indicate this allocation of coordinates, when working by hand, is to write at the top and on the right of the element stiffness matrix the coordinate numbers corresponding to the system nodal coordinates for the element as it is indicated in eq. (14.22) for element △. The stiffness matrices for elements △ and △ with the corresponding indication of system nodal coordinates are, respectively,

$$
[k^{(1)}] = 10^7
\begin{array}{cccc}
7 & 7 & 1 & 2 \\
\end{array}
\begin{bmatrix}
12 & 6 & -12 & 6 \\
6 & 4 & -6 & 2 \\
-12 & -6 & 12 & -6 \\
6 & 2 & -6 & 4
\end{bmatrix}
\begin{array}{c}
7 \\
7 \\
1 \\
2
\end{array}
\tag{14.23}
$$

and

$$
[k^{(3)}] = 10^7
\begin{array}{cccc}
3 & 4 & 5 & 6 \\
\end{array}
\begin{bmatrix}
12 & 6 & -12 & 6 \\
6 & 4 & -6 & 2 \\
-12 & -6 & 12 & -6 \\
6 & 2 & -6 & 4
\end{bmatrix}
\begin{array}{c}
3 \\
4 \\
5 \\
6
\end{array}.
\tag{14.24}
$$

Proceeding systematically to assemble the system stiffness matrix, we translate each entry in the element stiffness matrices, eqs. (14.22), (14.23), and (14.24), to the appropriate location in the system stiffness matrix. For instance, the stiffness coefficient for element △ $k_{13}^{(3)} = -12 \times 10^7$ should be translated to location at row three and column five since these are the coordinates indicated at right and top of matrix eq. (14.24) for this entry. Every element stiffness coefficient translated to its appropriate location in the system stiffness matrix is added to the other coefficients accumulated at that location. The stiffness coefficients corresponding to columns or rows carrying a label of a fixed system nodal coordinate (seven in the present example) are simply disregarded since the constrained nodal coordinates are not unknown quantities. The assemblage of the system matrix in the manner described results for this example in a 6 × 6 matrix, namely

$$[k] = 10^7 \begin{bmatrix} 24 & 0 & -12 & 6 & 0 & 0 \\ 0 & 8 & -6 & 2 & 0 & 0 \\ -12 & -6 & 24 & 0 & -12 & 6 \\ 6 & 2 & 0 & 8 & -6 & 2 \\ 0 & 0 & -12 & -6 & 12 & -6 \\ 0 & 0 & 6 & 2 & -6 & 4 \end{bmatrix}. \tag{14.25}$$

Equation (14.25) is thus the system stiffness matrix for the cantilever beam shown in Fig. 14.3 which has been segmented into three elements. As such, the system stiffness matrix relates the forces and the displacements at the nodal system coordinates in the same manner as the element stiffness matrix relates forces and displacements at the element nodal coordinates.

14.3 INERTIAL PROPERTIES—LUMPED MASS

The simplest method for considering the inertial properties for a dynamic system is to assume that the mass of the structure is lumped at the nodal coordinates where translational displacements are defined; hence the name *lumped mass method*. The usual procedure is to distribute the mass of each element to the nodes of the element. This distribution of the mass is determined by statics. Figure 14.4 shows for beam segments of length L and distributed mass $\overline{m}(x)$ per unit of length, the nodal allocation for uniform, triangular, and general mass distribution along the beam segment. The assemblage of the mass matrix for the entire structure will be a simple matter of adding the contributions of lumped masses at the nodal coordinates defined as translations.

In this method, the inertial effect associated with any rotational degree of freedom is usually assumed to be zero, although a finite value may be associated with rotational degrees of freedom by calculating the mass moment of inertia of a fraction of the beam segment about the nodal points. For example, for a uniform beam, this calculation would result in determining the mass moment of inertia of half of the beam segment about each node, that is

$$I_A = I_B = \frac{1}{3}\left(\frac{\overline{m}L}{2}\right)\left(\frac{L}{2}\right)^2,$$

where $\overline{m}$ is the mass per unit length along the beam. For the cantilever beam shown in Fig. 14.3 in which only translational mass effects are considered, the mass matrix of the system would be the diagonal matrix, namely

Mass distribution	Lumped mass
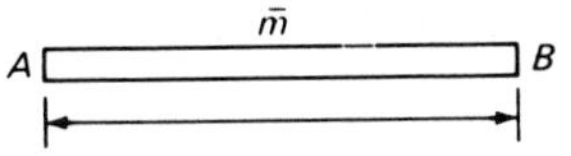 A $\overline{m}$ B Uniform	$m_A = \dfrac{\overline{m}L}{2}$ $m_B = \dfrac{\overline{m}L}{2}$
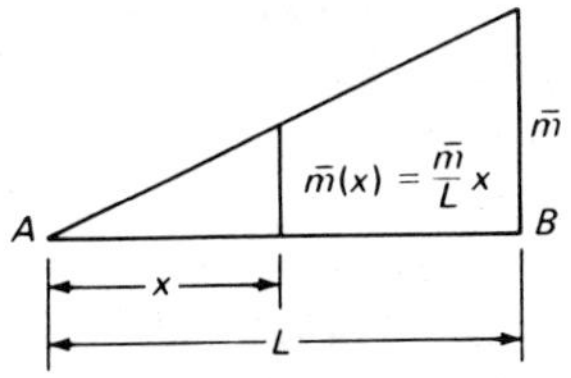 A $\overline{m}(x) = \dfrac{\overline{m}}{L}x$ $\overline{m}$ B x L Triangular	$m_A = \dfrac{\overline{m}L}{6}$ $m_B = \dfrac{\overline{m}L}{3}$
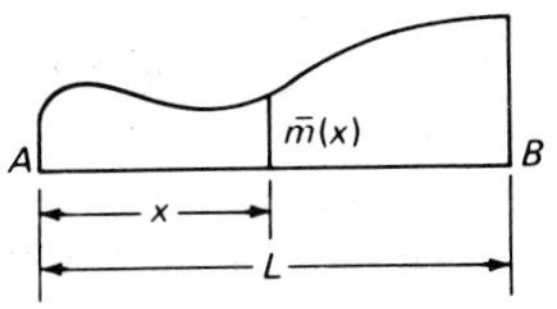 A $\overline{m}(x)$ B x L General	$m_A = \dfrac{\int_0^L (L-x)\overline{m}(x)dx}{\int_0^L \overline{m}(x)dx}$ $m_B = \dfrac{\int_0^L x\,\overline{m}(x)dx}{\int_0^L \overline{m}(x)dx}$

Fig. 14.4   Lumped masses for beam segments with distributed mass.

$$[M] = \begin{bmatrix} m_1 & & & & & \\ & 0 & & & & \\ & & m_3 & & & \\ & & & 0 & & \\ & & & & m_5 & \\ & & & & & 0 \end{bmatrix} \begin{matrix} 1 \\ 2 \\ 3 \\ 4 \\ 5 \\ 6 \end{matrix}$$

$$\begin{matrix} 1 & 2 & 3 & 4 & 5 & 6 \end{matrix}$$

(14.26)

in which

$$m_1 = \frac{\overline{m}L_1}{2} + \frac{\overline{m}L_2}{2},$$

$$m_3 = \frac{\overline{m}L_2}{2} + \frac{\overline{m}L_3}{2},$$

$$m_5 = \frac{\overline{m}L_3}{2}.$$

Using a special symbol ($\lceil$ $\rfloor$) for diagonal matrices, we may write eq. (14.26) as

$$[M] = \lceil m_1 \quad 0 \quad m_3 \quad 0 \quad m_5 \quad 0 \rfloor. \tag{14.27}$$

14.4   INERTIAL PROPERTIES–CONSISTENT MASS

It is possible to evaluate the mass coefficients corresponding to the nodal coordinates of a beam element by a procedure similar to the determination of element stiffness coefficients. First, we define the mass coefficient m_{ij} as the force at nodal coordinate i due to a unit acceleration at nodal coordinate j while all other nodal coordinates are maintained at zero acceleration.

Consider the beam segment shown in Fig. 14.5(a) which has distributed mass $\overline{m}(x)$ per unit of length. In the consistent mass method, it is assumed that the deflections resulting from unit dynamic displacements at the nodal coordinates of the beam element are given by the same functions $\psi_1(x)$, $\psi_2(x)$, $\psi_3(x)$, and $\psi_4(x)$ of eqs. (14.9) which were obtained from static considerations. If the beam segment is subjected to a unit nodal acceleration at one of the nodal coordinates, say $\ddot{\delta}_2 = 1$, the transverse acceleration developed along the length of the beam is given by the second derivative with respect to time of eq. (14.10). In this case, with $\ddot{\delta}_1 = \ddot{\delta}_3 = \ddot{\delta}_4 = 0$, we obtain

$$\ddot{y}_2(x) = \psi_2(x)\,\ddot{\delta}_2. \tag{14.28}$$

The inertial force per unit of length along the beam due to this acceleration is then given by

$$f_I(x) = \overline{m}(x)\,\ddot{y}_2(x)$$

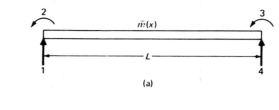

(a)

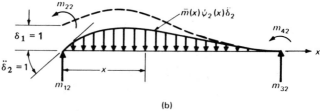

(b)

Fig. 14.5   (a) Beam element with distributed mass showing four nodal coordinates. (b) Beam element supporting inertial load due to acceleration $\ddot{\delta}_2 = 1$, undergoing virtual diaplacement $\delta_1 = 1$.

or using eq. (14.28) by

$$f_I(x) = \overline{m}(x)\, \psi_2(x)\, \ddot{\delta}_2$$

or, since $\ddot{\delta}_2 = 1$,

$$f_I(x) = \overline{m}(x)\, \psi_2(x). \tag{14.29}$$

Now to determine the mass coefficient m_{12}, we give to the beam in Fig. 14.5(b) a virtual displacement corresponding to a unit displacement at coordinate 1, $\delta_1 = 1$ and proceed to apply the principle of virtual work for an elastic system (external work equal to internal virtual work). The virtual work of the external force is simply

$$W_E = m_{12}\delta_1 = m_{12} \tag{14.30}$$

since the only external force undergoing virtual displacement is the inertial force reaction $m_{12}\delta_1$ with $\delta_1 = 1$. The virtual work of the internal forces per unit of length along the beam segment is

$$\delta W_I = f_I(x)\, \psi_1(x)$$

or by eq. (14.29),

$$\delta W_I = \overline{m}(x)\, \psi_2(x)\, \psi_1(x),$$

and for the entire beam

$$W_I = \int_0^L \overline{m}(x)\, \psi_2(x)\, \psi_1(x)\, dx. \tag{14.31}$$

Equating the external and internal virtual work given, respectively, by eqs. (14.30) and (14.31) results in

$$m_{12} = \int_0^L \overline{m}(x)\, \psi_2(x)\, \psi_1(x)\, dx \tag{14.32}$$

which is the expression for the consistent mass coefficient m_{12}. In general, a consistent mass coefficient may be calculated from

$$m_{ij} = \int_0^L \overline{m}(x)\, \psi_i(x)\, \psi_j(x)\, dx. \tag{14.33}$$

It may be seen from eq. (14.33) that $m_{ij} = m_{ji}$ since the interchange of the subindices only results in an interchange of the order of the factors $\psi_i(x)$ and $\psi_j(x)$ under the integral.

In practice, the cubic equations, eqs. (14.9), are used in calculating the mass coefficients of any straight beam element. For the special case of the beam with uniformly distributed mass, the use of eq. (14.33) gives the following relation between inertial forces and acceleration at the nodal coordinates:

$$
\begin{bmatrix} P_1 \\ P_2 \\ P_3 \\ P_4 \end{bmatrix} = \frac{\overline{m}L}{420} \begin{bmatrix} 156 & & & \text{symmetric} \\ 22L & 4L^2 & & \\ 54 & 13L & 156 & \\ -13L & -3L^2 & -22L & 4L^2 \end{bmatrix} \begin{bmatrix} \ddot{\delta}_1 \\ \ddot{\delta}_2 \\ \ddot{\delta}_3 \\ \ddot{\delta}_4 \end{bmatrix}. \qquad (14.34)
$$

When the mass matrix eq. (14.34) has been evaluated for each beam element of the structure, the mass matrix for the entire system is assembled by exactly the same procedure (direct method) as described in developing the stiffness matrix for the system. The resulting mass matrix will in general have the same arrangement of nonzero terms as the stiffness matrix.

The dynamic analysis using the lumped mass matrix requires considerable less computational effort than the analysis using the consistent mass method for the following reasons. The lumped mass matrix for the system results in a diagonal mass matrix while the consistent mass matrix has many off-diagonal terms which are called mass coupling. Also, the lumped mass matrix contains zeros in its main diagonal due to assumed zero rotational inertial forces. This fact permits the elimination by static condensation (Chapter 13) of the rotational degrees of freedom, thus reducing the dimension of the dynamic problem. Nevertheless, the dynamic analysis using the consistent mass matrix gives results which approximate better to the exact solution compared to the lumped mass method for the same element discretization.

Illustrative Example 14.1.   Determine the lumped mass and the consistent mass matrices for the cantilever beam in Fig. 14.6. Assume uniform mass, $\overline{m} = 420$ kg/m.

(a) Lumped Mass Matrix. The lumped mass at each node of any of the three beam segments, into which the cantilever beam has been divided, is simply half of the mass of the segment. In the present case, the lumped mass at each node is

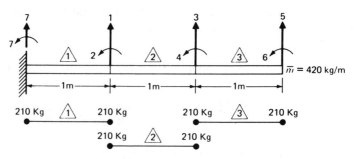

Fig. 14.6   Lumped masses for Example 14.1.

210 kg as shown in Fig. 14.6. The lumped mass matrix $[M_L]$ for this structure is a diagonal matrix of dimension 6 × 6, namely,

$$[M_L] = \lceil 420 \quad 0 \quad 420 \quad 0 \quad 210 \quad 0 \rfloor.$$

(b) Consistent Mass Matrix. The consistent mass matrix for a uniform beam segment is given by eq. (14.34). The substitution of numerical values for this example $L = 1m, \overline{m} = 420$ kg/m into eq. (14.34) gives the consistent mass matrix $[M_c]$ for any of three beam segments as

$$[M_c] = \begin{matrix} & 1 & 2 & 3 & 4 & \\ & \begin{bmatrix} 156 & 22 & 54 & -13 \\ 22 & 4 & 13 & -3 \\ 54 & 13 & 156 & -22 \\ -13 & -3 & -22 & 4 \end{bmatrix} & \begin{matrix} 1 \\ 2 \\ 3 \\ 4 \end{matrix} \end{matrix} \quad (a)$$

The assemblage of the system mass matrix from the element mass matrices is carried out in exactly the same manner as the assemblage of the system stiffness matrix from the element stiffness matrices, that is, the element mass matrices are allocated to appropriate entries in the system mass matrix. For the second beam segment, this allocation corresponds to the first four coordinates as indicated above and on the right of eq. (a). For the beam segment ⚠, the appropriate allocation is 3, 4, 5, 6 and for the beam segment ⚠, 7, 7, 1, 2 since these are the system nodal coordinates for these beam segments as indicated in Fig. 14.6. The consistent mass matrix $[M_c]$ for this example obtained in this manner is given by eq. (b) with all the entries left blank being equal to zero. Hence

$$[M_c] = \begin{matrix} & 1 & 2 & 3 & 4 & 5 & 6 & \\ & \begin{bmatrix} 312 & 0 & 54 & -13 & 0 & 0 \\ 0 & 8 & 13 & -3 & 0 & 0 \\ 54 & 13 & 312 & 0 & 54 & -13 \\ -13 & -3 & 0 & 8 & 13 & -3 \\ 0 & 0 & 54 & 13 & 156 & -22 \\ 0 & 0 & -13 & -3 & -22 & 4 \end{bmatrix} & \begin{matrix} 1 \\ 2 \\ 3 \\ 4 \\ 5 \\ 6 \end{matrix} \end{matrix} \quad (b)$$

We note that the mass matrix $[M_c]$ is symmetric and also banded as in the case of the stiffness matrix for this system. These facts are of great importance in developing computer programs for structural analysis, since it is possible to perform the necessary calculations storing in the computer only the diagonal elements and the elements to one of the sides of the main diagonal. The maximum number of nonzero elements in any row which are required to be stored is referred to as the bandwidth of the matrix. For the matrix eq. (b) the bandwidth is equal to four (NBW = 4). In this case, it is necessary to store a total of 6 × 4 = 24 coefficients, while if the square matrix were to be stored, it would require 6 × 6 = 36 storage spaces. This economy in storing spaces becomes more dramatic for structures with a large number of nodal coordinates. The dimension of the bandwidth is directly related to the largest difference of the nodal coordinates assigned to any of the elements of the structure. Therefore, it is important to number the system nodal coordinates so as to minimize this difference.

14.5   DAMPING PROPERTIES

Damping coefficients are defined in a manner entirely parallel to the definition of the stiffness coefficient or the mass coefficient. Specifically, the damping coefficient c_{ij} is defined as the force developed at coordinate i due to a unit velocity at j. If the damping forces distributed in the structure could be determined, the damping coefficients of the various structural elements would then be used in obtaining the damping coefficient corresponding to the system. For example, the damping coefficient c_{ij} for an element might be of the form

$$c_{ij} = \int_0^L c(x)\, \psi_i(x)\, \psi_j(x)\, dx, \qquad (14.35)$$

where $c(x)$ represents the distributed damping coefficient per unit length. If the element damping matrix could be calculated, the damping matrix for the entire structure could be assembled by a superposition process equivalent to the direct stiffness matrix. In practice, the evaluation of the damping property $c(x)$ is impracticable. For this reason, the damping is generally expressed in terms of damping ratios obtained experimentally rather than by a direct evaluation of the damping matrix using eq. (14.35). These damping ratios are evaluated or estimated for each natural mode of vibration. If the explicit expression of the damping matrix $[C]$ is needed, it may be computed from the specified relative damping coefficients by any of the methods described in Chapter 12.

14.6   EXTERNAL LOADS

When the dynamic loads acting on the structure consist of concentrated forces and moments applied at defined nodal coordinates, the load vector can be written directly. In general, however, loads are applied at points other than nodal coordinates. In addition, the external load may include the action of distributed forces. In this case, the load vector corresponding to the nodal coordinates consists of the equivalent or generalized forces. The procedure to deter-

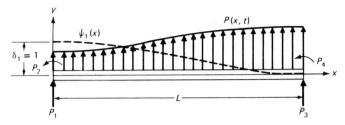

Fig. 14.7 Beam element supporting arbitrary distributed load undergoing virtual displacement $\delta_1 = 1$.

mine the equivalent nodal forces which is consistent with the derivation of the stiffness matrix and the consistent mass matrix is to assume the validity of the static deflection functions, eqs. (14.9), for the dynamic problem and to use the principle of virtual work.

Consider the beam element in Fig. 14.7 when subjected to an arbitrary distributed force $p(x, t)$ which is a function of position along the beam as well as a function of time. The equivalent force P_1 at coordinate 1 may be found by giving a virtual displacement $\delta_1 = 1$ at this coordinate and equating the resulting external work and internal work during this virtual displacement. In this case, the external work is

$$W_E = P_1 \delta_1 = P_1 \tag{14.36}$$

since $\delta_1 = 1$. The internal work per unit of length along the beam is $p(x, t)\psi_1(x)$ and the total internal work is then

$$W_I = \int_0^L p(x, t)\,\psi_1(x)\,dx. \tag{14.37}$$

Equating external work, eq. (14.36), and internal work, eq. (14.37), gives the equivalent nodal force as

$$P_1(t) = \int_0^L p(x, t)\,\psi_1(x)\,dx. \tag{14.38}$$

Thus the element equivalent nodal forces can be expressed in general as

$$P_i(t) = \int_0^L p(x, t)\,\psi_i(x)\,dx. \tag{14.39}$$

Illustrative Example 14.2.   Consider the beam segment in Fig. 14.8 and determine the element nodal forces for a uniform distributed force along the length

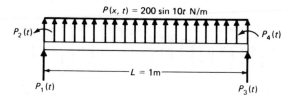

Fig. 14.8   Beam segment subjected to external distributed load showing equivalent nodal forces.

of the beam given by

$$p(x, t) = 200 \sin 10t \text{ N/m}.$$

Introduction of numerical values into the displacements functions, eqs. (14.9), and substitution in eq. (14.39) yield

$$P_1(t) = 200 \int_0^1 (1 - 3x^2 + 2x^3) \, dx \sin 10t = 100 \sin 10t,$$

$$P_2(t) = 200 \int_0^1 x(1 - x)^2 \, dx \sin 10t = 16.67 \sin 10t,$$

$$P_3(t) = 200 \int_0^1 (3x^2 - 2x^3) \, dx \sin 10t = 100 \sin 10t,$$

$$P_4(t) = 200 \int_0^1 x^2(x - 1) \, dx \sin 10t = -16.67 \sin 10t.$$

14.7   GEOMETRIC STIFFNESS

When a beam element is subjected to an axial force in addition to flexural loading, the stiffness coefficients are modified by the presence of the axial force. The modification corresponding to the stiffness coefficient k_{ij} is known as the geometric stiffness k_{Gij} which is defined as the force corresponding to the nodal coordinate i due to a unit displacement at coordinate j and resulting from the axial forces in the structure. These coefficients may be evaluated by application of the principle of virtual work. Consider a beam element as used previously but now subjected to a distributed axial force per unit of length $N(x)$, as depicted in Fig. 14.9(a). In the sketch in Fig. 14.9(b), the beam segment is subjected to a unit rotation of the left end, $\delta_2 = 1$. By definition, the nodal forces due to this

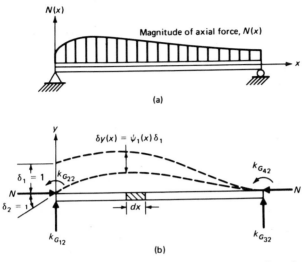

Fig. 14.9 (a) Beam element loaded with arbitrary distributed axial force. (b) Beam element acted on by nodal forces resulting from displacement $\delta_2 = 1$, undergoing a virtual displacement $\delta_1 = 1$.

displacement are the corresponding geometric stiffness coefficients, for example, k_{G12} is the vertical force at the left end. If we now give to this deformed beam a unit displacement $\delta_1 = 1$, the resulting external work is

$$W_e = k_{G12}\delta_1$$

or

$$W_e = k_{G12} \qquad (14.40)$$

since $\delta_1 = 1$.

The internal work during this virtual displacement is found by considering a differential element of length dx taken from the beam in Fig. 14.9(b) and shown enlarged in Fig. 14.10. The work done by the axial force $N(x)$ during the virtual displacement is

$$dW_I = N(x)\delta_e, \qquad (14.41)$$

where δ_e represents the relative displacement experienced by the normal force $N(x)$ acting on the differential element during the virtual displacement. From Fig. 14.10, by similar triangles (triangles I and II), we have

$$\frac{\delta_e}{d\psi_1(x)} = \frac{d\psi_2(x)}{dx}$$

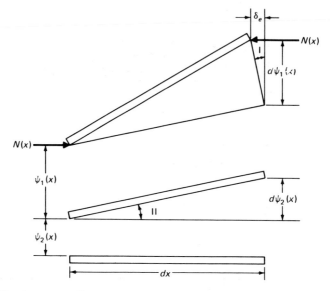

Fig. 14.10   Differential segment of deflected beam in Fig. 14.9.

or

$$\delta_e = \frac{d\psi_1}{dx} \cdot \frac{d\psi_2}{dx}\, dx,$$

$$\delta_e = \psi_1'(x) \cdot \psi_2'(x)\, dx,$$

in which $\psi_1'(x)$ and $\psi_2'(x)$ are the derivatives with respect to x of the corresponding displacement functions defined in eqs. (14.9).

Now, substituting δ_e into eq. (14.41), we have

$$dW_{\mathrm{I}} = N(x)\, \psi_1'(x)\, \psi_2'(x)\, dx. \tag{14.42}$$

Then integrating this expression and equating the result to the external work, eq. (14.40), finally give

$$k_{G12} = \int_0^L N(x)\, \psi_1'(x)\, \psi_2'(x)\, dx. \tag{14.43}$$

In general, any geometric stiffness coefficient may be expressed as

$$k_{Gij} = \int_0^L N(x)\, \psi_i'(x)\, \psi_j'(x)\, dx. \tag{14.44}$$

In the derivation of eq. (14.44), it is assumed that the normal force $N(x)$ is independent of time. When the displacement functions, eqs. (14.9), are used in eq.

(14.44) to calculate the geometric stiffness coefficients, the result is called the *consistent geometric stiffness* matrix. In the special case where the axial force is constant along the length of the beam, use of eqs. (14.44) and (14.9) gives the geometric stiffness equation as

$$
\begin{bmatrix} P_1 \\ P_2 \\ P_3 \\ P_4 \end{bmatrix} = \frac{N}{30L} \begin{bmatrix} 36 & \text{symmetric} & & \\ 3L & 4L^2 & & \\ -36 & -3L & 36 & \\ 3L & -L^2 & -3L & 4L^2 \end{bmatrix} \begin{bmatrix} \delta_1 \\ \delta_2 \\ \delta_3 \\ \delta_4 \end{bmatrix}.
\tag{14.45}
$$

The assemblage of the system geometric stiffness matrix can be carried out exactly in the same manner as for the assemblage of the elastic stiffness matrix. The resulting geometric stiffness matrix will have the same configuration as the elastic stiffness matrix. It is customary to define the geometric stiffness matrix for a compressive axial force. In this case, the combined stiffness matrix $[K_c]$ for the structure is given by

$$
[K_c] = [K] - [K_G]
\tag{14.46}
$$

in which $[K]$ is the assembled elastic stiffness matrix for the structure and $[K_G]$ the corresponding geometric stiffness matrix.

Illustrative Example 14.3. For the cantilever beam in Fig. 14.11, determine the system geometric matrix when an axial force of magnitude 30 N is applied at the free end as shown in this figure.

The substitution of numerical values into eq. (14.45) for any of the three beam segments in which the cantilever beam has been divided gives the element geometric matrix

$$
[K_G] = \begin{bmatrix} 36 & & \text{symmetric} & \\ 3 & 4 & & \\ -36 & -3 & 36 & \\ 3 & -1 & -3 & 4 \end{bmatrix}
$$

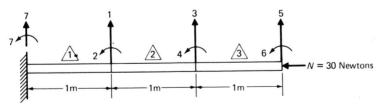

Fig. 14.11   Cantilever beam subjected to constant axial force (Example 14.3).

since in this example $L = 1m$ and $N = 30$ N. Use of the direct method gives the assembled system geometric matrix as

$$[K_G] = \begin{bmatrix} 72 & 0 & -36 & 3 & 0 & 0 \\ 0 & 8 & -3 & -1 & 0 & 0 \\ -36 & -3 & 72 & 0 & -36 & 3 \\ 3 & -1 & 0 & 8 & -3 & -1 \\ 0 & 0 & -36 & -3 & 36 & -3 \\ 0 & 0 & 3 & -1 & -3 & 4 \end{bmatrix}.$$

14.8 EQUATIONS OF MOTION

In the previous sections of this chapter, the distributed properties of a beam and its load were expressed in terms of discrete quantities at the nodal coordinates. The equations of motion as functions of these coordinates may then be established by imposing conditions of dynamic equilibrium between the inertial forces $\{F_I(t)\}$, damping forces $\{F_D(t)\}$, elastic forces $\{F_S(t)\}$ and the external forces $\{F(t)\}$, that is

$$\{F_I(t)\} + \{F_D(t)\} + \{F_S(t)\} = \{F(t)\}. \tag{14.47}$$

The forces on the left-hand side of eq. (14.47) are expressed in terms of the system mass matrix, the system damping matrix, and the system stiffness matrix as

$$\{F_I(t)\} = [M]\{\ddot{y}\}, \tag{14.48}$$

$$\{F_D(t)\} = [C]\{\dot{y}\}, \tag{14.49}$$

$$\{F_S(t)\} = [K]\{y\}. \tag{14.50}$$

Substitution of these equations into eq. (14.47) gives the differential equation of motion for a linear system as

$$[M]\{\ddot{y}\} + [C]\{\dot{y}\} + [K]\{y\} = \{F(t)\}. \tag{14.51}$$

In addition, if the effect of axial forces is considered in the analysis, eq. (14.51) is modified so that

$$[M]\{\ddot{y}\} + [C]\{\dot{y}\} + [K_c]\{y\} = \{F(t)\} \tag{14.52}$$

in which

$$[K_c] = [K] - [K_G]. \tag{14.53}$$

In practice, the solution of eq. (14.51) or eq. (14.52) is accomplished by standard methods of analysis and the assistance of appropriate computer programs as those described in this and the following chapters. We illustrate these methods by presenting here some simple problems for hand calculation.

Illustrative Example 14.4. Consider in Fig. 14.12 a uniform beam with the ends fixed against translation or rotation. In preparation for analysis, the beam has been divided into four equal segments. Determine the first three natural frequencies and corresponding modal shapes. Use the lumped mass method in order to simplify the numerical calculations.

We begin by numbering sequentially the nodal coordinates starting with the rotational coordinates which have to be condensed in the lumped mass method (no inertial effect associated with rotational coordinates), continuing to number the coordinates associated with translation and assigning the dummy last number 7 to any fixed nodal coordinates as shown in Fig. 14.12. The stiffness matrix for any of the beam segments for this example is obtained from eq. (14.20) as

$$
[K] = EI
\begin{array}{cccc}
4 & 1 & 5 & 2 \\
7 & 7 & 4 & 1 \\
\end{array}
\begin{bmatrix}
12 & 6 & -12 & 6 \\
6 & 4 & -6 & 2 \\
-12 & -6 & 12 & -6 \\
6 & 2 & -6 & 4
\end{bmatrix}
\begin{array}{cc}
7 & 4 \\
7 & 1 \\
4 & 5 \\
1 & 2
\end{array}
. \qquad (14.54)
$$

With the aid of the system nodal coordinates for each beam segment written at the top and on the right of the stiffness matrix, eq. (14.54), we proceed to assemble the system stiffness matrix using the direct method. For the beam segment △, the corresponding labels are 7, 7, 4, 1. Since the label 7 which corresponds to fixed coordinates should be ignored, we need to translate only the lowest 2 × 2 submatrix on the right to locations given by the combination of row indices 4, 1 and column indices 4, 1; for the beam segment △, we translate the 4 × 4 elements of matrix eq. (14.54) to the system stiffness matrix to rows

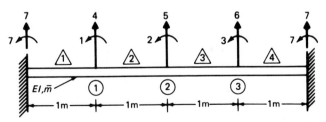

Fig. 14.12 Fixed beam divided in four elements with indication of system nodal coordinates (Example 14.4).

and columns designated by combination of indices 4, 1, 5, 2 as labeled for this element and so forth for the other two beam segments. The assembled system stiffness matrix obtained in this manner is

$$
[K] = EI
\begin{bmatrix}
8 & 2 & 0 & 0 & -6 & 0 \\
2 & 8 & 2 & 6 & 0 & -6 \\
0 & 2 & 8 & 0 & 6 & 0 \\
0 & 6 & 0 & 24 & -12 & 0 \\
-6 & 0 & 6 & -12 & 24 & -12 \\
0 & -6 & 0 & 0 & -12 & 24
\end{bmatrix}.
\tag{14.55}
$$

The reduction or condensation of eq. (14.55) is accomplished as explained in Chapter 13 by simply performing the Gauss-Jordan elimination of the first three rows since, in this case, we should condense these first three coordinates. This elimination reduces eq. (14.55) to the following matrix:

$$
[A] =
\left[
\begin{array}{ccc|ccc}
1 & 0 & 0 & -0.214 & -0.750 & 0.214 \\
0 & 1 & 0 & 0.858 & 0 & -0.858 \\
0 & 0 & 1 & -0.214 & 0.750 & 0.214 \\
\hline
0 & 0 & 0 & 18.86EI & -12.00EI & 5.14EI \\
0 & 0 & 0 & -12.00EI & 15.00EI & -12.00EI \\
0 & 0 & 0 & 5.14EI & -12.00EI & 18.86EI
\end{array}
\right].
\tag{14.56}
$$

Comparison of eq. (14.56) in partition form with eq. (13.1) permits the identification of the reduced stiffness matrix $[\overline{K}]$ and the transformation matrix $[\overline{T}]$, so that

$$
[\overline{K}] = EI
\begin{bmatrix}
18.86 & -12.00 & 5.14 \\
-12.00 & 15.00 & -12.00 \\
5.14 & -12.00 & 18.86
\end{bmatrix}
\tag{14.57}
$$

and

$$
[\overline{T}] =
\begin{bmatrix}
0.214 & 0.750 & -0.214 \\
-0.858 & 0 & 0.858 \\
0.214 & -0.750 & -0.214
\end{bmatrix}.
\tag{14.58}
$$

The general transformation matrix, eq. (13.6), is then

$$[T] = \begin{bmatrix} 0.214 & 0.750 & -0.214 \\ -0.858 & 0 & 0.858 \\ 0.214 & -0.750 & -0.214 \\ 1 & 0 & 0 \\ 0 & 1 & 0 \\ 0 & 0 & 1 \end{bmatrix}. \qquad (14.59)$$

As an exercise, the reader may check eq. (13.7) for this example by simply performing the matrix multiplications

$$[\bar{K}] = [T]^T [K] [T].$$

The lumped mass method applied to this example gives three equal masses of magnitude $\bar{m}$ at each of the three translatory coordinates as indicated in Fig. 14.13. Therefore, the reduced lumped mass matrix is

$$[\bar{M}] = \bar{m} \begin{bmatrix} 1 & 0 & 0 \\ 0 & 1 & 0 \\ 0 & 0 & 1 \end{bmatrix}. \qquad (14.60)$$

The natural frequencies and modal shapes are found by solving the undamped free vibration problem, that is

$$[\bar{M}] \{\ddot{y}\} + [\bar{K}] \{y\} = \{0\}. \qquad (14.61)$$

Assuming the harmonic solution ($\{y\} = \{a\} \sin \omega t$), we obtain

$$([\bar{K}] - \omega^2 [\bar{M}]) \{a\} = \{0\} \qquad (14.62)$$

requiring for a nontrivial solution that the determinant

$$|[\bar{K}] - \omega^2 [\bar{M}]| = 0. \qquad (14.63)$$

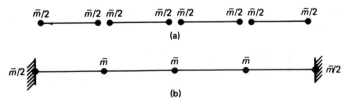

Fig. 14.13   (a) Lumped masses for uniform beam segments. (b) Lumped masses at system nodal coordinates.

Substitution in this last equation $[\bar{K}]$ and $[\bar{M}]$, respectively, from eqs. (14.57) and (14.60) yields

$$\begin{vmatrix} 18.86 - \lambda & -12.00 & 5.14 \\ -12.00 & 15.00 - \lambda & -12.00 \\ 5.14 & -12.00 & 18.86 - \lambda \end{vmatrix} = 0 \qquad (14.64)$$

in which

$$\lambda = \frac{\bar{m}\omega^2}{EI}. \qquad (14.65)$$

The roots of the cubic equation, eq. (14.64), are found to be

$$\lambda_1 = 1.943, \lambda_2 = 13.720, \text{ and } \lambda_3 = 37.057. \qquad (14.66)$$

Then from eq. (14.65)

$$\omega_1 = 1.393 \sqrt{EI/\bar{m}},$$
$$\omega_2 = 3.704 \sqrt{EI/\bar{m}},$$
$$\omega_3 = 6.087 \sqrt{EI/\bar{m}}. \qquad (14.67)$$

The first three natural frequencies for a uniform fixed beam of length $L = 4m$ determined by the exact analysis (Chapter 21) are

$$\omega_1 \text{ (exact)} = 1.398 \sqrt{EI/\bar{m}},$$
$$\omega_2 \text{ (exact)} = 3.857 \sqrt{EI/\bar{m}},$$
$$\omega_3 \text{ (exact)} = 7.540 \sqrt{EI/\bar{m}}. \qquad (14.68)$$

The first two natural frequencies determined using the three-degrees-of-freedom reduced system compare very well with the exact values. A practical rule in condensing degrees of freedom, is to reduce the system to a dimension of twice the number of required natural frequencies for the dynamic analysis.

The modal shapes are determined by solving two of the equations in eq. (14.62) after substituting successively values of ω_1, ω_2, ω_3 from eqs. (14.67) and letting the first element for each modal shape be equal to one. The resulting modal shapes are

$$\{a\}_1 = \begin{bmatrix} 1.00 \\ 1.84 \\ 1.00 \end{bmatrix}, \qquad \{a\}_2 = \begin{bmatrix} 1.00 \\ 0 \\ -1.00 \end{bmatrix}, \qquad \{a\}_3 = \begin{bmatrix} 1.00 \\ -1.08 \\ 1.00 \end{bmatrix}. \qquad (14.69)$$

The normalized modal shapes which are obtained by division of the elements of eqs. (14.69) by corresponding values of $\sqrt{\Sigma m_i a_{ij}^2}$ ($m_i = \bar{m} = 100$ kg/m for this example) are arranged in the columns of the modal matrix as

$$[\Phi] = \begin{bmatrix} 0.0431 & 0.0707 & 0.0562 \\ 0.0793 & 0 & -0.0607 \\ 0.0431 & -0.0707 & 0.0562 \end{bmatrix}. \tag{14.70}$$

Illustrative Example 14.5.   Determine the steady-state response for the beam of Example 14.4 when subjected to the harmonic forces

$$F_1 = F_{01} \sin \bar{\omega}t,$$

$$F_2 = F_{02} \sin \bar{\omega}t,$$

and

$$F_3 = F_{03} \sin \bar{\omega}t$$

acting, respectively, at joints 1, 2, and 3 of the beam in Fig. 14.12. Neglect damping and let $EI = 10^8 (\text{N} \cdot \text{m}^2)$, $m = 100$ kg/m, $\bar{\omega} = 3000$ rad/sec, $F_{01} = 2000$ N, $F_{02} = 3000$ N, and $F_{03} = 1000$ N.

The normal equations (uncoupled equations) can readily be written using the results of Example 14.4. In general the nth normal equation is given by

$$\ddot{z}_n + \omega_n^2 z_n = P_n \sin \bar{\omega}t \tag{14.71}$$

in which

$$P_n = \sum_{i=1}^{N} \phi_{in} F_{0i}.$$

The steady-state solution of eq. (14.71) is given by eq. (3.4) as

$$z_n = Z_n \sin \bar{\omega}t = \frac{P_n \sin \bar{\omega}t}{\omega_n^2 - \bar{\omega}^2}. \tag{14.72}$$

The calculations required in eq. (14.72) are conveniently arranged in Table 14.1.

TABLE 14.1   Modal Response for Example 14.5.

Mode n	ω_n^2	$P_n = \sum_i \phi_{in} F_{0i}$	$Z_n = \dfrac{P_n}{\omega_n^2 - \bar{\omega}^2}$
1	1.943×10^6	367.2	$-5.200 \ 10^{-5}$
2	13.720×10^6	70.7	$1.500 \ 10^{-5}$
3	37.057×10^6	-13.5	$-0.048 \ 10^{-5}$

The deflections at the nodal coordinates are found from the transformation

$$\{y\} = [\Phi]\{z\} \tag{14.73}$$

in which $[\Phi]$ is the modal matrix and

$$\{y\} = \{Y\} \sin \overline{\omega}t,$$

$$\{z\} = \{Z\} \sin \overline{\omega}t.$$

The substitution of the modal matrix from eq. (14.70) and the values of $\{Z\}$ from the last column of Table 14.1 into eq. (14.73) gives the amplitudes at nodal coordinates 4, 5, and 6 as

$$\begin{bmatrix} Y_4 \\ Y_5 \\ Y_6 \end{bmatrix} = \begin{bmatrix} 0.0431 & 0.0707 & 0.0562 \\ 0.0793 & 0 & -0.0607 \\ 0.0431 & -0.0707 & 0.0562 \end{bmatrix} \begin{bmatrix} -5.200 \\ 1.500 \\ -0.048 \end{bmatrix} 10^{-5}$$

or

$$Y_4 = -1.207 \times 10^{-6} \text{ m},$$

$$Y_5 = -4.094 \times 10^{-6} \text{ m},$$

$$Y_6 = -3.329 \times 10^{-6} \text{ m}.$$

Therefore, the motion at these nodal coordinates is given by

$$y_4 = -1.207 \times 10^{-6} \sin 3000t \text{ m},$$

$$y_5 = -4.094 \times 10^{-6} \sin 3000t \text{ m},$$

$$y_6 = -3.329 \times 10^{-6} \sin 3000t \text{ m}. \tag{14.74}$$

The minus sign in the resulting amplitudes of motion simply indicates that the motion is $180°$ out of phase with the applied harmonic forces.

14.9 ELEMENT FORCES AT NODAL COORDINATES

The central problem to be solved using the dynamic stiffness method is to determine the displacements at the nodal coordinates. Once these displacements have been determined, it is a simple matter of substituting the appropriate displacements in the condition of dynamic equilibrium for each element to calculate the forces at the nodal coordinates. The nodal element forces $\{P\}$ may be obtained by adding the inertial force $\{P_I\}$, the damping force $\{P_D\}$, the elastic force $\{P_S\}$, and subtracting the nodal equivalent forces $\{P_E\}$. Therefore, we may write

$$\{P\} = \{P_I\} + \{P_D\} + \{P_S\} - \{P_E\}$$

or

$${P} = [m] \{\ddot{\delta}\} + [c] \{\dot{\delta}\} + [k] \{\delta\} - \{P_E\}. \qquad (14.75)$$

In eq. (14.75) the inertial force, the damping force, and the elastic force are, respectively,

$${P_I} = [m] \{\ddot{\delta}\},$$

$${P_D} = [c] \{\dot{\delta}\},$$

$${P_S} = [k] \{\delta\}, \qquad (14.76)$$

where $[m]$ is the element mass matrix; $[c]$ the element damping matrix; $[k]$ the element stiffness matrix; and $\{\delta\}$, $\{\dot{\delta}\}$, $\{\ddot{\delta}\}$ represents, respectively, the displacement, velocity, and acceleration vectors at the nodal coordinates of the element.

The determination of the element nodal forces is illustrated in the following example.

Illustrative Example 14.6.   Determine the element nodal forces and moments for the four beam segments of Example 14.5. Since in this example damping is neglected and there are no external forces applied except those at the nodal coordinates, eq. (14.75) reduces to

$${P} = [m] \{\ddot{\delta}\} + [k] \{\delta\}. \qquad (14.77)$$

We begin by using the transformation eq. (13.2) to calculate the nodal displacements corresponding to the condensed coordinates. From eq. (13.2), we have

$$\{y_p\} = [\overline{T}] \{y_q\} \qquad (14.78)$$

where

$$\{y_p\} = \begin{bmatrix} z_1 \\ z_2 \\ z_3 \end{bmatrix}$$

are the condensed coordinates (angular displacement), and

$$\{y_q\} = \begin{bmatrix} z_4 \\ z_5 \\ z_6 \end{bmatrix}$$

the independent coordinates (translatory displacements). The substitution of $[\overline{T}]$ from eq. (14.58) and $\{y_q\}$ from eq. (14.74) into eq. (14.78) gives

$$\begin{bmatrix} y_1 \\ y_2 \\ y_3 \end{bmatrix} = \begin{bmatrix} 0.214 & 0.750 & -0.214 \\ -0.858 & 0 & 0.858 \\ 0.214 & -0.750 & -0.214 \end{bmatrix} \begin{bmatrix} -1.207 \\ -4.094 \\ -3.329 \end{bmatrix} 10^{-6} \sin 3000t$$

or

$$\begin{bmatrix} y_1 \\ y_2 \\ y_3 \end{bmatrix} = \begin{bmatrix} -2.616 \\ -1.818 \\ 3.524 \end{bmatrix} 10^{-6} \sin 3000t. \tag{14.79}$$

Consequently, the displacement functions for the six nodal coordinates of the beam in Fig. 14.12 are given by eqs. (14.74) and (14.79). These displacements are certainly also the displacements of the element nodal coordinates. The identification for this example of corresponding nodal coordinates between beam segments and system nodal coordinates is

$$\{\delta\}_1 = \begin{bmatrix} 0 \\ 0 \\ y_4 \\ y_1 \end{bmatrix}, \quad \{\delta\}_2 = \begin{bmatrix} y_4 \\ y_1 \\ y_5 \\ y_2 \end{bmatrix}, \quad \{\delta\}_3 = \begin{bmatrix} y_5 \\ y_2 \\ y_6 \\ y_3 \end{bmatrix}, \quad \{\delta\}_4 = \begin{bmatrix} y_6 \\ y_3 \\ 0 \\ 0 \end{bmatrix}$$

$$\tag{14.80}$$

where $\{\delta\}_i$ is the vector of nodal displacement for i beam segment. The substitution of appropriate quantities into eq. (14.77) for the first beam segment results in

$$\begin{bmatrix} P_1 \\ P_2 \\ P_3 \\ P_4 \end{bmatrix} = \begin{bmatrix} \dfrac{\bar{m}}{2} & 0 & & \\ 0 & \dfrac{\bar{m}}{2} & & \\ & & 0 \end{bmatrix} \begin{bmatrix} 0 \\ 0 \\ y_4 \\ y_1 \end{bmatrix} (-\bar{\omega}^2) \sin \bar{\omega} t$$

$$+ EI \begin{bmatrix} 12 & 6 & -12 & 6 \\ 6 & 4 & -6 & 2 \\ -12 & -6 & 12 & -6 \\ 6 & 2 & -6 & 4 \end{bmatrix} \begin{bmatrix} 0 \\ 0 \\ y_4 \\ y_1 \end{bmatrix} \sin \bar{\omega} t.$$

To complete this example, we substitute the numerical values of $\bar{\omega} = 3000$ rad/sec, $\bar{m} = 100$ kg/m, and $EI = 10^8 (\text{N} \cdot \text{m}^2)$ and obtain

$$\begin{bmatrix} P_1 \\ P_2 \\ P_3 \\ P_4 \end{bmatrix} = \begin{bmatrix} 7.5 & 6 & -12 & 6 \\ 6 & 4 & -6 & 2 \\ -12 & -6 & 7.5 & -6 \\ 6 & 2 & -6 & 4 \end{bmatrix} \begin{bmatrix} 0 \\ 0 \\ -1.207 \\ -2.616 \end{bmatrix} 10^2 \sin 3000t$$

TABLE 14.2   Element Nodal Forces (Amplitudes) for Example 14.6.

Force	Beam Segment				Units
	1	2	3	4	
P_1	-121.2	1347.1	1947.9	-382.3	N
P_2	201.0	322.2	-481.4	-587.0	N $\cdot$ m
P_3	664.3	1038.3	1392.4	1880.4	N
P_4	-322.2	481.4	587.0	-1292.6	N $\cdot$ m

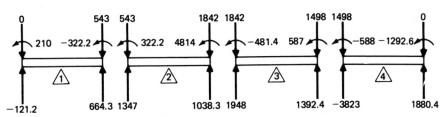

Fig. 14.14   Dynamic equilibrium for beam segments of Example 14.6.

which then gives

$$P_1 = -121.2 \sin 3000t \text{ N,}$$

$$P_2 = 201.0 \sin 3000t \text{ N} \cdot \text{m,}$$

$$P_3 = 664.3 \sin 3000t \text{ N,}$$

$$P_4 = -322.2 \sin 3000t \text{ N} \cdot \text{m.}$$

The nodal element forces found in this manner for all of the four beam segments
in this example are given in Table 14.2.

The results in Table 14.2 may be used to check that the dynamic conditions
of equilibrium are satisfied in each beam segment. The free body diagrams of the
four elements of this beam are shown in Fig. 14.14 with inclusion of nodal inertial forces. These forces are computed by multiplying the nodal mass by the corresponding nodal acceleration.

14.10   PROGRAM II—DYNAMIC ANALYSIS OF BEAMS (BEAM)

The computer program described in this section performs the analysis of beams
using the stiffness matrix method. The main section of the program loops over
the beam segments into which the beam has been divided. In this loop, the ele-

TABLE 14.3   Subroutine Programs.

Subroutine Name	Purpose	Chapter Presented
JACOBI	Determine natural frequencies and normal modes	10
CONDE	Condense the stiffness and mass matrices	13
HARMO	Determine steady-state response for harmonic excitation	11
STEPM	Determine response using step-by-step Wilson-θ method	19
MODAL	Determine response by modal superposition method	11
DAMP	Determine damping matrix from modal damping ratios	12

ment stiffness and mass matrices are computed and added to the proper entries of the system stiffness and mass matrices. After this loop has been completed for all the segments of the beam, concentrated masses at nodal coordinates, if present, are read and added to appropriate coefficients of the system mass matrix. The next portion of the program contains provisions to call, as required, subroutines to perform specific tasks as described in Table 14.3.

Table 14.3 gives a list of the subroutines which may be called by this program as well as by other computer programs described in the following chapters. The listing of these subroutines is given in the Appendix where the listings of all computer programs presented in this text are given.

A description of the principal variables used in Program 11 is given in Table 14.4 and a listing of the necessary input data and formats is given in Table 14.5.

The following are illustrative examples of the dynamics of beams using the computer program described in this section. These examples will serve to check the program as well as to illustrate the preparation of the required input data.

Illustrative Example 14.7.   Determine the natural frequencies and modal shapes of the fixed beam shown in Fig. 14.12. Proceed as in Example 14.4 to divide the beam into four segments and to condense the rotational degrees of freedom. Use the lumped mass method and neglect damping.

In the solution of this problem it is necessary to call subroutine CONDE to condense the rotational degrees of freedom and subroutine JACOBI to solve the eigenproblem. Hence "C" should be removed in the two statements of the program, calling these two subroutines while all other cards calling for other subroutines should remain as comment cards with "C" in the first column.

The listing of the required input data followed by the computer results is

TABLE 14.4 Description of Principal Variables for Program 11.

Variable	Symbol in Text	Description
NE		Number of elements
ND		Number of degrees of freedom
NCR	p	Number of degrees of freedom to condense
NCM		Number of concentrated masses
LOC		Mass index: 0 → lumped mass; 1 → consistent mass
IFPR		Print intermediate values in JACOBI: 0 → do not print, 1 → print
E	E	Modulus of elasticity
LE		Beam segment number
SL	L	Beam segment length
SMA	$\overline{m}$	Beam segment mass per unit of length
NC(L)		Beam segment nodal coordinates ($L = 1, 4$)
CM(L)		Concentrated masses ($L = 1$, NCM)
JC(L)		Nodal coordinate for CM(L)

TABLE 14.5 Input Data and Formats for Program 11.

Format	Variables
(6I5, E20.4)	NE ND NCR NCM LOC IFPR E
(I10, 3F10.0, 4I5)	LE SL SI SMA NC(L), ($L = 1, 4$) (one card for each beam segment)
(8 (I2, F8.2))	JC(L) CM(L), ($L = 1$, NCM)

given in Table 14.6. Computer results for this example which are shown in Table 14.6 check, as expected, the hand calculated values for the reduced stiffness matrix eq. (14.57), the transformation matrix eq. (14.59), the reduced mass matrix eq. (14.60), the eigenvalues which are shown in the second column of Table 14.1, and the eigenvectors shown in the columns of the modal matrix eq. (14.70).

TABLE 14.6   Data Input and Computer Results for Example 14.7.

INPUT DATA

4	6	3	0	0	0	1000000000000.			
	1	1.00	0.0001	100.00	7	7	4	1	
	2	1.00	0.0001	100.00	4	1	5	2	
	3	1.00	0.0001	100.00	5	2	6	3	
	4	1.00	0.0001	100.00	6	3	7	7	

THE REDUCED STIFFNESS MATRIX IS

0.1886D 10	-0.1200D 10	0.5143D 09
-0.1200D 10	0.1500D 10	-0.1200D 10
0.5143D 09	-0.1200D 10	0.1886D 10

THE TRANSFORMATION MATRIX IS

0.2143D 00	0.7500D 00	-0.2143D 00
-0.8571D 00	0.0000D 00	0.8571D 00
0.2143D 00	-0.7500D 00	-0.2143D 00
0.1000D 01	0.0000D 00	0.0000D 00
0.0000D 00	0.1000D 01	0.0000D 00
0.0000D 00	0.0000D 00	0.1000D 01

THE REDUCED MASS MATRIX IS

0.1000D 03	0.0000D 00	0.0000D 00
0.0000D 00	0.1000D 03	0.0000D 00
0.0000D 00	0.0000D 00	0.1000D 03

EIGENVALUES

0.37057D 08	0.19430D 07	0.13714D 08

EIGENVECTORS

0.56043D-01	0.43119D-01	-0.70711D-01
-0.60979D-01	0.79256D-01	0.48677D-17
0.56043D-01	0.43119D-01	0.70711D-01

Illustrative Example 14.8.   Determine the response of the fixed beam of Example 14.7 when subjected to a suddenly applied force of 5000 N at the center of the span. Use the lumped mass method and neglect damping.

The solution of this problem is obtained using Program 11 as a continuation

TABLE 14.7   Additional Data and Computer Results for Example 14.8.

0.0100	0.2000	0	2	0
0.00	5000.00		0.20	5000.00
0.000	0.000		0.000	

RESPONSE FOR ELASTIC SYSTEM

TIME	DISPLACEMENTS AT NODAL COORDINATES		
t	y_4	y_5	y_6
0.0000	0.0000D 00	0.0000D 00	0.0000D 00
0.0100	0.6428D-05	0.1367D-04	0.6428D-05
0.0200	0.1612D-04	0.3194D-04	0.1612D-04
0.0300	0.1368D-04	0.2526D-04	0.1368D-04
0.0400	0.2172D-05	0.5309D-05	0.2172D-05
0.0500	0.5451D-06	0.3613D-05	0.5451D-06
0.0600	0.1192D-04	0.2235D-04	0.1192D-04
0.0700	0.1717D-04	0.3235D-04	0.1717D-04
0.0800	0.8002D-05	0.1741D-04	0.8002D-05
0.0900	-0.1098D-06	0.7007D-06	-0.1098D-06
0.1000	0.5164D-05	0.9831D-05	0.5164D-05
0.1100	0.1513D-04	0.3037D-04	0.1513D-04
0.1200	0.1465D-04	0.2838D-04	0.1465D-04
0.1300	0.4063D-05	0.7533D-05	0.4063D-05
0.1400	-0.1726D-06	0.1901D-05	-0.1726D-06
0.1500	0.9595D-05	0.1961D-04	0.9595D-05
0.1600	0.1758D-04	0.3233D-04	0.1758D-04
0.1700	0.1019D-04	0.2046D-04	0.1019D-04
0.1800	-0.3563D-07	0.2337D-05	-0.3563D-07
0.1900	0.3597D-05	0.6783D-05	0.3597D-05

for the solution given in Example 14.7. In other words, after the computer has calculated the reduced stiffness, mass matrices, and has solved the corresponding eigenproblem (Table 14.6) subroutine MODAL is called to calculate the response at the three translatory nodal coordinates of the beam. Additional data input and computer results for this problem are shown in Table 14.7.

We can easily check the results given by the computer for this problem which are listed in Table 14.7. We proceed as follows.

In general the nth normal equation of motion is given by

$$\ddot{z}_n + \omega_n^2 z_n = \sum \phi_{in} F_{0i} = P_n$$

and the modal response for a constant force by

$$z_n = \frac{P_n}{\omega_n^2} (1 - \cos \omega_n t), \qquad n = 1, 2, 3.$$

TABLE 14.8   Modal Response Example 14.8.

Mode	ω_n (rad/sec)	$P_n = \sum \phi_{in} F_{0i}$	$z_n (t = 0.1 \text{ sec})$
1	1394	396.3	$1.243 \ 10^{-4}$
2	3703	0	0
3	6087	-304.9	$-2.316 \ 10^{-6}$

The calculations for this example are conveniently arranged in Table 14.8 in which the modal response is calculated for $t = 0.1$ sec.

Finally, the response in terms of the nodal coordinates is given by

$$\{y\} = [\Phi] \{z\}$$

$$\begin{bmatrix} y_4 \\ y_5 \\ y_6 \end{bmatrix} = \begin{bmatrix} 0.0431 & -0.0707 & 0.0560 \\ 0.0793 & 0 & 0.0610 \\ 0.0431 & 0.0707 & 0.0560 \end{bmatrix} \begin{bmatrix} 1.243 & 10^{-4} \\ 0 \\ -2.316 & 10^{-6} \end{bmatrix}$$

$$\begin{bmatrix} y_4 \\ y_5 \\ y_6 \end{bmatrix} = \begin{bmatrix} 0.523 & 10^{-5} \\ 0.971 & 10^{-5} \\ 0.523 & 10^{-5} \end{bmatrix}.$$

This last result checks closely the values as given in Table 14.7 for the response at time $t = 0.1$ sec.

14.11   SUMMARY

In this chapter, we have formulated the dynamic equations for beams in reference to a discrete number of nodal coordinates. These coordinates are translational and rotational displacements defined at joints between structural elements of the beam (beam segments). The dynamic equations for a linear system are conveniently written in matrix notation as

$$[M] \{\ddot{y}\} + [C] \{\dot{y}\} + [K] \{y\} = \{F(t)\}$$

where $F(t)$ is the force vector and $[M]$, $[C]$, and $[K]$ are, respectively, the mass, damping, and stiffness matrices of the structure. These matrices are assembled by the appropriate superposition (direct method) of the matrices determined for each beam segment of the structure.

The solution of the dynamic equations (i.e., the response) of a linear system may be found by the modal superposition method. This method requires the determination of the natural frequencies $\omega_n (n = 1, 2, 3, \ldots, N)$ and the corresponding normal modes which are conveniently written as the columns of the modal matrix $[\Phi]$. The linear transformation $\{y\} = [\Phi]\{z\}$ applied to the dynamic equations reduces them to a set of independent equations (uncoupled equations) of the form

$$\ddot{z}_n + 2\omega_n \xi_n \dot{z}_n + \omega_n^2 z_n = P_n(t),$$

where ξ_n is the modal damping ratio and $P_n(t) = \Sigma_i \, \phi_{in} F_i(t)$ is the modal force.

An alternate method for determining the response of linear systems (also valid for nonlinear systems) is the numerical integration of the dynamic equations. Chapter 19 presents the step-by-step linear acceleration method (with a modification introduced by Wilson) which is an efficient method for solving the dynamic equations.

A computer program is also described for the dynamic analysis of beams. This program performs the task of assembling the matrices of the system and then calling the necessary subroutines to perform specific calculations.

PROBLEMS

The following problems are intended for hand calculation, though it is recommended that, whenever possible, results should also be obtained using Program 11 BEAM.

14.1 A uniform beam of flexural stiffness $EI = 10^9$ (lb · in²) and length 300 in has one end fixed and the other simply supported. Determine the system mass matrix considering three beam segments and the nodal coordinates indicated in Fig. P14.1.

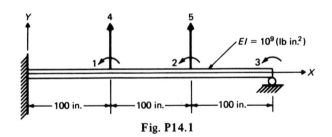

Fig. P14.1

14.2 Assuming that the beam shown in Fig. P14.1 carries a uniform weight per unit length $q = 386$ lb/in, determine the system mass matrix corresponding to the lumped mass formulation.

14.3 Determine the system matrix for Problem 14.2 using the consistent mass method.

14.4 For the beam in Problems 14.1 and 14.2, use static condensation to eliminate the massless degrees of freedom. Find the transformation matrix and the reduced stiffness and mass matrices.

14.5   For the beam in Problems 14.1 and 14.3 use static condensation to eliminate the rotational degrees of freedom. Find the transformation matrix
and the reduced stiffness and mass matrices.

14.6   Determine the natural frequencies and corresponding normal modes using the reduced stiffness and mass matrices obtained in Problem 14.4.

14.7   Determine the natural frequencies and corresponding normal modes using the reduced stiffness and mass matrices obtained in Problem 14.5.

14.8   Determine the geometric stiffness matrix for the beam of Problem 14.1
when it is subjected to a constant tensile force of 10,000 lb as shown in
Fig. P14.8.

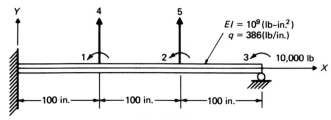

Fig. P14.8

14.9   Perform static condensation to reduce the geometric stiffness matrix obtained in Problem 14.8. Eliminate the rotational coordinates.

14.10   Use results from Problem 14.4 and 14.9 and determine the natural frequencies and corresponding normal modes for the beam shown in Fig.
P14.8.

14.11   Use the results of Problems 14.5 and 14.9 and determine the natural frequencies and corresponding normal modes for the beam shown in Fig.
P14.8.

14.12   Determine the stiffness matrix for a beam segment in which the flexural
stiffness has a linear variation as shown in Fig. P14.12.

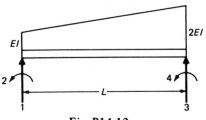

Fig. P14.12

14.13   Determine the lumped mass matrix for a beam segment in which the
mass has a linear distribution as shown in Fig. P14.13.

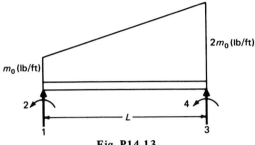

Fig. P14.13

14.14 Determine the consistent mass matrix for the beam segment shown in Fig. P14.13.

14.15 The uniform beam shown in Fig. P14.15 is subjected to a constant force of 5000 lb suddenly applied along the nodal coordinate 1. Use the results obtained in Problem 14.6 to determine the response by the modal superposition method. (Use only the two modes left by the static condensation.)

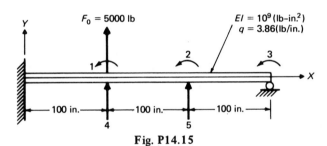

Fig. P14.15

14.16 Solve Problem 14.15 using the results obtained in Problem 14.7 which are based on the consistent mass formulation.

14.17 Solve Problem 14.15 using the results obtained in Problem 14.9 which includes the effect of the axial force in the stiffness of the system.

14.18 Determine the steady-state response for the beam shown in Fig. P14.18 which is acted upon by a hormonic force $F(t) = 5000 \sin 30t$ (lb) as

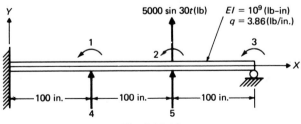

Fig. P14.18

shown in the figure. Eliminate the rotational coordinates by static condensation (Problem 14.5). Neglect damping in the system.

The following problems are intended for computer solution using Program 11 BEAM.

14.19   Determine the natural frequencies and corresponding normal modes for the beam shown in Fig. P14.1: (a) condensing the three rotational nodal coordinates, (b) no condensing coordinates. (Use consistent mass method.)

14.20   Determine the response for the beam shown in Fig. P14.15. Neglect damping. (a) Do not condense coordinates; (b) condense the three rotational coordinates.

14.21   Repeat Problem 14.20 assuming 10% damping in all the modes.

14.22   Determine the steady-state response for the beam shown in Fig. P14.18 when subjected to a harmonic force as shown in the figure. Do not condense coordinates and neglect damping in the system.

14.23   Repeat Problem 14.22 assuming that the damping is proportional to stiffness of the system where the constant of proportionality $a_0 = 0.2$.

14.24   Solve Problem 14.22 using subroutine CONDE to reduce the three rotational coordinates.

14.25   Repeat Problem 14.24 assuming 15% damping in all the modes.

14.26   Determine the steady-state response for the beam shown in Fig. P14.15. Do not condense coordinates and neglect damping in the system.

15

Dynamic analysis of plane frames

The dynamic analysis using the stiffness matrix method for structures modeled as beams was presented in Chapter 14. This method of analysis when applied to beams requires the calculation of element matrices (stiffness, mass, and damping matrices), the assemblage from these matrices of the corresponding system matrices, the formation of the force vector, and the solution of the resultant equations of motion. These equations, as we have seen, may be solved in general by the modal superposition method or by numerical integration of the differential equations of motion. In this chapter and in the following chapters, the dynamic analysis of structures modeled as frames is presented.

We begin in this chapter with the analysis of structures modeled as plane frames and with the loads acting in the plane of the frame. The dynamic analysis of such structures requires the inclusion of the axial effects in the stiffness and mass matrices. It also requires a coordinate transformation of the nodal coordinates from elements or local coordinates to system or global coordinates. Except for the consideration of axial effects and the need to transform these coordinates, the dynamic analysis by the stiffness method when applied to frames is identical to the analysis of beams as discussed in Chapter 14.

15.1   ELEMENT STIFFNESS MATRIX FOR AXIAL EFFECTS

The inclusion of axial forces in the stiffness matrix of a flexural beam segment requires the determination of the stiffness coefficients for axial loads. To derive the stiffness matrix for an axially loaded member, consider in Fig. 15.1 a beam segment acted on by the axial forces P_1 and P_2 producing axial displacements δ_1 and δ_2 at the nodes of the element. For a prismatic and uniform beam segment of length L and cross-sectional A, it is relatively simple to obtain the stiffness relation for axial effects by the application of Hooke's law. In relation to the beam shown in Fig. 15.1, the displacements δ_1 produced by the force P_1 acting at node 1 while node 2 is maintained fixed ($\delta_2 = 0$) is given by

$$\delta_1 = \frac{P_1 L}{AE}. \tag{15.1}$$

From eq. (15.1) and the definition of the stiffness coefficient k_{11} (force at node 1 to produce a unit displacement, $\delta_1 = 1$), we obtain

$$k_{11} = \frac{P_1}{\delta_1} = \frac{AE}{L}. \tag{15.2a}$$

The equilibrium of the beam segment acted upon by the force k_{11} requires a force k_{21} at the other end, namely

$$k_{21} = -k_{11} = -\frac{AE}{L} \tag{15.2c}$$

Analogously, the other stiffness coefficients are

$$k_{22} = \frac{AE}{L}. \tag{15.2c}$$

and

$$k_{12} = -\frac{AE}{L}. \tag{15.2d}$$

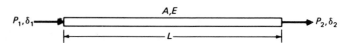

Fig. 15.1   Beam element with nodal axial loads P_1, P_2, and corresponding nodal displacements δ_1, δ_2.

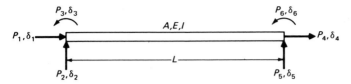

Fig. 15.2   Beam element showing flexural and axial nodal forces and displacements.

The stiffness coefficients as given by eqs. (15.2) are the elements of the stiffness matrix relating axial forces and displacements for a prismatic beam segment, that is,

$$\begin{bmatrix} P_1 \\ P_2 \end{bmatrix} = \frac{AE}{L} \begin{bmatrix} 1 & -1 \\ -1 & 1 \end{bmatrix} \begin{bmatrix} \delta_1 \\ \delta_2 \end{bmatrix}. \tag{15.3}$$

The stiffness matrix corresponding to the nodal coordinates for the beam segment shown in Fig. 15.2 is obtained by combining in a single matrix the stiffness matrix for axial effects, eq. (15.3), and the stiffness matrix for flexural effects, eq. (14.20). The matrix resulting from this combination relates the forces P_i and the displacements δ_i at the coordinates indicated in Fig. 15.2 as

$$\begin{bmatrix} P_1 \\ P_2 \\ P_3 \\ P_4 \\ P_5 \\ P_6 \end{bmatrix} = \frac{EI}{L^3} \begin{bmatrix} AL^2/I & & & & \text{symmetric} \\ 0 & 12 & & & & \\ 0 & 6L & 4L^2 & & & \\ -AL^2/I & 0 & 0 & AL^2/I & & \\ 0 & -12 & 6L & 0 & 12 & \\ 0 & 6L & 2L^2 & 0 & -6L & 4L^2 \end{bmatrix} \begin{bmatrix} \delta_1 \\ \delta_2 \\ \delta_3 \\ \delta_4 \\ \delta_5 \\ \delta_6 \end{bmatrix} \tag{15.4}$$

or, in concise notation,

$$\{P\} = [K] \{\delta\}. \tag{15.5}$$

15.2   ELEMENT MASS MATRIX FOR AXIAL EFFECTS

The determination of mass influence coefficients for axial effects of a beam element may be carried out by any of two methods indicated previously for the flexural effects: (1) the lumped mass method and (2) the consistent mass method. In the lumped mass method, the mass allocation to the nodes of the beam element is found from static considerations which for a uniform beam gives half of the total mass of the beam segment allocated at each node. Then for a prismatic beam segment, the relation between nodal axial forces and nodal accelerations is given by

$$\begin{bmatrix} P_1 \\ P_2 \end{bmatrix} = \frac{\overline{m}L}{2} \begin{bmatrix} 1 & 0 \\ 0 & 1 \end{bmatrix} \begin{bmatrix} \ddot{\delta}_1 \\ \ddot{\delta}_2 \end{bmatrix}, \tag{15.6}$$

where $\overline{m}$ is the mass per unit of length. The combination of the flexural lumped mass coefficient and axial mass coefficients gives, in reference to the nodal coordinates in Fig. 15.2, the following diagonal matrix:

$$\begin{bmatrix} P_1 \\ P_2 \\ P_3 \\ P_4 \\ P_5 \\ P_6 \end{bmatrix} = \frac{\overline{m}L}{2} \begin{bmatrix} 1 & & & & & \\ & 1 & & & & \\ & & 0 & & & \\ & & & 1 & & \\ & & & & 1 & \\ & & & & & 0 \end{bmatrix} \begin{bmatrix} \ddot{\delta}_1 \\ \ddot{\delta}_2 \\ \ddot{\delta}_3 \\ \ddot{\delta}_4 \\ \ddot{\delta}_5 \\ \ddot{\delta}_6 \end{bmatrix}. \tag{15.7}$$

To calculate the coefficients for the consistent mass matrix, it is necessary first to determine the displacement functions corresponding to a unit axial displacement at one of the nodal coordinates. Consider in Fig. 15.3 an axial unit displacement $\delta_1 = 1$ of node 1 while the other node 2 is kept fixed so that $\delta_2 = 0$. If $u = u(x)$ is the displacement at section x, the displacement at section $x + dx$ will be $u + du$. It is evident then that the element dx in the new position has changed in length by an amount du, and thus, the strain is du/dx. Since from Hooke's law, the ratio of stress to strain is equal to the modulus of elasticity E, we can write

$$\frac{du}{dx} = \frac{P}{AE}. \tag{15.8}$$

Integration with respect to x yields

$$u = \frac{P}{AE} x + C \tag{15.9}$$

in which C is a constant of integration. Introducing the boundary conditions, $u = 1$ at $x = 0$ and $u = 0$ at $x = L$, we obtain the displacement function $u_1(x)$ corresponding to a unit displacement δ_1 as

$$u_1(x) = 1 - \frac{x}{L}. \tag{15.10}$$

Analogously, the displacement function $u_2(x)$ corresponding to a unit displacement $\delta_2 = 1$ is

$$u_2(x) = \frac{x}{L}. \tag{15.11}$$

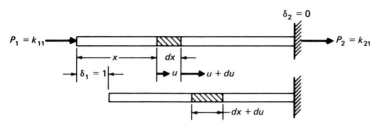

Fig. 15.3   Displacement at node 1 (δ_1 = 1) of a beam element.

The application of the principle of virtual work results in a general expression for the calculation of the stiffness coefficients. For example, consider the beam in Fig. 15.3 which is in equilibrium with the forces $P_1 = k_{11}$ and $P_2 = k_{21}$ at its two ends. Assume that a virtual displacement δ_2 = 1 takes place. Then, according to the principle of virtual work, during this virtual displacement, the work of the external and internal forces are equal. The external force k_{21} performs the work

$$W_E = k_{21}\delta_2$$

or

$$W_E = k_{21} \tag{15.12}$$

since δ_2 = 1. The internal force $P(x)$ at any section x is obtained from eq. (15.8) as

$$P(x) = AE\,u_1'(x) \tag{15.13}$$

in which $u_1'(x) = du_1/dx$. The relative displacement of element dx during this virtual displacement is

$$du_2 = \frac{du_2}{dx}\,dx \tag{15.14}$$

as shown in Fig. 15.4. Hence the internal work for element dx is obtained from eqs. (15.13) and (15.14) as

$$dW_I = AE\,u_1'(x)\,u_2'(x)\,dx$$

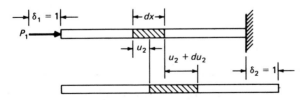

Fig. 15.4   Displacement at node 2 (δ_2 = 1) of a beam element subjected to axial displacement at node 1 (δ_1 = 1).

and for the beam segment of length L

$$W_I = \int_0^L AE\, u_1'(x)\, u_2'(x)\, dx. \tag{15.15}$$

Finally, equating $W_E = W_I$ from eqs. (15.12) and (15.15) gives the stiffness coefficient

$$k_{21} = \int_0^L AE\, u_1'(x)\, u_2'(x)\, dx. \tag{15.16}$$

In general, the stiffness coefficient k_{ij} for axial effects may be obtained from

$$k_{ij} = \int_0^L AE\, u_i'(x)\, u_j'(x)\, dx. \tag{15.17}$$

Using eq. (15.17), the reader may check the results obtained in eq. (15.3) for a uniform beam. However, eq. (15.17) could as well be used for nonuniform beams in which AE would in general be a function of x. In practice, the same displacement $u_1(x)$ and $u_2(x)$ obtained for a uniform beam are also used in eq. (15.17) for a nonuniform member. The displacement $y(x, t)$ at any section x of a beam element due to dynamic nodal displacements, $\delta_1(t)$ and $\delta_2(t)$, is obtained by superposition. Hence

$$y(x, t) = u_1(x)\,\delta_1(t) + u_2(x)\,\delta_2(t) \tag{15.18}$$

in which $u_1(x)$ and $u_2(x)$ are given by eqs. (15.10) and (15.11).

Now consider the beam of Fig. 15.5 while undergoing a unit acceleration, $\ddot{\delta}_1(t) = 1$ which by eq. (15.18) results in an acceleration at x given by

$$\ddot{u}_1(x, t) = u_1(x)\,\ddot{\delta}_1(t)$$

or

$$\ddot{u}_1(x, t) = u_1(x)$$

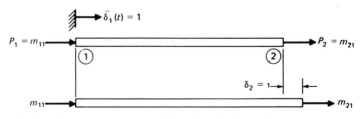

Fig. 15.5   Displacement at node 2 ($\delta_2 = 1$) of a beam element undergoing axial acceleration at node 1 ($\ddot{\delta}_1(t) = 1$).

since $\ddot{\delta}_1(t) = 1$. The inertial force per unit length along the beam resulting from this unit acceleration is

$$f_I = \overline{m}(x) u_1(x), \tag{15.19}$$

where $\overline{m}(x)$ is the mass per unit length along the beam. Now, to determine the mass coefficient m_{21}, we give to the beam shown in Fig. 15.5 a virtual displacement $\delta_2 = 1$. The only external force doing work during this virtual displacement is the reaction m_{21}. This work is then

$$W_E = m_{21} \delta_2$$

or

$$W_E = m_{21} \tag{15.20}$$

since $\delta_2 = 1$. The internal work per unit length along the beam performed by the inertial force f_I during this virtual displacement is

$$\delta W_I = f_I(x) u_2(x)$$

or, from eq. (15.19),

$$\delta W_I = \overline{m}(x) u_1(x) u_2(x).$$

Hence the total internal work is

$$W_I = \int_0^L \overline{m}(x) u_1(x) u_2(x) \, dx. \tag{15.21}$$

Finally, equating eqs. (15.20) and (15.21) yields

$$m_{21} = \int_0^L \overline{m}(x) u_1(x) u_2(x) \, dx \tag{15.22}$$

or, in general,

$$m_{ij} = \int_0^L \overline{m}(x) u_i(x) u_j(x) \, dx. \tag{15.23}$$

The application of eq. (15.23) to the special case of a uniform beam results in

$$m_{11} = \int_0^L \overline{m} \left(1 - \frac{x}{L}\right)^2 dx = \frac{\overline{m}L}{3}. \tag{15.24}$$

Similarly,

$$m_{22} = \frac{\overline{m}L}{3}$$

and

$$m_{12} = m_{21} = \int_0^L \overline{m}\left(1 - \frac{x}{L}\right)\left(\frac{x}{L}\right)dx = \frac{\overline{m}L}{6}. \tag{15.25}$$

In matrix form, the axial inertial force relationship for a uniform beam may be written as

$$\begin{bmatrix} P_1 \\ P_2 \end{bmatrix} = \frac{\overline{m}L}{6}\begin{bmatrix} 2 & 1 \\ 1 & 2 \end{bmatrix}\begin{bmatrix} \ddot{\delta}_1 \\ \ddot{\delta}_2 \end{bmatrix}. \tag{15.26}$$

Finally, combining the mass matrix eq. (14.34) for flexural effects with eq. (15.26) for the axial effects, we obtain the consistent mass matrix for a uniform beam element in reference to the nodal coordinates as shown in Fig. 15.2 as

$$\begin{bmatrix} P_1 \\ P_2 \\ P_3 \\ P_4 \\ P_5 \\ P_6 \end{bmatrix} = \frac{\overline{m}L}{420}\begin{bmatrix} 140 & & & & \text{Symmetric} & \\ 0 & 156 & & & & \\ 0 & 22L & 4L^2 & & & \\ 70 & 0 & 0 & 140 & & \\ 0 & 54 & 13L & 0 & 156 & \\ 0 & -13L & -3L^2 & 0 & -22L & 4L^2 \end{bmatrix}\begin{bmatrix} \ddot{\delta}_1 \\ \ddot{\delta}_2 \\ \ddot{\delta}_3 \\ \ddot{\delta}_4 \\ \ddot{\delta}_5 \\ \ddot{\delta}_6 \end{bmatrix} \tag{15.27}$$

or, in condensed notation,

$$\{P\} = [M_c]\{\ddot{\delta}\}$$

in which $[M_c]$ is the consistent mass matrix.

15.3   COORDINATE TRANSFORMATION

The stiffness matrix for the beam element in eq. (15.4) as well as the mass matrix in eq. (15.27) are in reference to nodal coordinates defined by coordinate axes fixed on the beam element. These axes are called *local* or *element axes* while the coordinate axes for the whole structure are known as *global* or *system axes*. Figure 15.6 shows a beam element with nodal forces $P_1, P_2, \ldots, P_6$ referred to the local coordinate axes x, y, z, and $\overline{P}_1, \overline{P}_2, \ldots, \overline{P}_6$ referred to global coordinate set of axes X, Y, Z. The objective is to transform the element matrices (stiffness, mass, etc.) from the reference of local coordinate axes to the global coordinate axes. This transformation is required in order that the matrices for all the elements refer to the same set of coordinates; hence, the matrices become compatible for assemblage into the system matrices for the structure. We begin by expressing the forces (P_1, P_2, P_3) in terms of the forces $(\overline{P}_1, \overline{P}_2, \overline{P}_3)$. Since these two sets of forces are equivalent, we obtain from Fig. 15.6, the following

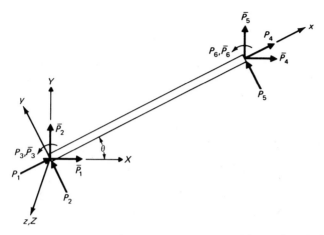

Fig. 15.6 **Beam element showing nodal forces in local (x, y, z) and global coordinate axes (X, Y, Z).**

relations:

$$P_1 = \overline{P}_1 \cos\theta + \overline{P}_2 \sin\theta,$$

$$P_2 = -\overline{P}_1 \sin\theta + \overline{P}_2 \cos\theta,$$

$$P_3 = \overline{P}_3. \tag{15.28}$$

The first two equations of eq. (15.28) may be written in matrix notation as

$$\begin{bmatrix} P_1 \\ P_2 \end{bmatrix} = \begin{bmatrix} \cos\theta & \sin\theta \\ -\sin\theta & \cos\theta \end{bmatrix} \begin{bmatrix} \overline{P}_1 \\ \overline{P}_2 \end{bmatrix}. \tag{15.29}$$

Analogously, we obtain for the forces on the other node the relations:

$$P_4 = \overline{P}_4 \cos\theta + \overline{P}_5 \sin\theta,$$

$$P_5 = -\overline{P}_4 \sin\theta + \overline{P}_5 \cos\theta,$$

$$P_6 = \overline{P}_6. \tag{15.30}$$

Equations (15.28) and (15.30) may conveniently be arranged in matrix form as

$$\begin{bmatrix} P_1 \\ P_2 \\ P_3 \\ P_4 \\ P_5 \\ P_6 \end{bmatrix} = \begin{bmatrix} \cos\theta & \sin\theta & 0 & 0 & 0 & 0 \\ -\sin\theta & \cos\theta & 0 & 0 & 0 & 0 \\ 0 & 0 & 1 & 0 & 0 & 0 \\ 0 & 0 & 0 & \cos\theta & \sin\theta & 0 \\ 0 & 0 & 0 & -\sin\theta & \cos\theta & 0 \\ 0 & 0 & 0 & 0 & 0 & 1 \end{bmatrix} \begin{bmatrix} \overline{P}_1 \\ \overline{P}_2 \\ \overline{P}_3 \\ \overline{P}_4 \\ \overline{P}_5 \\ \overline{P}_6 \end{bmatrix} \tag{15.31}$$

or in condensed notation

$$\{P\} = [T] \{\overline{P}\} \qquad (15.32)$$

in which $\{P\}$ and $\{\overline{P}\}$ are, respectively, the vectors of the element nodal forces in local and global coordinates and $[T]$ is the transformation matrix given by the square matrix in eq. (15.31).

Repeating the same procedure, we obtain the relation between nodal displacements $(\delta_1, \delta_2, \ldots, \delta_6)$ in local coordinates and the components of the nodal displacements in global coordinates $(\overline{\delta}_1, \overline{\delta}_2, \ldots, \overline{\delta}_6)$, namely

$$
\begin{bmatrix}
\delta_1 \\
\delta_2 \\
\delta_3 \\
\delta_4 \\
\delta_5 \\
\delta_6
\end{bmatrix}
=
\begin{bmatrix}
\cos\theta & \sin\theta & 0 & 0 & 0 & 0 \\
-\sin\theta & \cos\theta & 0 & 0 & 0 & 0 \\
0 & 0 & 1 & 0 & 0 & 0 \\
0 & 0 & 0 & \cos\theta & \sin\theta & 0 \\
0 & 0 & 0 & -\sin\theta & \cos\theta & 0 \\
0 & 0 & 0 & 0 & 0 & 1
\end{bmatrix}
\begin{bmatrix}
\overline{\delta}_1 \\
\overline{\delta}_2 \\
\overline{\delta}_3 \\
\overline{\delta}_4 \\
\overline{\delta}_5 \\
\overline{\delta}_6
\end{bmatrix}
\qquad (15.33)
$$

or

$$\{\delta\} = [T] \{\overline{\delta}\}. \qquad (15.34)$$

Now, the substitution of $\{P\}$ from eq. (15.32) and $\{\delta\}$ from eq. (15.34) into the stiffness equation referred to local axes $\{P\} = [K] \{\delta\}$ results in

$$[T] \{\overline{P}\} = [K] [T] \{\overline{\delta}\}$$

or

$$\{\overline{P}\} = [T]^{-1} [K] [T] \{\overline{\delta}\} \qquad (15.35)$$

where $[T]^{-1}$ is the inverse of matrix $[T]$. However, as the reader may verify, the transformation matrix $[T]$ in eq. (15.31) is an orthogonal matrix, that is, $[T]^{-1} = [T]^T$. Hence

$$\{\overline{P}\} = [T]^T [K] [T] \{\overline{\delta}\} \qquad (15.36)$$

or, in a more convenient notation,

$$\{\overline{P}\} = [\overline{K}] \{\overline{\delta}\} \qquad (15.37)$$

in which

$$[\overline{K}] = [T]^T [K] [T] \qquad (15.38)$$

is the stiffness matrix for a beam segment in reference to the global system of coordinates.

Repeating the procedure of transformation as applied to the stiffness matrix for the lumped mass, eq. (15.7), or consistent mass matrix, eq. (15.27), we obtain in a similar manner:

$$\{\bar{P}\} = [\bar{M}]\{\ddot{\bar{\delta}}\}$$

in which

$$[\bar{M}] = [T]^T [M] [T] \qquad (15.39)$$

is the mass matrix for a beam segment referred to global coordinates and $[T]$ is the matrix of the transformation given by the square matrix in eq. (15.33).

Illustrative Example 15.1. Consider in Fig. 15.7 a plane frame having two prismatic beam elements and three degrees of freedom as indicated in the figure. Using the consistent mass formulation, determine the three natural frequencies and normal modes corresponding to this discrete model of the frame.

The stiffness matrix for element $\triangle$ or $\triangle$ in local coordinates by eq. (15.4) is

$$[K_1] = [K_2] = 1000 \begin{bmatrix} 600 & & & & & \text{Symmetric} & \\ 0 & 12 & & & & & \\ 0 & 600 & 40{,}000 & & & & \\ -600 & 0 & 0 & 600 & & & \\ 0 & -12 & 600 & 0 & 12 & & \\ 0 & 600 & 20{,}000 & 0 & -600 & 40{,}000 \end{bmatrix}$$

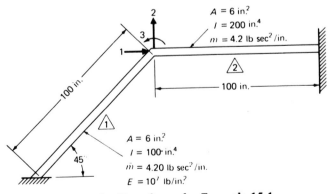

Fig. 15.7 Plane frame for Example 15.1.

The transformation matrix for element △ by eq. (15.31) with $\theta = 45°$ is

$$[T_1] = \frac{1}{\sqrt{2}} \begin{bmatrix} 1 & 1 & 0 & 0 & 0 & 0 \\ -1 & 1 & 0 & 0 & 0 & 0 \\ 0 & 0 & 1 & 0 & 0 & 0 \\ 0 & 0 & 0 & 1 & 1 & 0 \\ 0 & 0 & 0 & -1 & 1 & 0 \\ 0 & 0 & 0 & 0 & 0 & 1 \end{bmatrix}$$

and for element △ with $\theta = 0°$ is the identity matrix

$$[T_2] = [I].$$

The mass matrix in local coordinates for either of the two elements of this frame from eq. (15.27) is

$$[M_1] = [M_2] = \begin{bmatrix} 140 & & & & \text{Symmetric} & \\ 0 & 156 & & & & \\ 0 & 2200 & 40{,}000 & & & \\ 70 & 0 & 0 & 140 & & \\ 0 & 54 & 1300 & 0 & 156 & \\ 0 & -1300 & -30{,}000 & 0 & -2200 & 40{,}000 \end{bmatrix}$$

The element stiffness and mass matrices in reference to the global system of co-ordinates are, respectively, calculated by eqs. (15.38) and (15.39). For element △ the stiffness matrix is

$$[\overline{K}_1] = 10^6 \begin{bmatrix} 0.306 & & & & \text{Symmetric} & & \\ 0.294 & 0.306 & & & & & \\ -0.424 & 0.423 & 40.000 & & & & \\ -0.306 & -0.294 & 0.424 & 0.306 & & & \\ -0.294 & -0.306 & -0.424 & 0.294 & 0.306 & & \\ 0.424 & 0.424 & 20.000 & 0.424 & -0.424 & 40.000 \end{bmatrix} \begin{matrix} 4 \\ 4 \\ 4 \\ 1 \\ 2 \\ 3 \end{matrix}$$

with column headers $4 \quad 4 \quad 4 \quad 1 \quad 2 \quad 3$

and the mass matrix

$$
[\overline{M}_1] = \begin{array}{c c c c c c}
4 & 4 & 4 & 1 & 2 & 3 \\
\end{array}
$$

$$
[\overline{M}_1] = \left[\begin{array}{cccccc}
148 & & & \text{Symmetric} & & \\
-8 & 148 & & & & \\
-1556 & 1556 & 40{,}000 & & & \\
62 & 8 & -919 & 148 & & \\
8 & 62 & 919 & -8 & 148 & \\
909 & -909 & -30{,}000 & 1556 & -1556 & 40{,}000
\end{array}\right]
\begin{array}{c}
4 \\ 4 \\ 4 \\ 1 \\ 2 \\ 3
\end{array}
$$

For element △,

$$
\begin{array}{c c c c c c}
1 & 2 & 3 & 4 & 4 & 4
\end{array}
$$

$$
[\overline{K}_2] = 10^6 \left[\begin{array}{cccccc}
0.600 & & & \text{Symmetric} & & \\
0 & 0.012 & & & & \\
0 & 0.600 & 40.000 & & & \\
-0.600 & 0 & 0 & 0.600 & & \\
0 & -0.012 & -0.600 & 0 & 0.012 & \\
0 & 0.600 & 20.000 & 0 & -0.600 & 40.000
\end{array}\right]
\begin{array}{c}
1 \\ 2 \\ 3 \\ 4 \\ 4 \\ 4
\end{array}
$$

and

$$
\begin{array}{c c c c c c}
1 & 2 & 3 & 4 & 4 & 4
\end{array}
$$

$$
[\overline{M}_2] = \left[\begin{array}{cccccc}
140 & & & \text{Symmetric} & & \\
0 & 156 & & & & \\
0 & 2200 & 40{,}000 & & & \\
70 & 0 & 0 & 140 & & \\
0 & 54 & 1300 & 0 & 156 & \\
0 & -1300 & -30{,}000 & 0 & -2200 & 40{,}000
\end{array}\right]
\begin{array}{c}
1 \\ 2 \\ 3 \\ 4 \\ 4 \\ 4
\end{array}
$$

The system stiffness and mass matrices are assembled by the direct method. As was mentioned before, it is expedient for hand calculation of these matrices to indicate the corresponding system nodal coordinates at the top and right of the element matrices. We thus obtain the system stiffness matrix as:

$$
[K] = 10^6 \begin{bmatrix}
0.906 & 0.294 & 0.424 \\
0.294 & 0.318 & 0.176 \\
0.424 & 0.176 & 80.000
\end{bmatrix}
$$

and the system mass matrix as:

$$[M] = \begin{bmatrix} 288 & -8 & 1556 \\ -8 & 304 & 644 \\ 1556 & 644 & 80{,}000 \end{bmatrix}.$$

The natural frequencies are found as the roots of the characteristic equation

$$|[K] - \omega^2 [M]| = 0$$

which, upon substituting the values given for this example, yields

$$10^3 \begin{vmatrix} 906 - 0.288\omega^2 & 294 + 0.008\omega^2 & 424 - 1.556\omega^2 \\ 294 + 0.008\omega^2 & 318 - 0.304\omega^2 & 176 - 0.644\omega^2 \\ 424 - 1.556\omega^2 & 176 - 0.644\omega^2 & 80{,}000 - 80\omega^2 \end{vmatrix} = 0.$$

The roots then are

$$\omega_1^2 = 638, \qquad \omega_2^2 = 976, \qquad \omega_3^2 = 4212$$

and the natural frequencies are

$$\omega_1 = 25.26 \text{ rad/sec}, \qquad \omega_2 = 31.24 \text{ rad/sec}, \qquad \text{and} \quad \omega_3 = 64.90 \text{ rad/sec}$$

or

$$f_1 = 4.02 \text{ cps}, \qquad f_2 = 4.97 \text{ cps}, \qquad \text{and} \quad f_3 = 10.33 \text{ cps}.$$

The natural modes are given as the nontrivial solution of the eigenproblem

$$([K] - \omega^2 [M]) \{a\} = \{0\}.$$

Substituting $\omega_1^2 = 631.65$ and setting $a_{11} = 1.0$, we obtain the first mode shape as

$$\{a_1\} = \begin{bmatrix} a_{11} \\ a_{21} \\ a_{31} \end{bmatrix} = \begin{bmatrix} 1.00 \\ -2.38 \\ 0 \end{bmatrix}$$

which is normalized with the factor

$$\sqrt{\{a_1\}^T [M] \{a_1\}} = 45.81.$$

The normalized eigenvector is then

$$\{\phi_1\} = \begin{bmatrix} \phi_{11} \\ \phi_{21} \\ \phi_{31} \end{bmatrix} = \begin{bmatrix} 0.0218 \\ -0.0527 \\ 0 \end{bmatrix}.$$

Analogously, for the other two modes, we obtain

$$\{\phi_2\} = \begin{bmatrix} \phi_{12} \\ \phi_{22} \\ \phi_{32} \end{bmatrix} = \begin{bmatrix} 0.00498 \\ 0.00206 \\ 0.00341 \end{bmatrix} \quad \text{and} \quad \{\phi_2\} = \begin{bmatrix} \phi_{13} \\ \phi_{23} \\ \phi_{33} \end{bmatrix} = \begin{bmatrix} 0.0583 \\ 0.0241 \\ -0.0016 \end{bmatrix}.$$

From these vectors, we obtain the modal matrix

$$[\phi] = \begin{bmatrix} 0.0218 & 0.00498 & 0.0583 \\ -0.0527 & 0.00206 & 0.0241 \\ 0 & 0.00341 & -0.0016 \end{bmatrix}.$$

Illustrative Example 15.2. Determine the maximum displacement at the nodal coordinates of the frame in Fig. 15.7 when a force of magnitude 100,000 lb is suddenly applied at nodal coordinate 1. Neglect damping.

From Example 15.1, the natural frequencies are $\omega_1 = 25.26$ rad/sec, $\omega_2 = 31.24$ rad/sec, and $\omega_3 = 64.90$ rad/sec; and the modal matrix is

$$[\phi] = \begin{bmatrix} 0.0218 & 0.00498 & 0.0583 \\ -0.0527 & 0.00206 & 0.0241 \\ 0 & 0.00341 & -0.0016 \end{bmatrix}.$$

The modal equations have the form of

$$\ddot{z}_i + \omega_i^2 z_i = P_i \tag{a}$$

where

$$P_i = \sum_j \phi_{ji} F_j. \tag{b}$$

In this example, the nodal applied forces are

$$F_1 = 100,000 \text{ lb}, F_2 = 0, F_3 = 0.$$

We thus obtain the modal equations as

$$\ddot{z}_1 + 638 z_1 = 2180,$$
$$\ddot{z}_2 + 976 z_2 = 498,$$
$$\ddot{z}_3 + 4212 z_3 = 5830. \tag{c}$$

The solutions of these equations are of the form

$$z_i = \frac{P_i}{\omega_i^2} (1 - \cos \omega_i t).$$

Substitution for P_i and ω_i yields

$$z_1 = 3.417(1 - \cos 25.26t),$$
$$z_2 = 0.510(1 - \cos 31.24t),$$
$$z_3 = 1.384(1 - \cos 64.90t). \qquad \text{(d)}$$

The nodal displacements are obtained from

$$\{y\} = [\Phi]\,\{z\}$$

which results in

$$y_1 = \;\;\;0.1577 - 0.0745 \cos 25.26t - 0.00254 \cos 31.24t - 0.0807 \cos 64.9t,$$
$$y_2 = -0.1457 + 0.1800 \cos 25.26t - 0.00105 \cos 31.24t - 0.0333 \cos 64.9t,$$
$$y_3 = -0.0005 + \qquad 0 \cos 25.26t - 0.00174 \cos 31.24t + 0.0022 \cos 64.9t.$$

$$\text{(e)}$$

The maximum possible displacements at the nodal coordinates may then be estimated as the summation of the absolute values of the coefficients in the above expressions. Hence

$$y_{1_{max}} = 0.3154 \text{ in,}$$
$$y_{2_{max}} = 0.2914,$$
$$y_{3_{max}} = 0.0010. \qquad \text{(f)}$$

15.4   PROGRAM 12—DYNAMIC ANALYSIS OF PLANE FRAMES (FRAME)

The computer program described in this section performs the dynamic analysis of plane frames using the stiffness method. This program is organized in exactly the same form as Program 11 described in Chapter 14 for the analysis of beams. The two programs differ in the dimensions of the element matrix which are 4 X 4 in the beam program and 6 X 6 in the frame program. In the frame program, axial effects are considered in the analysis. Also in this program a rotation of coordinates is performed to refer all the element matrices to the global system of coordinates.

The subroutines listed in Table 14.3 for use in Program 11–Dynamic Analysis of Beams–should be available for Program 12–Dynamic Analysis of Plane Frames.

A description of the principal variables used in the program is given in Table 15.1 and a listing of the input data and corresponding formats required by the program is given in Table 15.2.

TABLE 15.1 Description of Principal Variables in Program 12.

Variable	Symbol in Text	Description
NE		Number of Elements
ND		Number of degrees of freedom
NCR		Number of coordinates to condense
NCM		Number of concentrates masses
LOC		Mass index: $0 \rightarrow$ lumped mass; $1 \rightarrow$ consistent mass
IFPR		Print intermediate values in JACOBI: $0 \rightarrow$ do not print, $1 \rightarrow$ print
E	E	Modulus of elasticity
LE		Element number
SL	L	Element length
AR	A	Element cross-sectional area
SI	I	Element cross-sectional moment of inertia
SMA	$\overline{m}$	Element mass per unit of length
TH	θ	Element slope
NC(L)		Element nodal coordinates $(L = 1, 6)$
CM(L)		Concentrated masses $(L = 1, NCM)$
JC(L)		Nodal coordinates for CM(L)

TABLE 15.2 Input Data and Formats for Program 12.

Format	Variables
(6I5, F10.0)	NE ND NCR NCM LOC IFPR E
(I10, 5F10.0, 6I2)	LE SL AR SI SMA TH NC(L), (L = 1, 6) (one card for each element)

Illustrative Example 15.3. Consider again the frame in Fig. 15.7 and use Program 12 to determine natural frequencies and normal modes. Also, determine the response to a force of magnitude 100,000 lb suddenly applied at nodal coordinate 1.

Subroutine JACOBI is called in Program 12 to solve for the natural frequencies and normal modes. The response could be obtained using either subroutine MODAL or subroutine STEPM. The former subroutine determines the response

TABLE 15.3   Input Data for Example 15.3.

						Data Listing							
2	3	0	0	1	1	10000000.							
1	100.	6.	100.	4.2	45.		4	4	4	1	2	3	
2	100.	6.	100.	4.2	0.		1	2	3	4	4	4	
1.4	0.001	0.02	2	0	0								
0.0	100000.	0.02		100000.									

TABLE 15.4   Computer Results for Example 15.3.

EIGENVALUES

SWEEP = 1

| 0.42116D 04 | 0.63850D 03 | 0.97658D 03 |

SWEEP = 2

| 0.42116D 04 | 0.63850D 03 | 0.97658D 03 |
| 0.42116D 04 | 0.63850D 03 | 0.97658D 03 |

EIGENVECTORS

0.58307D-01	-0.21830D-01	0.49789D-02
0.24152D-01	0.52702D-01	0.20623D-02
-0.16291D-02	-0.00000D 00	0.34093D-02

THE RESPONSE IS

CORD.	TIME	DISPL.	VELOC.	ACC.
1	0.001	0.000	0.389	389.140
2	0.001	0.000	0.027	26.717
3	0.001	-0.000	-0.008	-7.778
1	0.002	0.001	0.777	386.241
2	0.002	0.000	0.053	26.461
3	0.002	-0.000	-0.016	-7.707
1	0.003	0.002	1.161	381.427
2	0.003	0.000	0.080	26.035
3	0.003	-0.000	-0.023	-7.590
1	0.004	0.003	1.538	374.720
2	0.004	0.000	0.105	25.442
3	0.004	-0.000	-0.031	-7.427
1	0.005	0.005	1.909	366.154
2	0.005	0.000	0.130	24.685
3	0.005	-0.000	-0.038	-7.218

TABLE 15.4 (*Continued*)

	THE RESPONSE IS			
CORD.	TIME	DISPL.	VELOC.	ACC.
1	0.006	0.007	2.270	355.773
2	0.006	0.000	0.155	23.768
3	0.006	-0.000	-0.045	-6.965
1	0.007	0.009	2.619	343.627
2	0.007	0.001	0.178	22.696
3	0.007	-0.000	-0.052	-6.669
1	0.008	0.012	2.956	329.778
2	0.008	0.001	0.200	21.474
3	0.008	-0.000	-0.058	-6.332
1	0.009	0.015	3.277	314.295
2	0.009	0.001	0.221	20.109
3	0.009	-0.000	-0.065	-5.955
1	0.010	0.019	3.583	297.255
2	0.010	0.001	0.240	18.608
3	0.010	-0.000	-0.070	-5.541
1	0.011	0.022	3.871	278.742
2	0.011	0.002	0.258	16.980
3	0.011	-0.000	-0.076	-5.091
1	0.012	0.026	4.140	258.850
2	0.012	0.002	0.274	15.232
3	0.012	-0.001	-0.080	-4.609
1	0.013	0.031	4.388	237.678
2	0.013	0.002	0.288	13.374
3	0.013	-0.001	-0.085	-4.095
1	0.014	0.035	4.614	215.331
2	0.014	0.002	0.300	11.415
3	0.014	-0.001	-0.089	-3.554
1	0.015	0.040	4.818	191.920
2	0.015	0.003	0.311	9.367
3	0.015	-0.001	-0.092	-2.987
1	0.016	0.045	4.997	167.563
2	0.016	0.003	0.319	7.240
3	0.016	-0.001	-0.095	-2.398
1	0.017	0.050	5.152	142.382
2	0.017	0.003	0.325	5.045
3	0.017	-0.001	-0.097	-1.791
1	0.018	0.055	5.282	116.501
2	0.018	0.004	0.329	2.794
3	0.018	-0.001	-0.098	-1.167
1	0.019	0.060	5.329	90.142
2	0.019	0.004	0.327	0.508

TABLE 15.4   (Continued)

| | THE RESPONSE IS | | | |
CORD.	TIME	DISPL.	VELOC.	ACC.
3	0.019	-0.001	-0.098	-0.533
1	0.020	0.066	5.267	-326.348
2	0.020	0.004	0.317	-28.577
3	0.020	-0.001	-0.095	7.902

using the modal superposition method as described in Chapter 11, while subroutine STEPM uses the step-by-step Wilson-θ method to calculate the response as described in Chapter 19. In the solution that follows subroutine STEPM was used to obtain the response of this problem. Though damping could easily be considered in the analysis, it is neglected in the calculation of the response with the specific purpose of comparing the results with the solution previously obtained in Example 15.2.

The listing of the input data for this example and the computer results are given, respectively, in Tables 15.3 and 15.4.

Computer results for Example 15.3 consist of the eigenvalues ω_i^2, the modal matrix $[\Phi]$, and the time history of the response. These results compare very closely with the hand calculated values for the same structure in Example 15.2.

15.5   SUMMARY

The dynamic analysis of plane frames by the stiffness method requires the inclusion of the axial effects in the system matrices (stiffness, mass, etc. matrices). It also requires a transformation of coordinates in order to refer all the element matrices to the same coordinate system, so that the appropriate superposition can be applied to assemble the system matrices.

The required matrices for consideration of axial effects as well as the matrix required for the transformation of coordinates are developed in this chapter. A computer program for the dynamic analysis of plane frames is also presented. This program is organized following the pattern of the BEAM program of the preceding chapter. The same set of subroutines developed for the BEAM program should also be available for the FRAME program in order to perform the required calculations.

PROBLEMS

The following problems are intended for hand calculation, though it is recommended that whenever possible solutions should also be obtained using Program 12 FRAME.

15.1   For the plane frame shown in Fig. P15.1 determine the system stiffness and mass matrices. Base the analysis on the four nodal coordinates indicated in the figure. Use consistent mass method.

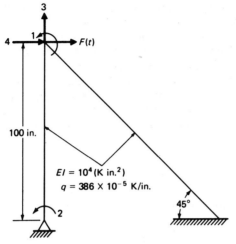

100 in.

$EI = 10^4 (K \text{ in.}^2)$
$q = 386 \times 10^{-5} \text{ K/in.}$

45°

Fig. P15.1

15.2   Use the results obtained in Problem 15.1 in performing the static condensation to eliminate the rotational degrees of freedom and determining the transformation matrix and the reduced stiffness and mass matrices.

15.3   Determine the natural frequencies and corresponding normal modes for the reduced system in Problem 15.2.

15.4   Determine the response of the frame shown in Fig. P15.1 when it is acted upon by a force $F(t) = 10$ kip suddenly applied at nodal coordinate 1 as shown in the figure. Use results of Problem 15.3 to obtain the modal equations. Neglect damping in the system.

15.5   Determine the maximum response of the frame shown in Fig. P15.1 when subjected to the triangular impulsive load (Fig. P15.5) along the

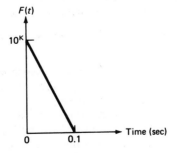

Fig. P15.5

nodal coordinate 1. Use results of Problem 15.3 to obtain the modal equations and use the appropriate response spectrum to find maximum modal response (Fig. 4.5). Neglect damping in the system.

15.6   Determine the steady-state response of the frame shown in Fig. P15.1 when subjected to harmonic force $F(t) = 10 \sin 30t$ (kip) along nodal coordinate 1. Use results of Problem 15.2 to obtain reduced equations of motion. Neglect damping in the system.

15.7   Repeat Problem 15.6 assuming that the damping is proportional to the stiffness of the system, $[C] = a_0 [K]$ where $a_0 = 0.2$.

15.8   The frame shown in Fig. P15.8 is acted upon by the dynamic forces shown in the figure. Determine the equivalent nodal forces corresponding to each member of the frame. (Hint: The equivalent nodal forces are equal to the negative value of the reactions of the members assumed to be fix-ended.)

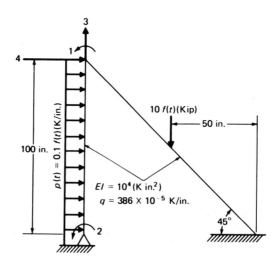

Fig. P15.8

15.9   Assemble the system equivalent nodal forces $\{F_e\}$ from equivalent member nodal forces which were calculated in Problem 15.8.

The following problems are intended for computer solution using Program 12 FRAME.

15.10   Determine the natural frequencies and corresponding normal modes for the frame shown in Fig. P15.1. (a) Condensing the rotational nodal coordinates, (b) no condensing coordinates.

15.11   Determine the response for the frame shown in Fig. P15.11 when subjected to the force $F(t)$ [Fig. P15.11(a)] acting along nodal coordinate 3. Condense the rotational coordinates and assume 5% damping in all the modes.

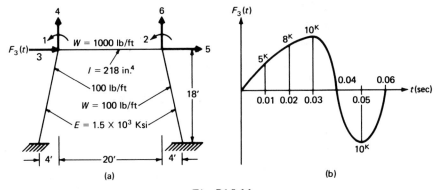

Fig. P15.11

15.12 Determine the steady-state response of the frame in Fig. P15.11 acted upon harmonic force $F_3(t) = 10 \cos 50t$ (kip) as indicated in the figure. Neglect damping in the system.

15.13 Solve Problem 15.12 using subroutine CONDE to condense the rotational nodal coordinates.

15.14 Determine the response of the frame shown in Fig. P15.1 when acted upon by the force $F(t)$ (depicted in Fig. P15.14) applied at nodal coordinate 1. Assume 10% damping in all the modes. Use modal superposition method (subroutine MODAL).

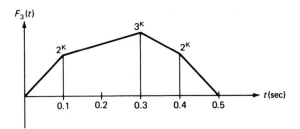

Fig. P15.14

15.15 Find the response in Problem 15.14 using step-by-step linear acceleration method (subroutine STEPM).

16

Dynamic analysis of grids

In Chapter 15 consideration was given to the dynamic analysis of the plane frame when subjected to forces acting on the plane of the structure. When the planar structural system is subjected to loads applied normally to its plane, the structure is referred to as a *grid*. This structure can also be treated as a special case of the three-dimensional frame to be presented in Chapter 17. The reason for considering the planar frame, whether loaded in its plane or normal to its plane, as a special case, is the immediate reduction of unknown nodal coordinates for a beam element; hence a considerable reduction in the number of unknown displacements for the structural system.

When analyzing the planar frame under action of loads in the plane, the possible components of joint displacements which had to be considered were translations in the X and Y directions and rotation about the Z axis. However, if a plane frame is loaded normal to the plane of the structure, the components of joint displacements required to describe the displacements of a joint are a translation in the Z direction and rotations about the X and Y axes. Thus treating the planar grid structure as a special case, it will be necessary to consider only three components of nodal displacements at each end of a typical grid member.

16.1 LOCAL AND GLOBAL COORDINATE SYSTEMS

For a beam element of a grid, the local orthogonal axes will be established such that the x defines the longitudinal centroidal axis of the member and the x-y plane will coincide with the plane of the structural system, which will be defined by the X-Y plane. In this case, the z axis will define the minor *principal axis* of the cross section while the y axis will define the *major axis* of the cross section. It will be assumed that the shear center of the cross section coincides with the centroid of the cross section. The grid member may have either a variable or constant cross section along its length.

The possible nodal displacements with respect to the local or to the global systems of coordinates are identified in Fig. 16.1. It can be seen that the trans-

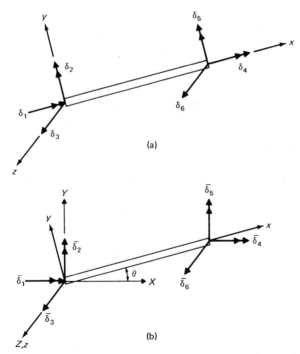

Fig. 16.1 Components of nodal displacements for a grid member. (a) Local coordinate system. (b) Global coordinate system.

latory displacements along the z direction for local axes and along the Z direction for the global system are identical since the two axes coincide. However, in general, rotational components at the nodal coordinates differ for these two coordinate systems. Hence, a transformation of coordinates will be required to transform the element matrices from the local to the global coordinates.

16.2   TORSIONAL EFFECTS

The dynamic analysis by the stiffness method for grid structures, that is, for plane frames subjected to normal loads, requires the determination of the torsional stiffness and mass coefficients for a typical member of the grid. The derivation of these coefficients is essentially identical to the derivation of the stiffness and mass coefficients for axial effects on a beam element. Similarity between these two derivations occurs because the differential equations for both problems have the same mathematical form. For the axial problem, the differential equation for the displacement function is given by eq. (15.8) as

$$\frac{du}{dx} = \frac{P}{AE}.$$

Likewise, the differential equation for torsional displacement is

$$\frac{d\theta}{dx} = \frac{T}{JG} \tag{16.1}$$

in which θ is the angular displacement, G is the modulus of elasticity in shear, and J is the torsional constant of the cross section (polar moment of inertia for circular sections).

As a consequence of the analogy between eqs. (15.8) and (16.1), we can write the following results already obtained for axial effects. The displacement functions for the torsional effects are the same as the corresponding functions giving the displacements for axial effects; hence by analogy to eqs. (15.10) and (15.11) and in reference to the nodal coordinates of Fig. 16.2, we obtain

$$\theta_1(x) = \left(1 - \frac{x}{L}\right) \tag{16.2}$$

and

$$\theta_2(x) = \frac{x}{L} \tag{16.3}$$

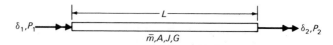

Fig. 16.2   Nodal torsional coordinates for a beam element.

in which the angular displacement function $\theta_1(x)$ corresponds to a unit angular displacement $\delta_1 = 1$ at nodal coordinate 1 and $\theta_2(x)$ corresponds to the displacement function resulting from a unit angular displacement $\delta_2 = 1$ at nodal coordinate 2. Analogous to eq. (15.17), the stiffness influence coefficients for torsional effects may be calculated from

$$k_{ij} = \int_0^L JG\theta_i'(x)\,\theta_j'(x)\,dx \qquad (16.4)$$

in which $\theta_1'(x)$ and $\theta_2'(x)$ are the derivatives with respect to x of the displacement functions $\theta_1(x)$ and $\theta_2(x)$. Also analogous to eq. (15.23), the consistent mass matrix coefficients for torsional effects are given by

$$m_{ij} = \int_0^L \frac{\overline{m}(x)J}{A}\,\theta_i(x)\,\theta_j(x)\,dx \qquad (16.5)$$

in which A is the cross-sectional area, J is the torsional constant, and $\overline{m}(x)$ is the mass per unit length along the beam element.

The application of eqs. (16.4) and (16.5) for a uniform beam yields the stiffness and mass matrices such that

$$\begin{bmatrix} T_1 \\ T_2 \end{bmatrix} = \frac{JG}{L} \begin{bmatrix} 1 & -1 \\ -1 & 1 \end{bmatrix} \begin{bmatrix} \delta_1 \\ \delta_2 \end{bmatrix} \qquad (16.6)$$

and

$$\begin{bmatrix} T_1 \\ T_2 \end{bmatrix} = \frac{J\overline{m}L}{6A} \begin{bmatrix} 2 & 1 \\ 1 & 2 \end{bmatrix} \begin{bmatrix} \ddot{\delta}_1 \\ \ddot{\delta}_2 \end{bmatrix} \qquad (16.7)$$

16.3 STIFFNESS MATRIX FOR A GRID ELEMENT

The torsional stiffness matrix, eq. (16.6), is combined with the flexural stiffness matrix, eq. (14.20), to obtain the stiffness matrix for a typical member of a grid structure. In reference to the local coordinate system indicated in Fig. 16.1(a), the stiffness equation for a uniform member is then

$$\begin{bmatrix} P_1 \\ P_2 \\ P_3 \\ P_4 \\ P_5 \\ P_6 \end{bmatrix} = \frac{EI}{L^3} \begin{bmatrix} JGL^2/EI & & & & \text{Symmetric} & \\ 0 & 4L^2 & & & & \\ 0 & 6L & 12 & & & \\ -JGL^2/EI & 0 & 0 & JGL^2/EI & & \\ 0 & 2L^2 & 6L & 0 & 4L^2 & \\ 0 & -6L & -12 & 0 & -6L & 12 \end{bmatrix} \begin{bmatrix} \delta_1 \\ \delta_2 \\ \delta_3 \\ \delta_4 \\ \delta_5 \\ \delta_6 \end{bmatrix}$$

$$(16.8)$$

or in condensed form

$$\{P\} = [K] \{\delta\}. \tag{16.9}$$

16.4   CONSISTENT MASS MATRIX FOR A GRID ELEMENT

The combination of the consistent mass matrix for flexural effects (14.34) with the consistent mass matrix for torsional effects (16.7) results in the consistent mass matrix for a typical member of a grid, namely

$$
\begin{bmatrix} P_1 \\ P_2 \\ P_3 \\ P_4 \\ P_5 \\ P_6 \end{bmatrix} = \frac{\overline{m}L}{420}
\begin{bmatrix}
140J/A & & & & \text{Symmetric} & \\
0 & 4L^2 & & & & \\
0 & 22L & 156 & & & \\
70J/A & 0 & 0 & 140J/A & & \\
0 & -3L^2 & -13L & 0 & 4L^2 & \\
0 & 13L & 54 & 0 & -22L & 156
\end{bmatrix}
\begin{bmatrix} \ddot{\delta}_1 \\ \ddot{\delta}_2 \\ \ddot{\delta}_3 \\ \ddot{\delta}_4 \\ \ddot{\delta}_5 \\ \ddot{\delta}_6 \end{bmatrix}
\tag{16.10}
$$

or in concise notation

$$\{P\} = [M_c] \{\delta\} \tag{16.11}$$

in which $[M_c]$ is the mass matrix for a typical uniform member of a grid structure.

16.5   LUMPED MASS MATRIX FOR A GRID ELEMENT

The lumped mass allocation to the nodal coordinates of a typical grid member is obtained from static considerations. For a uniform member having a distributed mass along its length, the nodal mass is simply one-half of the total rotational mass $J\overline{m}L/A$. The matrix equation for the lumped mass matrix corresponding to the· torsional effects is then

$$
\begin{bmatrix} P_1 \\ P_2 \end{bmatrix} = \frac{J\overline{m}L}{2A}
\begin{bmatrix} 1 & 0 \\ 0 & 1 \end{bmatrix}
\begin{bmatrix} \ddot{\delta}_1 \\ \ddot{\delta}_2 \end{bmatrix}.
\tag{16.12}
$$

The combination of the lumped torsional mass matrix from eq. (16.12) with the flexural mass matrix for a typical member of a grid results in the diagonal matrix which is the lumped mass matrix for the grid element. This matrix, relating forces and accelerations at nodal coordinates, is given by the following equation:

$$
\begin{bmatrix} P_1 \\ P_2 \\ P_3 \\ P_4 \\ P_5 \\ P_6 \end{bmatrix} = \frac{\overline{m}L}{2} \begin{bmatrix} J/A & & & & & \\ & 0 & & & & \\ & & 1 & & & \\ & & & J/A & & \\ & & & & 0 & \\ & & & & & 1 \end{bmatrix} \begin{bmatrix} \ddot{\delta}_1 \\ \ddot{\delta}_2 \\ \ddot{\delta}_3 \\ \ddot{\delta}_4 \\ \ddot{\delta}_5 \\ \ddot{\delta}_6 \end{bmatrix} \qquad (16.13)
$$

or briefly

$$
\{P\} = [M_L] \{\delta\} \qquad (16.14)
$$

in which $[M_L]$ is, in this case, the diagonal lumped mass matrix for a grid member.

16.6 TRANSFORMATION OF COORDINATES

The stiffness matrix, eq. (16.8), as well as the consistent and the lumped mass matrix in eqs. (16.10) and (16.13), respectively, are in reference to the local system of coordinates. Therefore, it is necessary to transform the reference of these matrices to the global system of coordinates before their assemblage in the corresponding matrices for the structure. As has been indicated, the z axis for the local coordinate system coincides with the Z axis for the global system. Therefore, the only step left to perform is a rotation of the coordinates in the x-y plane. The corresponding matrix for this transformation may be obtained by establishing the relations between components of the moments at the nodes expressed in these two systems of coordinates. In reference to Fig. 16.3, these

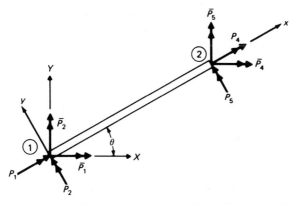

Fig. 16.3   Components of the nodal moments in local and global coordinates.

relations when written for node ① are

$$P_1 = \bar{P}_1 \cos\theta + \bar{P}_2 \sin\theta,$$
$$P_2 = -\bar{P}_1 \sin\theta + \bar{P}_2 \cos\theta,$$
$$P_3 = \bar{P}_3, \qquad\qquad (16.15a)$$

and for node ②

$$P_4 = \bar{P}_4 \cos\theta + \bar{P}_5 \sin\theta,$$
$$P_5 = -\bar{P}_4 \sin\theta + P_5 \cos\theta,$$
$$P_6 = \bar{P}_6. \qquad\qquad (16.15b)$$

The identical form of these equations with those derived for the transformation of coordinates for nodal forces of an element of a plane frame, eqs. (15.28) and (15.30), should be noted. Equations (16.15) may be written in matrix notation as

$$\begin{bmatrix} P_1 \\ P_2 \\ P_3 \\ P_4 \\ P_5 \\ P_6 \end{bmatrix} = \begin{bmatrix} \cos\theta & \sin\theta & 0 & 0 & 0 & 0 \\ -\sin\theta & \cos\theta & 0 & 0 & 0 & 0 \\ 0 & 0 & 1 & 0 & 0 & 0 \\ 0 & 0 & 0 & \cos\theta & \sin\theta & 0 \\ 0 & 0 & 0 & -\sin\theta & \cos\theta & 0 \\ 0 & 0 & 0 & 0 & 0 & 1 \end{bmatrix} \begin{bmatrix} \bar{P}_1 \\ \bar{P}_2 \\ \bar{P}_3 \\ \bar{P}_4 \\ \bar{P}_5 \\ \bar{P}_6 \end{bmatrix} \qquad (16.16)$$

or in short notation

$$\{P\} = [T]\,\{\bar{P}\} \qquad\qquad (16.17)$$

in which $\{P\}$ and $\{\bar{P}\}$ are, respectively, the vectors of the nodal forces of a typical grid member in local and global coordinates and $[T]$ the transformation matrix. The same transformation matrix $[T]$ serves also to transform the nodal components of the displacements from a global to a local system of coordinates. In condensed notation, this relation is given by

$$\{\delta\} = [T]\,\{\bar{\delta}\}, \qquad\qquad (16.18)$$

where $\{\delta\}$ and $\{\bar{\delta}\}$ are, respectively, the components of nodal displacement in local and global coordinates. The substitution of eqs. (16.17) and (16.18) in the stiffness relation eq. (16.9) yields the element stiffness matrix in reference to the global coordinate system, that is,

$$[T]\,\{\bar{P}\} = [K]\,[T]\,\{\bar{\delta}\}$$

or, since $[T]$ is an orthogonal matrix, it follows that

$$\{\bar{P}\} = [T]^T[K]\,[T]\,\{\bar{\delta}\}.$$

If we define $[\overline{K}]$ as

$$[\overline{K}] = [T]^T [K] [T], \qquad (16.19)$$

we obtain

$$\{\overline{P}\} = [\overline{K}] \{\overline{\delta}\}. \qquad (16.20)$$

Analogously, for the mass matrix, we find

$$\{\overline{P}\} = [\overline{M}] \{\overline{\ddot{\delta}}\} \qquad (16.21)$$

in which

$$[\overline{M}] = [T]^T [M] [T] \qquad (16.22)$$

is the transformed mass matrix.

Illustrative Example 16.1.   Figure 16.4 shows a grid structure in a horizontal plane consisting of two prismatic beam elements with a total of three degrees of freedom as indicated. Determine the natural frequencies and corresponding mode shapes. Use the consistent mass formulation.

The stiffness matrix for elements 1 or 2 of the grid in local coordinates by eq. (16.8) is

$$[K_1] = [K_2] = 10^6 \begin{bmatrix} 10 & 0 & 0 & -10 & 0 & 0 \\ 0 & 200 & 5 & 0 & 100 & -5 \\ 0 & 5 & 0.167 & 0 & 5 & -0.167 \\ -10 & 0 & 0 & 10 & 0 & 0 \\ 0 & 100 & 5 & 0 & 200 & -5 \\ 0 & -5 & -0.167 & 0 & -5 & 0.167 \end{bmatrix}$$

The transformation matrix for element △ with $\theta = 0°$ is simply the unit matrix $[T_1] = [I]$. Hence

$$[\overline{K}_1] = [T_1]^T [K_1] [T_1] = [K_1]$$

and for element △ with $\theta = 90°$

$$[T_2] = \begin{bmatrix} 0 & 1 & 0 & 0 & 0 & 0 \\ -1 & 0 & 0 & 0 & 0 & 0 \\ 0 & 0 & 1 & 0 & 0 & 0 \\ 0 & 0 & 0 & 0 & 1 & 0 \\ 0 & 0 & 0 & -1 & 0 & 0 \\ 0 & 0 & 0 & 0 & 0 & 1 \end{bmatrix}$$

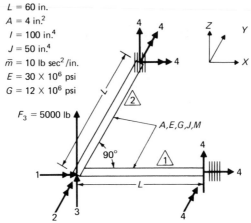

$L = 60$ in.
$A = 4$ in.^{2}
$I = 100$ in.^{4}
$J = 50$ in.^{4}
$\bar{m} = 10$ lb sec^2 /in.
$E = 30 \times 10^6$ psi
$G = 12 \times 10^6$ psi

$F_3 = 5000$ lb

A,E,G,J,M

$90°$

Fig. 16.4   Grid structure for Examples 16.1 and 16.2.

so that

$$[\overline{K}_2] = [T_2]^T [K_2] [T_2]$$

$$= 13,888 \begin{array}{cc} & \begin{array}{cccccc} 1 & \quad 2 & \quad 3 & \quad 4 & \quad 4 & \quad 4 \end{array} \\ \begin{bmatrix} 14,400 & 0 & -360 & 7200 & 0 & 360 \\ 0 & 720 & 0 & 0 & -720 & 0 \\ -360 & 0 & 12 & -360 & 0 & -12 \\ 7200 & 0 & -360 & 14,400 & 0 & 360 \\ 0 & -270 & 0 & 0 & 720 & 0 \\ 360 & 0 & -12 & 360 & 0 & 12 \end{bmatrix} & \begin{array}{c} 1 \\ 2 \\ 3 \\ 4 \\ 4 \\ 4 \end{array} \end{array}$$

The system matrix $[K_s]$ assembled from $[\overline{K}_1]$ and $[K_2]$ is

$$[K_s] = 10^6 \begin{bmatrix} 210 & 0 & -5 \\ 0 & 210 & 5 \\ -5 & 5 & 0.333 \end{bmatrix}.$$

Analogously, for the mass, we have from eq. (16.10)

$$[M_1] = [M_2] = \begin{bmatrix} 2500 & 0 & 0 & -1250 & 0 & 0 \\ 0 & 20,570 & 1886 & 0 & -15,430 & 1114 \\ 0 & 1886 & 93,600 & 0 & -1114 & 0 \\ -1250 & 0 & 0 & 2500 & 0 & 0 \\ 0 & -15,430 & -1114 & 0 & 20,520 & -1886 \\ 0 & 1114 & 0 & 0 & -1886 & 223 \end{bmatrix}.$$

We then calculate using eq. (16.22)

$$[\overline{M}_1] = [M_1]$$

since

$$[T_1] = [I]$$

and

$$
[\overline{M}_2] =
\begin{array}{ccccccc}
1 & 2 & 3 & 4 & 4 & 4 & \\
\begin{bmatrix}
20{,}570 & 0 & -1886 & 15{,}430 & 0 & 1114 \\
0 & 2500 & 0 & 0 & -1250 & 0 \\
1886 & 0 & 93{,}600 & 1114 & 0 & 0 \\
13{,}430 & 0 & 1114 & 20{,}570 & 0 & 1886 \\
0 & -1250 & 0 & 0 & 2500 & 0 \\
1114 & 0 & 0 & 1886 & 0 & 223 \\
\end{bmatrix}
&
\begin{matrix}
1 \\ 2 \\ 3 \\ 4 \\ 4 \\ 4
\end{matrix}
\end{array}
$$

From $[\overline{M}_1]$ and $[\overline{M}_2]$ we assemble the system mass matrix and obtain

$$
[M_s] =
\begin{bmatrix}
23{,}070 & 0 & -1886 \\
0 & 23{,}070 & 1886 \\
-1886 & 1886 & 187{,}200 \\
\end{bmatrix}.
$$

The natural frequencies and mode shapes are obtained from the solution of the eigenproblem

$$([K_s] - \omega^2[M_s])\{\phi\} = \{0\}$$

which gives the eigenvalues (squares of the natural frequencies)

$$\omega_1^2 = 0.509, \qquad \omega_2^2 = 9110, \qquad \text{and} \qquad \omega_3^2 = 9110$$

then

$$\omega_1 = 0.713 \text{ rad/sec}, \qquad \omega_2 = 95.45 \text{ rad/sec}, \qquad \text{and} \qquad \omega_3 = 94.45 \text{ rad/sec}$$

and the eigenvectors (modal matrix)

$$
[a] =
\begin{bmatrix}
1.00 & 1.00 & 1.00 \\
-1.00 & 1.00 & -1.00 \\
42.04 & 0 & 0 \\
\end{bmatrix}.
$$

The eigenvectors are conveniently normalized by dividing the columns of the modal matrix, respectively, by the factors

$$\sqrt{\{a_1\}^T[M_s]\{a_1\}} = 1.818,$$

$$\sqrt{\{a_2\}^T [M_s] \{a_2\}} = 0.0215,$$

$$\sqrt{\{a_3\}^T [M_s] \{a_3\}} = 0.0215.$$

The normalized eigenvectors are arranged in columns of the modal matrix, so that

$$[\Phi] = 10^{-4} \begin{bmatrix} 0.55 & 46.55 & 46.55 \\ -0.55 & 46.55 & -46.55 \\ 23.32 & 0 & 0 \end{bmatrix}.$$

Illustrative Example 16.2. Determine the response of the grid shown in Fig. 16.4 when subjected to a suddenly applied force $F_3 = 5000$ lb as indicated in the figure.

The natural frequencies and modal shapes for this structure were calculated in Example 16.1. The modal equation is given in general as

$$\ddot{z}_n + \omega_n^2 z_n = P_n$$

where

$$P_n = \sum_i \phi_{in} F_i$$

and F_i the external forces at the nodal coordinates which for this example are $F_1 = F_2 = 0$ and $F_3 = 5000$ lb. Hence, we obtain

$$\ddot{z}_1 + 0.509 z_1 = 11.66$$

$$\ddot{z}_2 + 9110 z_2 = 0$$

$$\ddot{z}_3 + 9110 z_3 = 0.$$

The solution of these equations for zero initial conditions is

$$z_1 = \frac{11.66}{0.509} (1 - \cos 0.713t)$$

$$z_2 = z_3 = 0.$$

The displacements at the nodal coordinates are calculated from

$$\{y\} = [\Phi] \{z\},$$

$$\begin{bmatrix} y_1 \\ y_2 \\ y_3 \end{bmatrix} = 10^{-4} \begin{bmatrix} 0.55 & 46.55 & 46.55 \\ -0.55 & 46.55 & -46.55 \\ 23.32 & 0 & 0 \end{bmatrix} \begin{bmatrix} 22.9 \\ 0 \\ 0 \end{bmatrix} (1 - \cos 0.713t)$$

and finally

$$y_1 = 0.001260 (1 - \cos 0.713t)$$

$$y_2 = 0.001260 (1 - \cos 0.713t)$$

$$y_3 = 0.05340 (1 - \cos 0.713t).$$

16.7 PROGRAM 13–DYNAMIC ANALYSIS OF GRIDS (GRID)

The computer program described in this section for the dynamic analysis of a plane frame with normal loads (grids) follows along much of the same pattern of organization as the programs described in the preceding chapters for the

TABLE 16.1 Description of Principal Variables for Program 13.

Variable	Symbol in Text	Description
NE		Number of elements
ND		Number of degrees of freedom
NCR		Number of degrees of freedom condensed
NCM		Number of concentrated masses
LOC		Mass index:
		0 → lumped mass;
		1 → consistent mass
IFPR		Print intermediate values in JACOBI:
		0 → do not print,
		1 → print
E	E	Modulus of elasticity
G	G	Modulus of rigidity
LE		Beam segment number
AR	A	Beam cross-sectional area
SL	L	Beam segment length
SI	I	Beam moment of inertia
SJ	J	Beam torsional constant
SMA	$\overline{m}$	Beam mass per unit length
TH	θ	Slope of beam segment
NC(L)		Beam segment nodal coordinates ($L = 1, 6$)
CM(L)		Concentrated masses ($L = 1$, NCM)
JC(L)		Nodal coordinates for CM(L)
BK(I, J)	$[K]$	Element stiffness matrix
BM(I, J)	$[M]$	Element mass matrix
T(I, J)	$[T]$	Element transformation matrix
SK(I, J)	$[K_s]$	System stiffness matrix
SM(I, J)	$[M_s]$	System mass matrix

TABLE 16.2   Input Data and Formats for Program 13.

Format	Variables
(6I5, 3F10.0)	NE ND NCR NCM LOC IFPR F G
(I2, 2F8.2, 4F10.2, 6I2)	LE AR SL SI SJ SMA TH (NC(L), L = 1, 6) (one card for each beam segment)
(8(I2, F8.2))	JC(L) CM(L) (L = 1, NCM)

dynamic analysis of beams or plane frames. Essentially, the major difference with those programs is the need to consider torsional effects in the analysis of the structures designated as grids.

The description of the important variables used in the program is given in Table 16.1 and the listing of input variables and formats is given in Table 16.2. The listing of the computer program itself is given in the Appendix. As in the other structural programs presented in this text, the various subroutines indicated in Table 14.3 should be available as required in the calling statements of the main program.

Illustrative Example 16.3.   For the structure shown in Fig. 16.4 and analyzed in the previous examples, use Program 13 to calculate the natural frequencies and mode shapes, and to determine the response to a constant force of 5000 lb suddenly applied in the direction of nodal coordinate 3. Compare these results with those obtained by hand calculations in Examples 16.1 and 16.2.

In the solution of this problem it is required to call subroutines JACOBI and MODAL. Hence "C" should be removed from the two statements in the program calling these subroutines. The listing of the required input data for this example and the computer results giving the eigenvalues and eigenvectors, are given, respectively, in Tables 16.3 and 16.4. Table 16.5 gives the additional data required by subroutine MODAL followed by the computer results given the response at the three nodal coordinates of this structure. The results given by the computer compare very closely to those obtained in Examples 16.1 and 16.2 by hand calculations. For example, the displacement given by the computer

TABLE 16.3   Data Input for Example 16.3.

			Data Listing								
	2	3	0	0	1	0	30000000.		12000000.		
1	4.00	60.00	100.00	50.00	10.00	0.00	1	2	3	4	4	4
2	4.00	60.00	100.00	50.00	10.00	90.00	1	2	3	4	4	4

TABLE 16.4   Computer Results for Example 16.3.

EIGENVALUES

| 0.91022D 04 | 0.91097D 04 | 0.50917D 00 |

EIGENVECTORS

0.46553D-02	-0.46588D-02	0.55045D-04
0.46553D-02	0.46588D-02	-0.55045D-04
0.42320D-18	-0.66553D-04	0.23122D-02

TABLE 16.5   Additional Data Input and Computer Results for Example 16.3.

0.0100	0.2000	0	0	2
0.00	5000.00	0.20	5000.00	
0.000	0.000	0.000		

RESPONSE FOR ELASTIC SYSTEM

TIME	DISPLACEMENTS AT NODAL COORDINATES		
t	y_1	y_2	y_3
0.000	0.0000D 00	0.0000D 00	0.0000D 00
0.010	0.1036D-06	-0.1036D-06	0.1338D-05
0.020	0.3539D-06	-0.3539D-06	0.5349D-05
0.030	0.6202D-06	-0.6202D-06	0.1203D-04
0.040	0.8120D-06	-0.8120D-06	0.2139D-04
0.050	0.9554D-06	-0.9554D-06	0.3341D-04
0.060	0.1171D-05	-0.1171D-05	0.4811D-04
0.070	0.1572D-05	-0.1572D-05	0.6548D-04
0.080	0.2169D-05	-0.2169D-05	0.8552D-04
0.090	0.2861D-05	-0.2861D-05	0.1082D-03
0.100	0.3520D-05	-0.3520D-05	0.1336D-03
0.110	0.4099D-05	-0.4099D-05	0.1616D-03
0.120	0.4674D-05	-0.4674D-05	0.1923D-03
0.130	0.5376D-05	-0.5376D-05	0.2257D-03
0.140	0.6282D-05	-0.6282D-05	0.2617D-03
0.150	0.7353D-05	-0.7353D-05	0.3004D-03
0.160	0.8461D-05	-0.8461D-05	0.3418D-03
0.170	0.9502D-05	-0.9502D-05	0.3858D-03
0.180	0.1048D-04	-0.1048D-04	0.4325D-03
0.190	0.1151D-04	-0.1151D-04	0.4818D-03

in Table 16.5 for nodal coordinate 3, a time $t = 0.1$ sec, is

$$y_3(t = 0.1) = 0.1336 \ 10^{-3} \text{ in}$$

while the hand calculation from Example 16.2 is

$$y_3(t = 0.1) = 0.534 \ (1 - \cos 0.0713)$$

$$= 0.1356 \ 10^{-3} \text{ in.}$$

16.8   SUMMARY

This chapter has presented the dynamic analysis of plane frames (grids) support-ing loads applied normally to its plane. The dynamic analysis of grids requires the inclusion of torsional effects in the element stiffness and mass matrices. It also requires a transformation of coordinates of the element matrices previous to the assembling of the system matrix. The required matrices for torsional effects are developed and a computer program for the dynamic analysis of grids is presented. This program is also organized along the same pattern of the programs in the two preceding chapters for the dynamic analysis of beams and plane frames.

PROBLEMS

The following problems are intended for hand calculation, though it is recom-mended that whenever possible solutions should also be obtained using Program 13 GRID.

16.1   For the grid shown in Fig. P16.1 determine the system stiffness and mass matrices. Base the analysis on the three nodal coordinates indicated in the figure. Use consistent mass method.

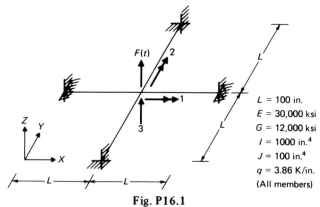

$L = 100$ in.
$E = 30{,}000$ ksi
$G = 12{,}000$ ksi
$I = 1000$ in.^{4}
$J = 100$ in.^{4}
$q = 3.86$ K/in.
(All members)

Fig. P16.1

16.2   Use static condensation to eliminate the rotational degrees of freedom and determine the transformation matrix and the reduced stiffness and mass matrices in Problem 16.1.

16.3   Determine the natural frequency for the reduced system in Problem 16.1.

16.4   Determine the natural frequencies and corresponding normal modes for the grid analyzed in Problem 16.1.

16.5   Determine the response of the grid shown in Fig. P16.1 when acted upon by a force $F(t)$ = 10 kip suddenly applied at the nodal coordinate 3 as shown in the figure. Use results of Problem 16.2 to obtain the equation of motion for the condensed system. Assume 10% modal damping.

16.6   Use results from Problem 16.4 to solve Problem 16.5 on the basis of the three nodal coordinates as indicated in Fig. P16.1.

16.7   Determine the steady-state response of the grid shown in Fig. P16.1 when subjected to harmonic force $F(t)$ = 10 sin 50t (kip) along nodal coordinate 3. Neglect damping in the system.

16.8   Repeat Problem 16.7 assuming that the damping is proportional to the stiffness of the system, $[C] = a_0[K]$ where a_0 = 0.3.

16.9   Determine the equivalent nodal forces for a member of a grid loaded with a dynamic force, $P(t) = P_0 f(t)$, uniformly distributed along its length.

16.10   Determine the equivalent nodal forces for a member of a grid supporting a concentrated dynamic force $F(t)$ as shown in Fig. P16.10.

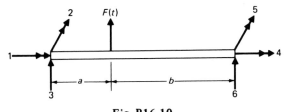

Fig. P16.10

The following problems are intended for computer solution using Program 13 GRID.

16.11   Determine the natural frequencies and corresponding normal modes for the grid shown in Fig. P16.1.

16.12   Determine the response of the grid shown in Fig. P16.1 when acted upon by the force depicted in Fig. P16.12 acting along nodal coordinate 3. Neglect damping in the system.

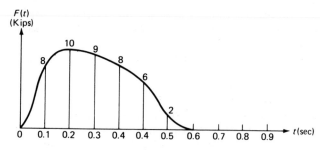

Fig. P16.12

16.13   Repeat Problem 16.12 for 15% damping in all the modes. Use modal
        superposition method.

16.14   Repeat Problem 16.12 for 15% damping in all the modes. Use step-by-
        step linear acceleration method (subroutine STEPM).

17

Three-dimensional frames

The stiffness method for dynamic analysis of frames presented in Chapter 15 for plane frames and in Chapter 16 for grids can readily be expanded for the analysis of three-dimensional space frames. While for the plane frame or for the grid, there were only three nodal coordinates at each joint, the three-dimensional frame has a total of six possible nodal displacements at each unconstrained joint: three translations along the x, y, z axes and three rotations about these axes. Consequently, a beam element of a space frame has for its two joints a total of 12 nodal coordinates; hence the resulting element matrices will be of dimension 12×12.

The dynamic analysis of three-dimensional frames results in a comparatively longer computer program in general requiring substantially more input data and the availability of a computer with larger storage memory. However, except for size, the analysis of three-dimensional frames by the stiffness method of dynamic analysis is identical to the analysis of plane frames or plane grids.

17.1 ELEMENT STIFFNESS MATRIX

Figure 17.1 shows a beam segment of a three-dimensional frame with its 12 nodal coordinates numbered consecutively. The convention adopted is to label first the three translatory displacements of the first joint followed by the three rotational displacements of the same joint; then to continue with the three translatory displacements of the second joint and finally the three rotational displacements of this second joint. The double arrows used in Fig. 17.1 serve to indicate rotational nodal coordinates; hence, these are distinguished from translational nodal coordinates for which single arrows are used.

The stiffness matrix for a three-dimensional uniform beam segment is readily written by the superposition of the axial stiffness matrix from eq. (15.3), the torsional stiffness matrix from eq. (16.6), and the flexural stiffness matrix from eq. (14.20). The flexural stiffness matrix is used twice in forming the stiffness matrix of a three-dimensional beam segment to account for the flexural effects in the two principal planes of the cross section. Proceeding to combine in an appropriate manner these matrices, we obtain in eq. (17.1) the stiffness equation for a uniform beam segment of a three-dimensional frame, namely

$$
\begin{Bmatrix} P_1 \\ P_2 \\ P_3 \\ P_4 \\ P_5 \\ P_6 \\ P_7 \\ P_8 \\ P_9 \\ P_{10} \\ P_{11} \\ P_{12} \end{Bmatrix} =
\begin{bmatrix}
\frac{EA}{L} & & & & & & & & & & & \\
0 & \frac{12EI_z}{L^3} & & & & & \text{Symmetric} & & & & & \\
0 & 0 & \frac{12EI_y}{L^3} & & & & & & & & & \\
0 & 0 & 0 & \frac{GJ}{L} & & & & & & & & \\
0 & 0 & \frac{-6EI_y}{L^2} & 0 & \frac{4EI_y}{L} & & & & & & & \\
0 & \frac{6EI_z}{L^3} & 0 & 0 & 0 & \frac{4EI_z}{L} & & & & & & \\
\frac{-EA}{L} & 0 & 0 & 0 & 0 & 0 & \frac{EA}{L} & & & & & \\
0 & \frac{-12EI_z}{L^2} & 0 & 0 & 0 & \frac{-6EI_z}{L^2} & 0 & \frac{12EI_z}{L^3} & & & & \\
0 & 0 & \frac{-12EI_y}{L^3} & 0 & \frac{6EI_y}{L^2} & 0 & 0 & 0 & \frac{12EI_y}{L^3} & & & \\
0 & 0 & 0 & \frac{-GJ}{L^2} & 0 & 0 & 0 & 0 & 0 & \frac{GJ}{L} & & \\
0 & 0 & \frac{-6EI_y}{L^2} & 0 & \frac{2EI_y}{L} & 0 & 0 & 0 & \frac{6EI_y}{L^2} & 0 & \frac{4EI_y}{L} & \\
0 & \frac{6EI_z}{L^2} & 0 & 0 & 0 & \frac{2EI_z}{L} & 0 & \frac{-6EI_z}{L^2} & 0 & 0 & 0 & \frac{4EI_z}{L}
\end{bmatrix}
\begin{Bmatrix} \delta_1 \\ \delta_2 \\ \delta_3 \\ \delta_4 \\ \delta_5 \\ \delta_6 \\ \delta_7 \\ \delta_8 \\ \delta_9 \\ \delta_{10} \\ \delta_{11} \\ \delta_{12} \end{Bmatrix}
$$

(17.1)

or in condensed notation

$$\{P\} = [K]\,\{\delta\} \tag{17.2}$$

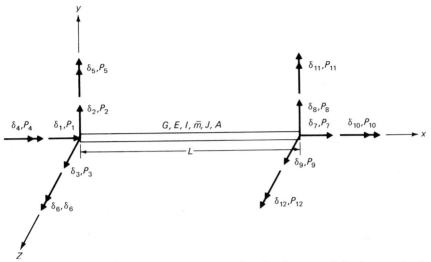

Fig. 17.1   Beam segment of a space frame showing forces and displacements at the nodal coordinates.

in which I_y and I_z are, respectively, the cross-sectional moments of inertia with respect to the principal axes labeled as y and z in Fig. 17.1, and L, A and J are respectively the length, cross-sectional area and torsional constant of the beam element.

17.2   ELEMENT MASS MATRIX

The lumped mass matrix for the uniform beam segment of a three-dimensional frame is simply a diagonal matrix in which the coefficients corresponding to translatory and torsional displacements are equal to one-half of the total inertia of the beam segment while the coefficients corresponding to flexural rotations are assumed to be zero. The diagonal lumped mass matrix for the uniform beam of distributed mass $\overline{m}$ per unit of length may be written conveniently as

$$M_l = \frac{\overline{m}L}{2} \begin{bmatrix} 1 & 1 & 1 & J/A & 0 & 0 & 1 & 1 & 1 & J/A & 0 & 0 \end{bmatrix}. \quad (17.3)$$

The consistent mass matrix for a uniform beam segment of a three-dimensional frame is readily obtained combining the consistent mass matrices, eq. (15.36) for axial effects, eq. (16.7) for torsional effects, and eq. (14.34) for flexural effects. Appropriate combination of these matrices results in the consistent mass matrix for the uniform beam segment of a three-dimensional frame, namely

$$
\begin{Bmatrix} P_1 \\ P_2 \\ P_3 \\ P_4 \\ P_5 \\ P_6 \\ P_7 \\ P_8 \\ P_9 \\ P_{10} \\ P_{11} \\ P_{12} \end{Bmatrix} = \frac{\overline{m}L}{420}
\begin{bmatrix}
140 & & & & & & & & & & & \\
0 & 156 & & & & & \text{Symmetric} & & & & & \\
0 & 0 & 156 & & & & & & & & & \\
0 & 0 & 0 & \dfrac{140J}{A} & & & & & & & & \\
0 & 0 & -2L & 0 & 4L^2 & & & & & & & \\
0 & -2L & 0 & 0 & 0 & 4L^2 & & & & & & \\
70 & 0 & 0 & 0 & 0 & 0 & 140 & & & & & \\
0 & 54 & 0 & 0 & 0 & 13L & 0 & 156 & & & & \\
0 & 0 & 54 & 0 & -13L & 0 & 0 & 0 & 156 & & & \\
0 & 0 & 0 & \dfrac{70J}{A} & 0 & 0 & 0 & 0 & 0 & \dfrac{140J}{A} & & \\
0 & 0 & 13L & 0 & -3L^2 & 0 & 0 & 0 & 22L & 0 & 4L^2 & \\
0 & -13L & 0 & 0 & 0 & -3L^2 & 0 & -22L & 0 & 0 & 0 & 4L^2
\end{bmatrix}
\begin{Bmatrix} \delta_1 \\ \delta_2 \\ \delta_3 \\ \delta_4 \\ \delta_5 \\ \delta_6 \\ \delta_7 \\ \delta_8 \\ \delta_9 \\ \delta_{10} \\ \delta_{11} \\ \delta_{12} \end{Bmatrix}
$$

$$(17.4)$$

or in condensed notation

$$\{P\} = [M]\{\delta\}. \tag{17.5}$$

17.3 ELEMENT DAMPING MATRIX

The damping matrix for a uniform beam segment of a three-dimensional frame may be obtained in a manner entirely parallel to those of the stiffness, eq. (17.1), and mass, eq. (17.4), matrices. Nevertheless, as was discussed in Section 14.5, in practice, damping is generally expressed in terms of damping ratios for each mode of vibration. Therefore, if the response is sought using the modal superposition method, these damping ratios are directly introduced in the modal equations. When the damping matrix is required explicitly, it may be determined from given values of damping ratios by the methods presented in Chapter 12.

17.4 TRANSFORMATION OF COORDINATES

The stiffness and the mass matrices given by eqs. (17.1) and (17.4), respectively, are referred to local coordinate axes fixed on the beam segment. Inasmuch as the elements of these matrices corresponding to the same nodal coordinates of the structure should be added to obtain the system stiffness and mass matrices, it is necessary first to transform these matrices to the same reference system, the global system of coordinates. Figure 17.2 shows these two reference systems, the x, y, z axes representing the local system of coordinates and the X, Y, Z axes representing the global system of coordinates. Also shown in this figure is a general vector A with its components along the axes X, Y, and Z. This vector A with its components may represent any force or displacement at the nodal coordi-

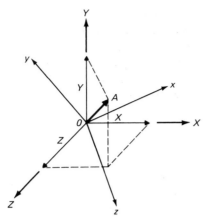

Fig. 17.2   Components of a general vector A in local and global coordinates.

nates of one of the joints of the structure. To obtain the components of vector A along one of the local axes x, y, or z, it is necessary to add the projections of X, Y, and Z components along that axis. For example, the component x of vector A along the x coordinate is given by

$$x = X \cos xX + Y \cos xY + Z \cos xZ \tag{17.6a}$$

in which $\cos xX$ is the cosine of the angle between axes x and X and corresponding definitions for other cosines. Similarly, the y and z components of A are

$$y = X \cos yX + Y \cos yY + Z \cos yZ, \tag{17.6b}$$

$$z = X \cos zX + Y \cos zY + Z \cos zZ. \tag{17.6c}$$

These equations are conveniently written in matrix notation as

$$\begin{bmatrix} x \\ y \\ z \end{bmatrix} = \begin{bmatrix} \cos xX & \cos xY & \cos xZ \\ \cos yX & \cos yY & \cos yZ \\ \cos zX & \cos zY & \cos zZ \end{bmatrix} \begin{bmatrix} X \\ Y \\ Z \end{bmatrix} \tag{17.7}$$

or in short notation

$$\{A\} = [T_1] \{\overline{A}\} \tag{17.8}$$

in which $\{A\}$ and $\{\overline{A}\}$ are, respectively, the components in the local and global systems of the general vector A and $[T_1]$ the transformation matrix given by

$$[T_1] = \begin{bmatrix} \cos xX & \cos xY & \cos xZ \\ \cos yX & \cos yY & \cos yZ \\ \cos zX & \cos zY & \cos zZ \end{bmatrix}. \tag{17.9}$$

For the beam segment of a three-dimensional frame, the transformation of the nodal displacement vectors involve the transformation of linear and angular displacement vectors at each joint of the segment. Therefore, a three-dimensional beam element requires for the two joints, the transformation of a total of four displacement vectors. The transformation of the 12 nodal displacements $\{\bar{\delta}\}$ in global coordinates to the displacements $\{\delta\}$ in local coordinates may be written in abbreviated form as

$$\{\delta\} = [T]\{\bar{\delta}\} \tag{17.10}$$

in which

$$[T] = \begin{bmatrix} [T_1] & & & \\ & [T_1] & & \\ & & [T_1] & \\ & & & [T_1] \end{bmatrix}. \tag{17.11}$$

Analogously, the transformation from nodal forces $\{\bar{P}\}$ in global coordinates to nodal forces $\{P\}$ in local coordinates is given by

$$\{P\} = [T]\{\bar{P}\}. \tag{17.12}$$

Finally, to obtain the stiffness matrix $[\bar{K}]$ and the mass matrix $[\bar{M}]$ in reference to the global system of coordinates, we simply substitute into eq. (17.2), $\{\delta\}$ from eq. (17.10) and $[P]$ from eq. (17.12) to obtain

$$[T]\{\bar{P}\} = [K][T]\{\bar{\delta}\}$$

or

$$\{\bar{P}\} = [T]^T[K][T]\{\bar{\delta}\} \tag{17.13}$$

since $[T]$ is an orthogonal matrix. From eq. (17.13), we may write

$$\{\bar{P}\} = [\bar{K}]\{\bar{\delta}\} \tag{17.14}$$

in which $[\bar{K}]$ is defined as

$$[\bar{K}] = [T]^T[K][T]. \tag{17.15}$$

Analogously, the mass matrix in eq. (17.5) is transformed from local to global coordinates by

$$[\bar{M}] = [T]^T[M][T] \tag{17.16}$$

and the damping matrix $[C]$ by

$$[\bar{C}] = [T]^T[C][T]. \tag{17.17}$$

17.5 DIFFERENTIAL EQUATION OF MOTION

The direct method which was explained in detail in Chapter 14 may also be used to assemble the stiffness, mass, and damping matrices from the corresponding matrices for a three-dimensional beam segment, eqs. (17.15), (17.16), and (17.17), which are referred to the global system of coordinates. The differential equations of motion which are obtained by establishing the dynamic equilibrium among the inertial, damping, and elastic forces with the external forces may be expressed in matrix notation as

$$[M] \{\ddot{y}\} + [C] \{\dot{y}\} + [K] \{y\} = \{F(t)\} \qquad (17.18)$$

in which $[M]$, $[C]$, and $[K]$ are, respectively, the system mass, damping, and stiffness matrices; $\{\ddot{y}\}$, $\{\dot{y}\}$, and $\{y\}$ the system acceleration, velocity, and displacement vectors; and $F(t)$ the force vector which includes the forces applied directly to the joints of the structure and the equivalent nodal forces for the forces not applied at the joints.

17.6 DYNAMIC RESPONSE

The integration of the differential equations of motion, eq. (17.18), may be accomplished by any of the methods presented in previous chapters to obtain the response of structures modeled as beams, plane frames, or grids. The selection of the particular method of solution depends, as discussed previously, on the linearity of the differential equation, that is, whether the stiffness matrix $[K]$ or any other coefficient matrix is constant and also depends on the complexity of the excitation as a function of time. When the differential equations of motion, eq. (17.18), are linear, the modal superposition method is applicable. This method as discussed in the preceding chapters, requires the solution of an eigenproblem to uncouple the differential equations and to give as a result the modal equations of motion.

If the structure is assumed to follow an elastoplastic behavior or any other form of nonlinearity, it is necessary to resort to some kind of numerical integration in order to solve the differential equations of motion, eq. (17.18). In Chapter 19, the linear acceleration method with a modification known as the Wilson-θ method is presented together with a computer program for the seismic analysis of structures with an elastic or elastoplastic behavior.

17.6 SUMMARY

The dynamic analysis of space frames by the stiffness method requires a substantially greater computational effort than the corresponding analysis for the beams, plane frames, or grids presented in preceding chapters. A member of a space frame has a total of 12 nodal coordinates; hence, the element matrices are of dimension 12 X 12. The dynamic analysis of space frames is entirely analogous to the analysis of the special framed structures.

This chapter presented the necessary matrices for the dynamic analysis of space frames. A computer program could easily be developed as an extension of the special programs for framed structures presented in previous chapters. However, such a program would require a computer with a very large memory capacity for the solution of almost any three-dimensional frame. It is then of particular importance, in this case, to develop such programs employing highly efficient computational techniques in order to reduce both computer memory requirements and computational time.

18

Dynamic analysis of trusses

The static analysis of trusses whose members are pin-connected reduces to the problem of determining the bar forces due to a set of loads applied at the joints. When the same trusses are subjected to the action of dynamics forces, the simple situation of only axial stresses in the members is no longer present. The inertial forces developed along the members of the truss will, in general, produce flexural bending in addition to axial forces. The bending moments at the ends of the truss members will still remain zero in the absence of external joint moments. The dynamic stiffness method for the analysis of trusses is developed as in the case of framed structures by establishing the basic relations between elastic forces, damping forces, inertial forces, and the resulting displacements, velocities, and accelerations at the nodal coordinates; that is, by determining the stiffness, damping, and mass matrices for a member of the truss. The assemblage of system stiffness, damping, and mass matrices of the truss as well as the solution for the displacements at the nodal coordinates follows along the standard method presented in the preceding chapters for framed structures.

18.1   STIFFNESS AND MASS MATRICES FOR THE PLANE TRUSS

A member of a plane truss has two nodal coordinates at each joint, that is, a total of four nodal coordinates (Fig. 18.1). For small deflections, it may be assumed that the force-displacement relationship for the nodal coordinates along the axis of the member (coordinates 1 and 3 in Fig. 18.1) are independent of the transverse displacements along nodal coordinates 2 and 4. This assumption is equivalent to stating that a displacement along nodal coordinates 1 or 3 does not produce forces along nodal coordinates 2 or 4 and vice versa.

The stiffness and mass coefficients corresponding to the axial nodal coordinates were derived in Chapter 15 and are given, in general, by eq. (15.17) for the stiffness coefficients and by eq. (15.23) for consistent mass coefficients. Applying these equations to a uniform beam element, we obtain using the notation of Fig. 18.1 the following coefficients:

$$k_{11} = k_{33} = \frac{AE}{L}, \qquad k_{13} = k_{31} = -\frac{AE}{L}, \qquad (18.1)$$

$$m_{11} = m_{33} = \frac{\overline{m}L}{3}, \qquad m_{13} = m_{31} = \frac{\overline{m}L}{6} \qquad (18.2)$$

in which $\overline{m}$ is the mass per unit length, A is the cross-sectional area, and L is the length of the element.

The stiffness coefficients, for pin-ended element, corresponding to the nodal coordinates 1 and 4 are all equal to zero, since a force is not required to produce displacements at these coordinates. Therefore, arranging the coefficients given by eq. (18.1), we obtain the stiffness matrix for a uniform member of a truss as

$$\begin{bmatrix} P_1 \\ P_2 \\ P_3 \\ P_4 \end{bmatrix} = \frac{AE}{L} \begin{bmatrix} 1 & 0 & -1 & 0 \\ 0 & 0 & 0 & 0 \\ -1 & 0 & 1 & 0 \\ 0 & 0 & 0 & 0 \end{bmatrix} \begin{bmatrix} \delta_1 \\ \delta_2 \\ \delta_3 \\ \delta_4 \end{bmatrix} \qquad (18.3)$$

or in condensed notation

$$\{P\} = [K]\{\delta\} \qquad (18.4)$$

in which $[K]$ is the element stiffness matrix.

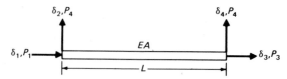

Fig. 18.1   Member of a plane truss showing nodal displacements and forces.

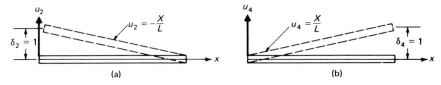

Fig. 18.2   Displacements functions. (a) For a unit displacement $\delta_2 = 1$. (b) For a unit displacement $\delta_4 = 1$.

The consistent mass matrix is obtained, as previously demonstrated, using expressions for static displacement functions in the application of the principle of virtual work. The displacement functions corresponding to a unit deflection at nodal coordinates 2 and 4 indicated in Fig. 18.2 are given by

$$u_2 = 1 - \frac{x}{L} \tag{18.5}$$

and

$$u_4 = \frac{x}{L}. \tag{18.6}$$

The consistent mass coefficients are given by the general expression, eq. (15.23), which is repeated here for convenience, namely

$$m_{ij} = \int_0^L \overline{m}(x)\, u_i(x)\, u_j(x)\, dx. \tag{18.7}$$

For a uniform member of mass $\overline{m}$ per unit length, the substitution of eqs. (18.5) and (18.6) into eq. (18.7) yields

$$m_{22} = m_{44} = \frac{\overline{m}L}{3}$$

$$m_{24} = m_{42} = \frac{\overline{m}L}{6}. \tag{18.8}$$

Finally, the combination of the mass coefficients from eqs. (18.2) and (18.8) forms the consistent mass matrix relating forces to accelerations at the nodal coordinates for a uniform member of plane truss, namely

$$
\begin{bmatrix} P_1 \\ P_2 \\ P_3 \\ P_4 \end{bmatrix} = \frac{\overline{m}L}{6}
\begin{bmatrix} 2 & 0 & 1 & 0 \\ 0 & 2 & 0 & 1 \\ 1 & 0 & 2 & 0 \\ 0 & 1 & 0 & 2 \end{bmatrix}
\begin{bmatrix} \ddot{\delta}_1 \\ \ddot{\delta}_2 \\ \ddot{\delta}_3 \\ \ddot{\delta}_4 \end{bmatrix} \tag{18.9}
$$

or in concise notation

$$\{P\} = [M] \{\ddot{\delta}\}. \qquad (18.10)$$

18.2   TRANSFORMATION OF COORDINATES

The stiffness matrix, eq. (18.3), and the mass matrix, eq. (18.9), were derived in reference to nodal coordinates associated with the local or element system of coordinates. As discussed before in the chapters on framed structures, it is necessary to transform these matrices to a common system of reference, the global coordinate system. The transformation of displacements and forces at the nodal coordinates is accomplished, as was demonstrated in Chapter 15, performing a rotation of coordinates. Deleting the angular coordinates in eq. (15.31) and relabeling the remaining coordinates result in the following transformation for the nodal forces:

$$
\begin{bmatrix} P_1 \\ P_2 \\ P_3 \\ P_4 \end{bmatrix} =
\begin{bmatrix}
\cos\theta & \sin\theta & 0 & 0 \\
-\sin\theta & \cos\theta & 0 & 0 \\
0 & 0 & \cos\theta & \sin\theta \\
0 & 0 & -\sin\theta & \cos\theta
\end{bmatrix}
\begin{bmatrix} \bar{P}_1 \\ \bar{P}_2 \\ \bar{P}_3 \\ \bar{P}_4 \end{bmatrix}
\qquad (18.11)
$$

or in condensed notation

$$\{P\} = [T] \{\bar{P}\} \qquad (18.12)$$

in which $\{P\}$ and $\{\bar{P}\}$ are the nodal forces in reference to local and global coordinates, respectively, and $[T]$ the transformation matrix defined in eq. (18.11).

The same transformation matrix $[T]$ also serves to transform the nodal displacement vector $\{\bar{\delta}\}$ in the global coordinate system to the nodal displacement vector $\{\delta\}$ in local coordinates:

$$\{\delta\} = [T] \{\bar{\delta}\}. \qquad (18.13)$$

The substitution of eqs. (18.12) and (18.13) into the stiffness equation, eq. (18.4), gives

$$[T] [\bar{P}] = [K] [T] \{\bar{\delta}\}.$$

Since $[T]$ is an orthogonal matrix ($[T]^{-1} = [T]^T$), it follows that

$$\{\bar{P}\} = [T]^T [K] [T] \{\bar{\delta}\}$$

or

$$\{\bar{P}\} = [\bar{K}] \{\bar{\delta}\} \qquad (18.14)$$

in which

$$[\overline{K}] = [T]^T [K] [T] \tag{18.15}$$

is the element stiffness matrix in the global coordinate system. Analogously, substituting eqs. (18.12) and (18.13) into eq. (18.10) results in

$$\{\overline{P}\} = [T]^T [M] [T] \{\ddot{\overline{\delta}}\} \tag{18.16}$$

or

$$\{\overline{P}\} = [\overline{M}] \{\ddot{\overline{\delta}}\} \tag{18.17}$$

$$[\overline{M}] = [T]^T [M] [T] \tag{18.18}$$

in which $[\overline{M}]$ is the element mass matrix referred to the global system of coordinates. A similar relation is also obtained for the element damping matrix, namely

$$[\overline{C}] = [T]^T [C] [T] \tag{18.19}$$

in which $[\overline{C}]$ and $[C]$ are the damping matrices referred, respectively, to the global and the local systems of coordinates.

Illustrative Example 18.1. The plane truss shown in Fig. 18.3 which has only three members is used to illustrate the application of the stiffness method for trusses. For this truss determine the system stiffness and the system consistent mass matrices.

The stiffness matrix, eq. (18.3), the mass matrix, eq. (18.9), and the transfor-

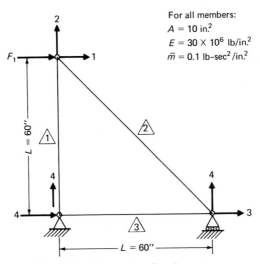

For all members:
$A = 10$ in.^{2}
$E = 30 \times 10^6$ lb/in.^{2}
$\overline{m} = 0.1$ lb-sec^2/in.^{2}

Fig. 18.3   Example of a plane truss.

mation matrix, eq. (18.11), are applied to the three members of this truss. For member $\triangle$, $\theta = 90°$,

$$[K_1] = \frac{AE}{L} \begin{bmatrix} 1 & 0 & -1 & 0 \\ 0 & 0 & 0 & 0 \\ -1 & 0 & 1 & 0 \\ 0 & 0 & 0 & 0 \end{bmatrix}, \quad [M_1] = \frac{\overline{m}L}{6} \begin{bmatrix} 2 & 0 & 1 & 0 \\ 0 & 2 & 0 & 1 \\ 1 & 0 & 2 & 0 \\ 0 & 1 & 0 & 2 \end{bmatrix},$$

and

$$[T_1] = \begin{bmatrix} 0 & 1 & 0 & 0 \\ -1 & 0 & 0 & 0 \\ 0 & 0 & 0 & 1 \\ 0 & 0 & -1 & 0 \end{bmatrix}.$$

Then from eqs. (18.15) and (18.18)

$$[\overline{K}_1] = [T_1]^T [K_1] [T_1] = \frac{AE}{L} \begin{bmatrix} 0 & 0 & 0 & 0 \\ 0 & 1 & 0 & -1 \\ 0 & 0 & 0 & 0 \\ 0 & -1 & 0 & 1 \end{bmatrix}$$

and

$$[\overline{M}_1] = [T_1]^T [M_1] [T_1] = \frac{\overline{m}L}{6} \begin{bmatrix} 2 & 0 & 1 & 0 \\ 0 & 2 & 0 & 1 \\ 1 & 0 & 2 & 0 \\ 0 & 1 & 0 & 2 \end{bmatrix}.$$

For member $\triangle$, $\theta = 135°$,

$$[K_2] = \frac{AE}{\sqrt{2}L} \begin{bmatrix} 1 & 0 & -1 & 0 \\ 0 & 0 & 0 & 0 \\ -1 & 0 & 1 & 0 \\ 0 & 0 & 0 & 0 \end{bmatrix}, \quad [M_2] = \frac{\overline{m}\sqrt{2}L}{6} \begin{bmatrix} 2 & 0 & 1 & 0 \\ 0 & 2 & 0 & 1 \\ 1 & 0 & 2 & 0 \\ 0 & 1 & 0 & 2 \end{bmatrix},$$

and

$$[T_2] = \frac{1}{\sqrt{2}} \begin{bmatrix} -1 & 1 & 0 & 0 \\ -1 & -1 & 0 & 0 \\ 0 & 0 & -1 & 1 \\ 0 & 0 & -1 & -1 \end{bmatrix}.$$

Then from eqs. (18.15) and (18.18)

$$[\bar{K}_2] = [T_2]^T [K_2] [T_2] = \frac{AE}{2\sqrt{2L}} \begin{bmatrix} 1 & -1 & -1 & 1 \\ -1 & 1 & 1 & -1 \\ -1 & 1 & 1 & -1 \\ 1 & -1 & -1 & 1 \end{bmatrix}$$

and

$$[\bar{M}_2] = [T_2]^T [M_2] [T_2] = \frac{\bar{m}L}{6\sqrt{2}} \begin{bmatrix} 4 & 0 & 2 & 0 \\ 0 & 4 & 0 & 2 \\ 2 & 0 & 4 & 0 \\ 0 & 2 & 0 & 4 \end{bmatrix}$$

For member △, $\theta = 0°$,

$$[\bar{K}_3] = [K_3] = \frac{AE}{L} \begin{bmatrix} 1 & 0 & -1 & 0 \\ 0 & 0 & 0 & 0 \\ -1 & 0 & 1 & 0 \\ 0 & 0 & 0 & 0 \end{bmatrix}$$

$$[\bar{M}_3] = [M_3] = \frac{\bar{m}L}{6} \begin{bmatrix} 2 & 0 & 1 & 0 \\ 0 & 2 & 0 & 1 \\ 1 & 0 & 2 & 0 \\ 0 & 1 & 0 & 2 \end{bmatrix}$$

substituting the proper numerical values for this example; $L = 60$ in, $A = 10$ in^2, $\bar{m} = 0.1$ lb · sec^2/in^2, $E = 30 \times 10^6$ lb/in^2, and following the rules of the direct method of assembling the system stiffness and mass matrix from the above element matrices, we obtain

$$[K_S] = 10^6 \begin{bmatrix} 1.768 & -1.768 & -1.768 \\ -1.768 & 6.768 & 1.768 \\ -1.768 & 1.768 & 6.768 \end{bmatrix}$$

$$[M_S] = \begin{bmatrix} 4.828 & 0 & 1.414 \\ 0 & 4.828 & 0 \\ 1.414 & 0 & 4.828 \end{bmatrix}$$

where $[K_S]$ and $[M_S]$ are, respectively, the system stiffness and mass matrices for the truss shown in Fig. 18.3

Illustrative Example 18.2. Determine the natural frequencies and normal modes for the truss of Example 18.1. The differential equations of motion for this system are

$$[M_S]\,\{\ddot{y}\} + [K_S]\,\{y\} = 0. \tag{a}$$

Substituting $\{y\} = \{a\}\sin\omega t$, we obtain

$$([K_S] - \omega^2\,[M_S])\,\{a\} = \{0\}. \tag{b}$$

For the nontrivial solution, we require

$$\left|[K_S] - \omega^2[M_S]\right| = 0. \tag{c}$$

Substituting from Example 18.1 $[K_S]$ and $[M_S]$ and expanding the above determinant give a cubic equation in $\lambda = \omega^2\overline{m}L^2/6AE$, which has the following roots

$$\lambda_1 = 0.00344 \quad\text{or}\quad \omega_1 = 415\ \text{rad/sec},$$

$$\lambda_2 = 0.0214 \quad\text{or}\quad \omega_2 = 1034\ \text{rad/sec},$$

$$\lambda_3 = 0.0466 \quad\text{or}\quad \omega_3 = 1526\ \text{rad/sec}.$$

Substituting in turn ω_1, ω_2, and ω_3 into eq. (b), setting $a_1 = 1$, and solving for a_2 and a_3 give the modal vectors

$$\{a_1\} = \begin{bmatrix} 1.000 \\ 0.216 \\ 0.274 \end{bmatrix}, \quad \{a_2\} = \begin{bmatrix} 1.000 \\ 5.488 \\ -4.000 \end{bmatrix}, \quad \{a_3\} = \begin{bmatrix} 1.000 \\ -1.000 \\ -1.524 \end{bmatrix}$$

which may be normalized using the factors

$$\sqrt{\{a_1\}^T[M_S]\,\{a_1\}} = 2.489, \qquad \sqrt{\{a_2\}^T[M_S]\,\{a_2\}} = 14.695,$$

$$\sqrt{\{a_3\}^T[M_S]\,\{a_3\}} = 4.066.$$

This normalization results in

$$\{\phi_1\} = \begin{bmatrix} 0.402 \\ 0.087 \\ 0.110 \end{bmatrix}, \quad \{\phi_2\} = \begin{bmatrix} 0.068 \\ 0.373 \\ -0.272 \end{bmatrix}, \quad \{\phi_3\} = \begin{bmatrix} 0.246 \\ -0.246 \\ -0.375 \end{bmatrix}.$$

These normalized eigenvectors form the modal matrix:

$$[\phi] = \begin{bmatrix} 0.402 & 0.068 & 0.246 \\ 0.087 & 0.373 & -0.246 \\ 0.110 & -0.272 & -0.375 \end{bmatrix}.$$

Illustrative Example 18.3.   Determine the response of the truss in Examples 18.1 and 18.2 when a constant force $F_1 = 5000$ lb is suddenly applied along coordinate 1 as shown in Fig. 18.3.

The modal equations are given in general by

$$\ddot{z}_n + \omega_n^2 z_n = P_n$$

in which the modal force

$$P_n = \sum_i \phi_{in} F_i.$$

Hence using the results which were calculated in Example 18.2 we obtain

$$\ddot{z}_1 + (415)^2 z_1 = 2010$$
$$\ddot{z}_2 + (1034)^2 z_2 = 340$$
$$\ddot{z}_3 + (1526)^2 z_3 = 1230.$$

The solution of the above equations for zero initial conditions ($z_n = 0, \dot{z}_n = 0$) is given by eqs. (4.5) as

$$z_1 - \frac{2010}{(415)^2} (1 - \cos 415t)$$

$$z_2 = \frac{340}{(1034)^2} (1 - \cos 1034t)$$

$$z_3 = \frac{1230}{(1526)^2} (1 - \cos 1526t).$$

The response at the nodal coordinates is then calculated from

$$\{y\} = [\Phi] \{z\},$$

$$\begin{bmatrix} y_1 \\ y_2 \\ y_3 \end{bmatrix} = \begin{bmatrix} 0.402 & 0.068 & 0.246 \\ 0.087 & 0.373 & -0.246 \\ 0.110 & -0.272 & -0.375 \end{bmatrix} \begin{bmatrix} z_1 \\ z_2 \\ z_3 \end{bmatrix}$$

or

$$y_1 = 10^{-3}[4.843 - 4.692 \cos 415t - 0.022 \cos 1034t - 0.130 \cos 1526t]$$
$$y_2 = 10^{-3}[1.004 - 1.015 \cos 415t - 0.119 \cos 1034t + 0.130 \cos 1526t]$$
$$y_3 = 10^{-3}[0.999 - 1.284 \cos 415t + 0.087 \cos 1034t + 0.198 \cos 1526t].$$

18.3   PROGRAM 14–DYNAMIC ANALYSIS OF PLANE TRUSSES (TRUSS)

In this section we present a computer program for the dynamic analysis of plane trusses. The listing of the program is given in the Appendix, the description of the principal variables used in the program is given in Table 18.1, and the listing of the input variables and required formats are shown in Table 18.2.

TABLE 18.1   Description of Principal Variables for Program 14.

Variable	Symbol in Text	Description
NE		Number of elements
ND		Number of degrees of freedom
NCR		Number of degrees of freedom to condense
NCM		Number of concentrated masses
LOC		Mass index:
		$\quad 0 \rightarrow$ lumped mass;
		$\quad 1 \rightarrow$ consistent mass
IFPR		Print intermediate values in JACOBI:
		$\quad 0 \rightarrow$ do not print,
		$\quad 1 \rightarrow$ print
E	E	Modulus of elasticity
LE		Element number
SL	L	Element length
AR	A	Cross-sectional area
SMA	$\overline{m}$	Distributed mass
TH	θ	Slope element
NC(L)		Element nodal coordinates ($L = 1, 4$)
MC(L)		Concentrated masses ($L = 1$, NCM)
JC(L)		Nodal coordinates for MC(L)
BK(I, J)	$[K]$	Element stiffness matrix
BM(I, J)	$[M]$	Element mass matrix
T(I, J)	$[T]$	Element transformation matrix
SK(I, J)	$[K_s]$	System stiffness matrix
SM(I, J)	$[M_s]$	System mass matrix

TABLE 18.2   Input Data and Formats for Program 14.

Format	Variables
(6I5, 2F10.0)	NE ND NCR NCM LOC IFPR E
(I10, 4F10.2, 4I2)	LE SL AR SMA TH (NC(L), $L = 1, 4$)
(8(I2, F8.2))	JC(L) CM(L) ($L = 1$, NCM)

TABLE 18.3 Data Input for Example 18.4.

						Input Data				
3	3	0	0	1	1	30000000.				
	1	60.00		10.	0.1	90.	4	4	0	2
	2	84.85		10.	0.1	135.	3	4	1	2
	3	60.00		10.	0.1	0.	4	4	3	4

TABLE 18.4 Computer Results for Example 18.4.

EIGENVALUES

0.17258D 06   0.10685D 07 0.23288D 07

EIGENVECTORS

0.40178D 00	0.68052D-01	-0.24594D 00
0.86811D-01	0.37346D 00	0.24516D 00
0.11034D 00	-0.27173D 00	0.37487D 00

Program 14 is organized along the same general outline of the programs which were previously presented for the beam, plane frame, and grid. The same set of subroutines associated with those earlier programs should also be available for calling by Program 14 to perform specific calculations.

Illustrative Example 18.4. Use Program 14 for the dynamic analysis of the truss shown in Fig. 18.3. Determine the natural frequencies, modal shapes, and the response to a constant force of $F_1 = 5000$ lb suddenly applied at nodal coordinate 1 as shown in the figure.

In the solution of this problem it is necessary for Program 14 to call subroutine JACOBI to solve the eigenproblem (natural frequencies and modal shapes) and subroutine MODAL to determine the response. The listing of the required input data for this example and the computer solution of the eigenproblem are given, respectively, in Tables 18.3 and 18.4. The additional data required by subroutine MODAL followed by the computer output giving the response at the three nodal coordinates of the structure are shown in Table 18.5.

18.4 STIFFNESS AND MASS MATRICES FOR SPACE TRUSSES

The stiffness matrix and the mass matrix for a space truss can be obtained as an extension of the corresponding matrices for the plane truss. Figure 18.4 shows

TABLE 18.5   Additional Data Input and Computer Results for Example 18.4.

0.0100	0.2000	2	0	0
0.00	5000.00	2.00	5000.00	
0.000	0.000	0.000		

RESPONSE FOR ELASTIC SYSTEM

TIME	DISPLACEMENTS AT NODAL COORDINATES		
t	y_1	y_2	y_3
0.000	0.0000D 00	0.0000D 00	0.0000D 00
0.010	0.7436D-02	0.1491D-02	0.1449D-02
0.020	0.6806D-02	0.1554D-02	0.1666D-02
0.030	0.1861D-03	-0.1437D-03	-0.2427D-03
0.040	0.7750D-02	0.1697D-02	0.1671D-02
0.050	0.6352D-02	0.1410D-02	0.1577D-02
0.060	0.3534D-03	-0.1872D-03	-0.3750D-03
0.070	0.7961D-02	0.1948D-02	0.2002D-02
0.080	0.6079D-02	0.1067D-02	0.1181D-02
0.090	0.2864D-03	0.7820D-04	-0.6767D-04
0.100	0.8452D-02	0.1855D-02	0.1852D-02
0.110	0.5507D-02	0.1020D-02	0.1213D-02
0.120	0.4694D-03	0.1602D-03	-0.5848D-04
0.130	0.8827D-02	0.1810D-02	0.1819D-02
0.140	0.4863D-02	0.1053D-02	0.1334D-02
0.150	0.8934D-03	0.5672D-04	-0.3334D-03
0.160	0.8872D-02	0.2025D-02	0.2219D-02
0.170	0.4540D-02	0.7877D-03	0.9550D-03
0.180	0.1070D-02	0.2425D-03	-0.1494D-03
0.190	0.9075D-02	0.2019D-02	0.2303D-02

the nodal coordinates in the local system (unbarred) and in the global system (barred) for a member of a space truss. The local x axis is directed along the longitudinal axes of the member while the y and z axes are set to agree with the principal directions of the cross section of the member. The following matrices may then be written for a uniform member of a space truss as an extension of the stiffness, eq. (18.3), and the mass, eq. (18.9), matrices for a member of a plane truss.

Stiffness matrix:

$$[K] = \frac{AE}{L} \begin{bmatrix} 1 & 0 & 0 & -1 & 0 & 0 \\ 0 & 0 & 0 & 0 & 0 & 0 \\ 0 & 0 & 0 & 0 & 0 & 0 \\ -1 & 0 & 0 & 1 & 0 & 0 \\ 0 & 0 & 0 & 0 & 0 & 0 \\ 0 & 0 & 0 & 0 & 0 & 0 \end{bmatrix} \qquad (18.20)$$

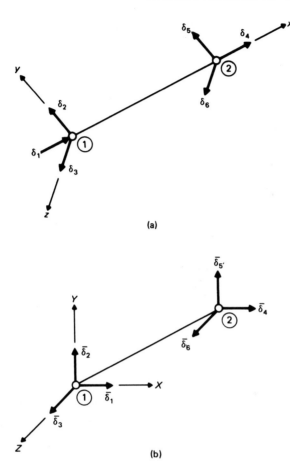

(a)

(b)

Fig. 18.4 Member of a space truss showing nodal coordinates. (a) In the local system (unbarred). (b) In the global system (barred).

Consistent mass matrix:

$$[M_c] = \frac{\overline{m}L}{6} \begin{bmatrix} 2 & 0 & 0 & 1 & 0 & 0 \\ 0 & 2 & 0 & 0 & 1 & 0 \\ 0 & 0 & 2 & 0 & 0 & 1 \\ 1 & 0 & 0 & 2 & 0 & 0 \\ 0 & 1 & 0 & 0 & 2 & 0 \\ 0 & 0 & 1 & 0 & 0 & 2 \end{bmatrix}. \tag{18.21}$$

Lumped mass matrix:

$$[M_L] = \frac{\overline{m}L}{2} \begin{bmatrix} 1 & 1 & 1 & 1 & 1 & 1 \end{bmatrix}. \tag{18.22}$$

The transformation matrix $[T_1]$ corresponding to three nodal coordinates at a joint is given by eq. (17.9). It is repeated here for convenience.

$$[T_1] = \begin{bmatrix} \cos xX & \cos xY & \cos xZ \\ \cos yX & \cos yY & \cos yZ \\ \cos zX & \cos zY & \cos zZ \end{bmatrix} \qquad (18.23)$$

in which, $\cos xY$ is the cosine of the angle between the local axis x and the global axis Y. The transformation matrix for the nodal coordinates at the two ends of a truss member is then given by

$$[T] = \begin{bmatrix} [T_1] & [0] \\ [0] & [T_1] \end{bmatrix} \qquad (18.24)$$

in which $[T_1]$ is given by eq. (18.23). The following transformations are then required to obtain the member stiffness matrix $[\overline{K}]$ and the member mass matrix $[\overline{M}]$ in reference to the global system of coordinates:

$$[\overline{K}] = [T]^T [K] [T] \qquad (18.25)$$

and

$$[\overline{M}] = [T]^T [M] [T] \qquad (18.26)$$

where $[K]$ and $[M]$ are the stiffness and mass matrices in local coordinates.

18.5   EQUATION OF MOTION FOR SPACE TRUSS

The dynamic equilibrium conditions at the nodes of the space truss result in the differential equations of motion which in matrix notation may be written as follows:

$$[M] \{\ddot{y}\} + [C] \{\dot{y}\} + [K] \{y\} = \{F(t)\} \qquad (18.27)$$

in which $\{y\}$, $\{\dot{y}\}$, and $\{\ddot{y}\}$ are, respectively, the displacement, velocity, and acceleration vectors at the nodal coordinates, $\{F(t)\}$ the vector of external nodal forces, and $[M]$, $[C]$, and $[K]$ the system mass, damping, and stiffness matrices.

In the stiffness method of analysis the system matrices in eq. (18.27) are obtained by appropriate superposition of the corresponding member matrices using the direct method as we have shown previously for the framed structures. As was discussed in the preceding chapters, the practical way of evaluating damping is to prescribe damping ratios relative to the critical damping for each mode. Consequently, when eq. (18.27) is solved using the modal superposition method, the specified modal damping ratios are introduced directly into the modal equations. In this case, there is no need for explicitly obtaining the system damping matrix $[C]$. However, this matrix is required when the solution of eq. (18.27) is sought by other methods of solution, such as the step-by-step integration method. In this case, the system damping matrix $[C]$ can be obtained from the specified modal damping ratios by any of the methods discussed in Chapter 12.

18.6 SUMMARY

The dynamic analysis of trusses by the stiffness matrix method was presented in this chapter. As in the case of framed structures, discussed in the preceding chapters, the stiffness and mass matrices for a member of a truss are developed. Though the bending moments at the ends of truss members are zero, the inertial force does produce flexural bending along the members in addition to the axial forces. The system matrices for a truss are assembled as explained for framed structures by the appropriate superposition of the element matrices.

The analysis of trusses is presented for both plane and space trusses, although the computer program (Program 14) described in this chapter is limited to the dynamic analysis of plane trusses.

PROBLEMS

The following problems are intended for hand calculation, though it is recommended that whenever possible solutions should also be obtained using Program 14 TRUSS.

18.1 For the plane truss shown in Fig. P18.1 determine the system stiffness and mass matrices corresponding to the two nodal coordinates indicated in the figure.

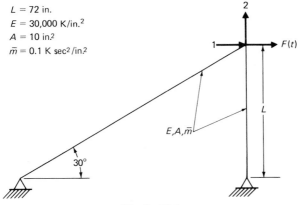

$L = 72$ in.
$E = 30,000$ K/in.^{2}
$A = 10$ in.^{2}
$\bar{m} = 0.1$ K sec^2/in.^{2}

30°

$E, A, \bar{m}$

$F(t)$

L

1

2

Fig. P. 18.1

18.2 Determine the natural frequencies and corresponding normal modes for truss shown in Fig. P18.1.

18.3 Determine the response of the truss shown in Fig. P18.1 when subjected to a force $F(t) = 10$ kip suddenly applied at nodal coordinate 1. Use the results of Problem 18.2 to obtain the modal equations. Neglect damping in the system.

18.4 Solve Problem 18.3 assuming 10% damping in all the modes.

18.5 Determine the maximum response of the truss shown in Fig. P18.1 when subjected to a rectangular pulse of magnitude $F_0 = 10$ kip and duration $t_d = 0.1$ sec. Use the appropriate response spectrum to determine the maximum modal response (Fig. 4.4). Neglect damping in the system.

18.6 Determine the dynamic response of the frame shown in Fig. P18.1 when subjected to a harmonic force $F(t) = 10 \sin 10t$ (kips) along nodal coordinate 1. Neglect damping in the system.

18.7 Repeat Problem 18.6 assuming that the damping in the system is proportional to the stiffness, $[C] = a_0[K]$ where $a_0 = 0.1$.

18.8 Determine the response of the truss shown in Fig. P18.8 when acted upon by the forces $F_1(t) = 10t$ and $F_2(t) = 5t^2$ during 1 sec. Neglect damping.

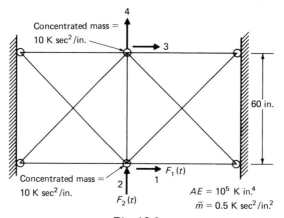

Fig. 18.8

18.9 Solve Problem 18.8 assuming 10% modal damping in all the modes.

19

Nonlinear response of multidegree-of-freedom systems

In Chapter 7 the nonlinear analysis of a single degree-of-freedom system using the step-by-step linear acceleration method was presented. The extension of this method, with a modification known as the Wilson-θ method, for the solution of structures modeled as multidegree-of-freedom systems is developed in this chapter. The modification introduced in the method by Wilson[1] serves to assure the numerical stability of the solution process regardless of the magnitude selected for the time step; for this reason, such a method is said to be *unconditionally stable*. On the other hand, without Wilson's modification, the step-by-step linear acceleration method is only conditionally stable and for numerical stability of the solution may require such an extremely small time step as to make the method impractical if not impossible. The development of the necessary algorithm for the linear and nonlinear

[1] Wilson, E. L., Farhoomand, I., and Bathe, K. J., "Nonlinear dynamic analysis of complex structures," *Int. J. Earthquake Engineering and Structural Dynamics*, Vol. 1, pp. 241–252, 1973.

multidegree-of-freedom systems by the step-by-step linear acceleration method parallels the presentation for the single degree-of-freedom system in Chapter 7.

19.1   INCREMENTAL EQUATIONS OF MOTION

The basic assumption of the Wilson-θ method is that the acceleration varies linearly over the time interval from t to $t + \theta \Delta t$ where $\theta \geqslant 1.0$. The value of the factor θ is determined to obtain optimum stability of the numerical process and accuracy of the solution. It has been shown by Wilson that, for $\theta \geqslant 1.38$, the method becomes unconditionally stable.

The equations expressing the incremental equilibrium conditions for a multidegree-of-freedom system can be derived as the matrix[2] equivalent of the incremental equation of motion for the single degree-of-freedom system, eq. (7.8). Thus taking the difference between dynamic equilibrium conditions defined at time t_i and at $t_i + \tau$, where $\tau = \theta \Delta t$, we obtain the incremental equations

$$M\hat{\Delta}\ddot{y}_i + C(\dot{y})\,\hat{\Delta}\dot{y}_i + K(y)\,\hat{\Delta}y_i = \hat{\Delta}F_i \tag{19.1}$$

in which the circumflex over Δ indicates that the increments are associated with the extended time step $\tau = \theta \Delta t$. Thus

$$\hat{\Delta}y_i = y(t_i + \tau) - y(t_i), \tag{19.2}$$

$$\hat{\Delta}\dot{y}_i = \dot{y}(t_i + \tau) - \dot{y}(t_i), \tag{19.3}$$

$$\hat{\Delta}\ddot{y}_i = \ddot{y}(t_i + \tau) - \ddot{y}(t_i), \tag{19.4}$$

and

$$\hat{\Delta}F_i = F(t_i + \tau) - F(t_i). \tag{19.5}$$

In writing eq. (19.1), we assumed, as explained in Chapter 7 for single degree-of-freedom systems, that the stiffness and damping are obtained for each time step as the initial values of the tangent to the corresponding curves as shown in Fig. 19.1 rather than the slope of the secant line which requires iteration. Hence the stiffness coefficient is defined as

$$k_{ij} = \frac{dF_{si}}{dy_j} \tag{19.6}$$

and the damping coefficient as

$$c_{ij} = \frac{dF_{Di}}{d\dot{y}_j} \tag{19.7}$$

---

[2]Matrices and vectors are denoted with boldface lettering throughout this chapter.

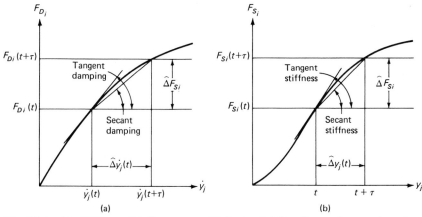

Fig. 19.1   Definition of influence coefficients. (a) Nonlinear viscous damping, c_{ij}. (b) Nonlinear stiffness, k_{ij}.

in which F_{si} and F_{Di} are, respectively, the elastic and damping forces at nodal coordinate i; y_j and $\dot{y}_j$ are, respectively, the displacement and velocity at nodal coordinate j.

19.2   THE WILSON-θ METHOD

The integration of the nonlinear equations of motion by the step-by-step linear acceleration method with the extended step introduced by Wilson is based, as it has already been mentioned, on the assumption that the acceleration may be represented by a linear function during the time step $\tau = \theta \Delta t$ as shown in Fig. 19.2. From this figure we can write the linear expression for the accelera-

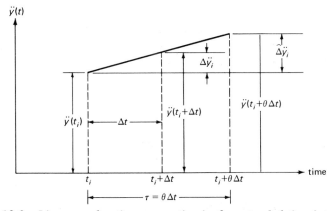

Fig. 19.2   Linear acceleration assumption in the extended time interval.

tion during the extended time step as

$$\ddot{y}(t) = \ddot{y}_i + \frac{\hat{\Delta}\ddot{y}_i}{\tau}(t - t_i) \tag{19.8}$$

in which $\hat{\Delta}\ddot{y}_i$ is given by eq. (19.4). Integrating eq. (19.8) twice yields

$$\dot{y}(t) = \dot{y}_i + \ddot{y}_i(t - t_i) + \frac{1}{2}\frac{\hat{\Delta}\ddot{y}_i}{\tau}(t - t_i)^2 \tag{19.9}$$

and

$$y(t) = y_i + \dot{y}(t - t_i) + \frac{1}{2}\ddot{y}_i(t - t_i)^2 + \frac{1}{6}\frac{\hat{\Delta}\ddot{y}_i}{\tau}(t - t_i)^3. \tag{19.10}$$

Evaluation of eqs. (19.9) and (19.10) at the end of the extended interval $t = t_i + \tau$ gives

$$\hat{\Delta}\dot{y}_i = \ddot{y}_i\tau + \frac{1}{2}\hat{\Delta}\ddot{y}_i\tau \tag{19.11}$$

and

$$\hat{\Delta}y_i = \dot{y}_i\tau + \frac{1}{2}\ddot{y}_i\tau^2 + \frac{1}{6}\hat{\Delta}\ddot{y}_i\tau^2 \tag{19.12}$$

in which $\hat{\Delta}y_i$ and $\hat{\Delta}\dot{y}_i$ are defined by eqs. (19.2) and (19.3), respectively. Now eq. (19.12) is solved for the incremental acceleration $\hat{\Delta}\ddot{y}_i$ and substituted in eq. (19.11). We obtain

$$\hat{\Delta}\ddot{y}_i = \frac{6}{\tau^2}\hat{\Delta}y_i - \frac{6}{\tau}\dot{y}_i - 3\ddot{y}_i \tag{19.13}$$

and

$$\hat{\Delta}\dot{y}_i = \frac{3}{\tau}\hat{\Delta}y_i - 3\dot{y}_i - \frac{\tau}{2}\ddot{y}_i. \tag{19.14}$$

Finally, substituting eqs. (19.13) and (19.14) into the incremental equation of motion, eq. (19.1), results in an equation for the incremental displacement $\hat{\Delta}y_i$ which may be conveniently written as

$$\overline{K}_i\hat{\Delta}y_i = \overline{\hat{\Delta}F}_i \tag{19.15}$$

where

$$\overline{K}_i = K_i + \frac{6}{\tau^2}M + \frac{3}{\tau}C_i \tag{19.16}$$

and

$$\overline{\hat{\Delta}F}_i = \hat{\Delta}F_i + M\left(\frac{6}{\tau}\dot{y}_i + 3\ddot{y}_i\right) + C_i\left(3\dot{y}_i + \frac{\tau}{2}\ddot{y}_i\right). \tag{19.17}$$

Equation (19.15) has the same form as the static incremental equilibrium equation and may be solved for the incremental displacements $\hat{\Delta y}_i$ by simply solving a system of linear equations.

To obtain the incremental accelerations $\hat{\Delta \ddot{y}}_i$ for the extended time interval, the value of $\hat{\Delta y}_i$ obtained from the solution of eq. (19.15) is substituted into eq. (19.13). The incremental acceleration $\Delta \ddot{y}_i$ for the normal time interval Δt is then obtained by a simple linear interpolation. Hence

$$\Delta \ddot{y} = \frac{\hat{\Delta \ddot{y}}}{\theta}.$$

(19.18)

To calculate the incremental velocity $\Delta \dot{y}_i$ and incremental displacement Δy_i corresponding to the normal interval Δt, use is made of eqs. (19.11) and (19.12) with the extended time interval parameter τ substituted for Δt, that is

$$\Delta \dot{y}_i = \ddot{y}_i \Delta t + \tfrac{1}{2} \Delta \ddot{y}_i \Delta t$$

(19.19)

and

$$\Delta y_i = \dot{y}_i \Delta t + \tfrac{1}{2} \ddot{y}_i \Delta t^2 + \tfrac{1}{6} \Delta \ddot{y}_i \Delta t^2.$$

(19.20)

Finally, the displacement y_{i+1} and velocity $\dot{y}_{i+1}$ at the end of the normal time interval are calculated by

$$y_{i+1} = y_i + \Delta y_i$$

(19.21)

and

$$\dot{y}_{i+1} = \dot{y}_i + \Delta \dot{y}_i.$$

(19.22)

As mentioned in Chapter 7 for the single degree-of-freedom system, the initial acceleration for the next step should be calculated from the condition of dynamic equilibrium at the time $t + \Delta t$; thus

$$\ddot{y}_{i+1} = M^{-1}[F_{i+1} - C_{i+1}\dot{y}_{i+1} - K_{i+1}y_{i+1}]$$

(19.23)

in which the products $C_{i+1}\dot{y}_{i+1}$ and $K_{i+1}y_{i+1}$ represent, respectively, the damping force and stiffness force vectors evaluated at the end of the time step $t_{i+1} = t_i + \Delta t$. Once the displacement, velocity, and acceleration vectors have been determined at time $t_{i+1} = t_i + \Delta t$, the outlined procedure is repeated to calculate these quantities at the next time step $t_{i+2} = t_{i+1} + \Delta t$ and the process is continued to any desired final time.

The step-by-step linear acceleration, as indicated in the discussion for the single degree-of-freedom system, involves two basic approximations: (1) the acceleration is assumed to vary linearly during the time step, and (2) the damping and stiffness characteristics of the structure are evaluated at the initiation of the time step and are assumed to remain constant during this time interval. The algorithm for the integration process of a linear system by the Wilson-θ method

is outlined in the next section. The application of this method to linear structures is then developed in the following section.

19.3   ALGORITHM FOR STEP-BY-STEP SOLUTION OF A LINEAR SYSTEM USING THE WILSON-θ INTEGRATION METHOD

19.3.1   Initialization

(1) Assemble the system stiffness matrix K, mass matrix M, and damping matrix C.
(2) Set initial values for displacement y_0, velocity $\dot{y}_0$, and forces F_0.
(3) Calculate initial acceleration $\ddot{y}_0$ from

$$M\ddot{y}_0 = F_0 - C\dot{y}_0 - Ky_0.$$

(4) Select a time step Δt, the factor θ (usually taken as 1.4), and calculate the
constants τ, a_1, a_2, a_3, and a_4 from the relations

$$\tau = \theta \Delta t; \qquad a_1 = \frac{3}{\tau}; \qquad a_2 = \frac{6}{\tau}; \qquad a_3 = \frac{\tau}{3}; \qquad a_4 = \frac{6}{\tau^2}.$$

(5) Form the effective stiffness matrix $\overline{K}$, namely

$$\overline{K} = K + a_4 M + a_1 C.$$

19.3.2   For Each Time Step

(1) Calculate by linear interpolation the incremental load $\hat{\Delta}F_i$ for the time
interval t_i to $t_i + \tau$, from the relation

$$\hat{\Delta}F_i = F_{i+1} + (F_{i+2} - F_{i+1})(\theta - 1) - F_i.$$

(2) Calculate the effective incremental load $\overline{\hat{\Delta}F_i}$ for the time interval t_i to $t_i + \tau$,
from the relation

$$\overline{\hat{\Delta}F_i} = \hat{\Delta}F_i + (a_2 M + 3C)\dot{y}_i + (3M + a_3 C)\ddot{y}_i.$$

(3) Solve for the incremental displacement $\hat{\Delta}y_i$ from

$$\overline{K}\hat{\Delta}y_i = \overline{\hat{\Delta}F_i}.$$

(4) Calculate the incremental acceleration for the extended time interval τ, from
the relation

$$\hat{\Delta}\ddot{y}_i = a_4 \hat{\Delta}y_i - a_2 \dot{y}_i - 3\ddot{y}_i.$$

(5) Calculate the incremental acceleration for the normal interval from

$$\Delta\ddot{y} = \frac{\hat{\Delta\ddot{y}}}{\theta}.$$

(6) Calculate the incremental velocity $\Delta\dot{y}_i$ and the incremental displacement Δy_i from time t_i to $t_i + \Delta t$ from the relations

$$\Delta\dot{y}_i = \ddot{y}_i\Delta t + \tfrac{1}{2}\,\Delta\ddot{y}_i\Delta t,$$
$$\Delta y_i = \dot{y}_i\Delta t + \tfrac{1}{2}\ddot{y}_i\Delta t^2 + \tfrac{1}{6}\Delta\ddot{y}_i\Delta t^2.$$

(7) Calculate the displacement and velocity at time $t_{i+1} = t_i + \Delta t$ using

$$y_{i+1} = y_i + \Delta y_i,$$
$$\dot{y}_{i+1} = \dot{y}_i + \Delta\dot{y}_i.$$

(8) Calculate the acceleration $\ddot{y}_{i+1}$ at time $t_{i+1} = t_i + \Delta t$ directly from the equilibrium equation of motion, namely

$$M\ddot{y}_{i+1} = F_{i+1} - C\dot{y}_{i+1} - Ky_{i+1}.$$

Illustrative Example 19.1.   Calculate the displacement response for a two-story shear building of Fig. 19.3 subjected to a suddenly applied force of 10 kip at the level of the second floor. Neglect damping and assume elastic behavior.
The equations of motion for this structure are:

$$\begin{bmatrix} 0.136 & 0 \\ 0 & 0.066 \end{bmatrix}\begin{bmatrix} \ddot{y}_1 \\ \ddot{y}_2 \end{bmatrix} + \begin{bmatrix} 75.0 & -44.3 \\ -44.3 & 44.3 \end{bmatrix}\begin{bmatrix} y_1 \\ y_2 \end{bmatrix} = \begin{bmatrix} 0 \\ 10 \end{bmatrix}$$

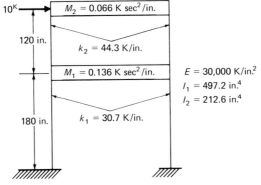

Fig. 19.3   Two-story shear building for Examples 19.1 and 19.2.

which, for free vibration, become

$$\begin{bmatrix} 0.136 & 0 \\ 0 & 0.066 \end{bmatrix} \begin{bmatrix} \ddot{y}_1 \\ \ddot{y}_2 \end{bmatrix} + \begin{bmatrix} 75.0 & -44.3 \\ -44.3 & 44.3 \end{bmatrix} \begin{bmatrix} y_1 \\ y_2 \end{bmatrix} = \begin{bmatrix} 0 \\ 0 \end{bmatrix}.$$

Substitution of $y_i = a_i \sin \omega t$ results in the eigenproblem

$$\begin{bmatrix} 75.0 - 0.136\omega^2 & -44.3 \\ -44.3 & 44.3 - 0.066\omega^2 \end{bmatrix} \begin{bmatrix} a_1 \\ a_2 \end{bmatrix} = \begin{bmatrix} 0 \\ 0 \end{bmatrix}$$

which requires a nontrivial solution

$$\begin{vmatrix} 75.0 - 0.136\omega^2 & -44.3 \\ -44.3 & 44.3 - 0.066\omega^2 \end{vmatrix} = 0.$$

Expansion of this determinant yields

$$\omega^4 - 1222.68\omega^2 + 151516 = 0$$

which has the roots

$$\omega_1^2 = 139.94 \quad \text{and} \quad \omega_2^2 = 1082.0.$$

Hence, the natural frequencies are

$$\omega_1 = 11.83 \text{ rad/sec}, \qquad \omega_2 = 32.90 \text{ rad/sec}$$

or

$$f_1 = 1.883 \text{ cps}, \qquad f_2 = 5.237 \text{ cps}$$

and the natural periods are

$$T_1 = 0.531 \text{ sec}, \qquad T_2 = 0.191 \text{ sec}.$$

The initial acceleration is calculated from

$$\begin{bmatrix} 0.136 & 0 \\ 0 & 0.066 \end{bmatrix} \begin{bmatrix} \ddot{y}_{10} \\ \ddot{y}_{20} \end{bmatrix} + \begin{bmatrix} 75.0 & -44.3 \\ -44.3 & 44.3 \end{bmatrix} \begin{bmatrix} 0 \\ 0 \end{bmatrix} = \begin{bmatrix} 0 \\ 10 \end{bmatrix}$$

giving

$$\ddot{y}_{10} = 0$$

$$\ddot{y}_{20} = 151.51 \text{ in/sec}^2.$$

If we select $\Delta t = 0.02$ and $\theta = 1.4, \tau = \theta \Delta t = 0.028$, and calculate the constants, we obtain

$$a_1 = \frac{3}{\tau} = 107.14, \qquad a_3 = \frac{\tau}{2} = 0.014,$$

$$a_2 = \frac{6}{\tau} = 214.28, \qquad a_4 = \frac{6}{\tau^2} = 7653.$$

The effective stiffness is then

$$\overline{K} = K + a_4 M + a_1 C \quad (C = 0 \text{ undamped system}),$$

$$\overline{K} = \begin{bmatrix} 75.0 & -44.3 \\ -44.3 & 44.3 \end{bmatrix} + 7653 \begin{bmatrix} 0.136 & 0 \\ 0 & 0.066 \end{bmatrix},$$

$$\overline{K} = \begin{bmatrix} 1115.8 & -44.3 \\ -44.3 & 549.4 \end{bmatrix},$$

and the effective force

$$\overline{\Delta F} = \hat{\Delta} F + (a_2 M + 3C)\dot{y} + (3M + a_3 C)\ddot{y}$$

$$\overline{\Delta F} = \begin{bmatrix} 0 \\ 0 \end{bmatrix} + 214.28 \begin{bmatrix} 0.136 & 0 \\ 0 & 0.066 \end{bmatrix} \begin{bmatrix} 0 \\ 0 \end{bmatrix} + 3 \begin{bmatrix} 0.136 & 0 \\ 0 & 0.066 \end{bmatrix} \begin{bmatrix} 0 \\ 151.51 \end{bmatrix},$$

$$\overline{\Delta F} = \begin{bmatrix} 0 \\ 30 \end{bmatrix}.$$

Solving for $\hat{\Delta}y$ from $\overline{K}\hat{\Delta}y = \overline{\Delta F}$ yields

$$\begin{bmatrix} 1115.8 & -44.3 \\ -44.3 & 549.4 \end{bmatrix} \begin{bmatrix} \hat{\Delta}y_1 \\ \hat{\Delta}y_2 \end{bmatrix} = \begin{bmatrix} 0 \\ 30 \end{bmatrix}, \qquad \hat{\Delta}y = \begin{bmatrix} 0.002175 \\ 0.054780 \end{bmatrix}.$$

Solving for $\hat{\Delta}\ddot{y}$ from eq. (19.13), we obtain

$$\hat{\Delta}\ddot{y} = 7653 \begin{bmatrix} 0.002175 \\ 0.054780 \end{bmatrix} - 214.28 \begin{bmatrix} 0 \\ 0 \end{bmatrix} - 3 \begin{bmatrix} 0 \\ 151.51 \end{bmatrix}, \qquad \hat{\Delta}\ddot{y} = \begin{bmatrix} 16.645 \\ -35.299 \end{bmatrix}.$$

Then

$$\Delta\ddot{y} = \frac{\hat{\Delta}\ddot{y}}{\theta} = \frac{1}{1.4} \begin{bmatrix} 16.647 \\ -35.299 \end{bmatrix} = \begin{bmatrix} 11.891 \\ -25.21 \end{bmatrix}.$$

From eq. (19.19), it follows that

$$\Delta\dot{y}_1 = \begin{bmatrix} 0 \\ 151.51 \end{bmatrix} (0.02) + \frac{0.02}{2} \begin{bmatrix} 11.891 \\ -25.21 \end{bmatrix} = \begin{bmatrix} 0.1189 \\ 2.7781 \end{bmatrix}.$$

From eq. (19.20),

$$\Delta y_1 = \begin{bmatrix} 0 \\ 0 \end{bmatrix} (0.02) + \frac{(0.02)^2}{2} \begin{bmatrix} 0 \\ 151.51 \end{bmatrix} + \frac{(0.02)^2}{6} \begin{bmatrix} 11.891 \\ -25.21 \end{bmatrix} = \begin{bmatrix} 0.0008 \\ 0.0286 \end{bmatrix}.$$

From eqs. (19.21) and (19.22),

$$\{y_2\} = \begin{bmatrix} 0 \\ 0 \end{bmatrix} + \begin{bmatrix} 0.0008 \\ 0.0286 \end{bmatrix} = \begin{bmatrix} 0.0008 \\ 0.0286 \end{bmatrix} \tag{a}$$

and

$$\{\dot{y}_2\} = \begin{bmatrix} 0 \\ 0 \end{bmatrix} + \begin{bmatrix} 0.1189 \\ 2.7781 \end{bmatrix} = \begin{bmatrix} 0.1189 \\ 2.7781 \end{bmatrix}. \tag{b}$$

From eq. (19.23),

$$\begin{bmatrix} 0.136 & 0 \\ 0 & 0.066 \end{bmatrix} \{\ddot{y}_2\} = \begin{bmatrix} 0 \\ 10 \end{bmatrix} - \begin{bmatrix} 75.0 & -44.3 \\ -44.3 & 44.3 \end{bmatrix} \begin{bmatrix} 0.0008 \\ 0.0286 \end{bmatrix},$$

which gives

$$\{\ddot{y}_2\} = \begin{bmatrix} 8.875 \\ 132.85 \end{bmatrix}. \tag{c}$$

The results given in eqs. (a), (b), and (c) for the displacement, velocity, and acceleration, respectively, at time $t_2 = t_1 + \Delta t$ complete a first cycle of the integration process. The continuation in determining the response for this structure is given in Example 19.2 with the use of the computer program described in the next section.

19.4 PROGRAM 15—TESTING SUBROUTINE STEPM

Subroutine STEPM performs the step-by-step integration of the equations of motion for a linear system using the linear acceleration method with the Wilson-θ modification. The main program makes available to this subroutine the stiffness matrix, the mass matrix, and the damping matrix of the system. Input data to subroutine STEPM include the value of the parameter θ, the time step Δt, the maximum time computation TMAX, and a listing of time-force values for each load applied at the nodal coordinates of the structure.

The statements in this subroutine follow the algorithm outlined in the pre-

TABLE 19.1 Variables and Symbols for Subroutine STEPM.

Variable	Symbol in Text	Description
SK(I, J)	$[K]$	System stiffness matrix
SM(I, J)	$[M]$	System mass matrix
SC(I, J)	$[C]$	System damping matrix
ND	N	Number of degrees of freedom
THETA	θ	Wilson-θ factor
DT	Δt	Time step of integration
TMAX		Maximum time of integration
NEQ(L)		Number of data points for excitation at nodal coordinates ($L = 1$, ND)
TC(I), P(I)	$t_i, F_i(t)$	Time-force values

TABLE 19.2 Input Data and Formats for Program 15.

Format	Variables	Program
(I10)	ND	
(8F10.0)	SK(I, J) (read by rows)	Testing
(8F10.0)	SM(I, J) (read by rows)	
(8F10.0)	SC(I, J) (read by rows)	
(3F10.3, 24I2)	THETA DT TMAX NEQ(L)	Subroutine
(8F10.2)	TC(I) P(I) (I = 1, NEQ(L)) (one card and continuation cards as need to input the time-force values for every external load)	STEPM

ceding section. A list of the principal variables and symbols used in the program is given in Table 19.1. The corresponding algebraic symbols that were used in the equations are also given. Input data cards and corresponding formats are indicated in Table 19.2.

The program performs a linear interpolation between the excitation data points, which result in a table giving the magnitude of the applied forces at each nodal coordinate calculated at increments of time equal to the time step Δt. Comments are interspersed throughout the program to facilitate the understanding of what is being done in each segment of the program. The output consists of a table giving the response for each nodal coordinate in terms of displacement, velocity, and acceleration at increments of time Δt up to the maximum time specified, TMAX. The computer program for subroutine STEPM and a main testing program are listed in the Appendix as Program 15.

Illustrative Example 19.2. Use the testing program associated with subroutine STEPM to determine the response of the two-story shear building shown in Fig. 19.3. The first cycle of the integration process for this structure has been hand calculated in Example 19.1. The necessary input data for this example are given in Table 19.3 and the corresponding output results in Table 19.4. It

TABLE 19.3 Input Data Listing for Example 19.2.

2			
75.	-44.3		
-44.3	44.3		
0.136	0.		
0.	0.066		
0.	0.		
0.	0.		
1.400	0.020	0.400 0 2	
0.000	10.000	0.400	10.000

TABLE 19.4   Computer Results for Example 19.2.

THE RESPONSE IS CORD.	TIME	DISPL.	VELOC.	ACC.
1	0.020	0.001	0.119	8.886
2	0.020	0.029	2.778	132.836
1	0.040	0.007	0.533	31.343
2	0.040	0.107	4.911	83.897
1	0.060	0.025	1.405	57.243
2	0.060	0.218	5.978	21.894
1	0.080	0.066	2.685	74.236
2	0.080	0.339	5.923	-31.939
1	0.100	0.134	4.111	72.678
2	0.100	0.449	5.064	-60.432
1	0.120	0.229	5.297	49.555
2	0.120	0.539	3.946	-56.959
1	0.140	0.342	5.875	9.871
2	0.140	0.609	3.124	-27.793
1	0.160	0.458	5.635	-35.066
2	0.160	0.668	2.938	10.584
1	0.180	0.562	4.601	-71.805
2	0.180	0.731	3.377	38.161
1	0.200	0.639	3.031	-89.763
2	0.200	0.806	4.090	39.457
1	0.220	0.682	1.321	-85.142
2	0.220	0.893	4.532	9.826
1	0.240	0.693	-0.133	-62.247
2	0.240	0.982	4.188	-42.495
1	0.260	0.680	-1.077	-31.685
2	0.260	1.054	2.795	-99.378
1	0.280	0.653	-1.483	-6.153
2	0.280	1.087	0.465	-139.725
1	0.300	0.623	-1.535	4.540
2	0.300	1.069	-2.333	-147.611
1	0.320	0.592	-1.546	-2.702
2	0.320	0.995	-4.932	-118.509
1	0.340	0.560	-1.814	-23.111
2	0.340	0.876	-6.702	-61.154
1	0.360	0.517	-2.490	-45.996
2	0.360	0.735	-7.273	5.693
1	0.380	0.458	-3.505	-59.051
2	0.380	0.593	-6.666	60.333
1	0.400	0.376	-4.602	-54.135
2	0.400	0.471	-5.667	88.042
1	0.420	0.275	-5.446	-31.829
2	0.420	0.368	-4.996	-62.325

may be seen from this last table that as expected the computer results for the first cycle check with the hand calculations in Example 19.1.

19.5 PROGRAM 16—SEISMIC RESPONSE OF SHEAR BUILDINGS (SRSB)

A computer program for the analysis of a multidegree-of-freedom shear building with elastoplastic behavior, linear viscous damping, and subjected to an arbitrary acceleration at the foundation, is presented in this section. This program may be conceived as a combination of three computer programs already presented: (1) the elastoplastic single degree-of-freedom system (Program 4) of Chapter 7; (2) the seismic response of elastic shear buildings using the modal superposition method (Program 7) of Chapter 11; and (3) the subroutine STEPM using the Wilson-θ integration method for linear systems in this chapter (Program 15).

The listing of Program 16 is given in the Appendix. The program calls subroutine JACOBI (Chapter 10) to solve the eigenproblem of the system in the linear range and then calls subroutine DAMP (Chapter 12) to determine from specified modal damping ratios, the damping matrix of the system. A listing of the principal variables used in the program is given in Table 19.5. Input data cards and corresponding formats are indicated in Table 19.6.

TABLE 19.5   Variables and Symbols for Program 16.

Variable	Symbol in Text	Description
SK(I, J)	$[K]$	Stiffness matrix
SM(I, J)	$[M]$	Mass matrix
SC(I, J)	$[C]$	Damping matrix
THETA	θ	Wilson-θ factor
DT	Δt	Time step
E	E	Modulus of elasticity
GR	g	Acceleration of gravity
TMAX		Maximum time of calculation
NEQ	NT	Number of data points for the excitation
ND	N	Number of degrees of freedom
IFPR		Printing index of subroutine JACOBI: 1 → print eigenvalues during iteration, 0 → do not print
SI	I	Moment of inertia of story columns
SL	L	Height of story
SM(I, I)	M	Mass at floor level
PM	M_p	Plastic moment of story
TC(I), P(I)	t_i, F_i	Time-acceleration values (acceleration in g's)
XIS(I)	ξ_i	Modal damping ratios

TABLE 19.6   Input Data and Formats for Program 16.

Format	Variables
(2F10.2, 3F10.0, 3I5)	THETA DT E GR TMAX NEQ ND IFPR
(8F10.0)	SI SL SM(I, I) PM (one card per degree of freedom)
(8F10.2)	TC(L) P(L) (L = 1, NEQ)
(8F10.3)	XIS(L) (L = 1, ND)

Illustrative Example 19.3.   Use Program 16 to determine the response of the two-story shear building of Example 19.2, subjected to a constant acceleration of 0.28 g applied suddenly at the foundation. The plastic moment for the columns on the first or second story is M_p = 254,942 lb · in.

The listing of the input data followed by the computer results (up to t = 0.02 sec) is given in Table 19.7

TABLE 19.7   Input Data and Computer Results for Example 19.3.

Input Data

1.400	0.00130000000.000		386.000	0.047	2	2	0
497.20	180.00	136.00	254942.
212.60	120.00	66.00	254942.

EIGENVALUES

0.81323D-01   0.92402D-01

EIGENVECTORS

0.64370E-01   -0.56652E-01
0.81323E-01   0.92402E-01

THE DAMPING MATRIX IS

0.0000D 00   0.0000D 00
0.0000D 00   0.0000D 00
0.000   0.280   0.047   0.280

THE RESPONSE IS				
CORD.	TIME	DISPL.	VELOC.	ACC.
1	0.001	-0.0001	-0.1081	-108.0678
2	0.001	-0.0001	-0.1081	-108.0800
1	0.002	-0.0002	-0.2161	-108.0312
2	0.002	-0.0002	-0.2162	-108.0800
1	0.003	-0.0005	-0.3241	-107.9703
2	0.003	-0.0005	-0.3242	-108.0799
1	0.004	-0.0009	-0.4320	-107.8850

TABLE 19.7   (*Continued*)

CORD.	TIME	DISPL.	VELOC.	ACC.
2	0.004	-0.0009	-0.4323	-108.0798
1	0.005	-0.0014	-0.5399	-107.7755
2	0.005	-0.0014	-0.5404	-108.0795
1	0.006	-0.0019	-0.6476	-107.6417
2	0.006	-0.0019	-0.6485	-108.0791
1	0.007	-0.0026	-0.7551	-107.4838
2	0.007	-0.0026	-0.7566	-108.0783
1	0.008	-0.0035	-0.8625	-107.3019
2	0.008	-0.0035	-0.8646	-108.0771
1	0.009	-0.0044	-0.9697	-107.0959
2	0.009	-0.0044	-0.9727	-108.0754
1	0.010	-0.0054	-1.0767	-106.8662
2	0.010	-0.0054	-1.0808	-108.0731
1	0.011	-0.0065	-1.1834	-106.6127
2	0.011	-0.0065	-1.1889	-108.0699
1	0.012	-0.0078	-1.2899	-106.3356
2	0.012	-0.0078	-1.2969	-108.0657
1	0.013	-0.0091	-1.3961	-106.0351
2	0.013	-0.0091	-1.4050	-108.0604
1	0.014	-0.0106	-1.5020	-105.7113
2	0.014	-0.0106	-1.5130	-108.0537
1	0.015	-0.0121	-1.6075	-105.3645
2	0.015	-0.0122	-1.6211	-108.0455
1	0.016	-0.0138	-1.7127	-104.9947
2	0.016	-0.0138	-1.7291	-108.0354
1	0.017	-0.0155	-1.8175	-104.6023
2	0.017	-0.0156	-1.8372	-108.0233
1	0.018	-0.0174	-1.9219	-104.1873
2	0.018	-0.0175	-1.9452	-108.0089
1	0.019	-0.0194	-2.0258	-103.7501
2	0.019	-0.0195	-2.0532	-107.9919
1	0.020	-0.0215	-2.1294	-103.2909
2	0.020	-0.0216	-2.1612	-107.9721

19.6   ELASTOPLASTIC BEHAVIOR OF FRAMED STRUCTURES

The dynamic analysis of frames having linear elastic behavior was presented in Chapter 15. To extend this analysis to frames whose members may be strained beyond the yield point of the material, it is necessary to develop the member stiffness matrix for the assumed elastoplastic behavior. The analysis is then carried out by a step-by-step numerical integration of the differential equations of motion. Within each short time interval Δt, the structure is assumed to behave in a linear elastic manner, but the elastic properties of the structure are changed from one interval to another as dictated by the response. Consequently, the nonlinear response is obtained as a sequence of linear responses of different elastic systems. For each successive interval, the stiffness of the structure is evaluated based on the moments in the members at the beginning of the time increment.

The changes in displacements of the linear system are computed by integration of the differential equations of motion over the finite interval and the total displacements by addition of the incremental displacement to the displacements calculated in the previous time step. The incremental displacements are also used to calculate the increment in member end forces and moments from the member stiffness equation. The magnitude of these end moments relative to the yield conditions (plastic moments) determine the characteristics of the stiffness matrix to be used in the next time step.

19.7 MEMBER STIFFNESS MATRIX

If only bending deformations are considered, the moment-displacement (rotational displacement) relationship for a uniform beam segment (Fig. 19.4) with elastic behavior (no hinges) is given by eq. (14.20). This equation may be written in incremental quantities as follows:

$$\begin{bmatrix} \Delta P_1 \\ \Delta P_2 \\ \Delta P_3 \\ \Delta P_4 \end{bmatrix} = \frac{2EI}{L^3} \begin{bmatrix} 6 & 3L & -6 & 3L \\ 3L & 2L^2 & -3L & L^2 \\ -6 & -3L & 6 & -3L \\ 3L & L^2 & -3L & 2L^2 \end{bmatrix} \begin{bmatrix} \Delta\delta_1 \\ \Delta\delta_2 \\ \Delta\delta_3 \\ \Delta\delta_4 \end{bmatrix} \qquad (19.24)$$

in which ΔP_i and $\Delta \delta_i$ are, respectively, the incremental forces and the incremental displacements at the nodal coordinates of the beam segment. When the moment at one end of the beam reaches the value of the plastic moment M_p, a hinge is formed at that end. Under the assumption of an elastoplastic relation between the bending moment and the angular displacement as depicted in

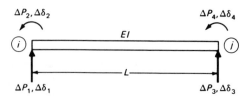

Fig. 19.4   Beam segment indicating incremental end forces and corresponding incremental displacements.

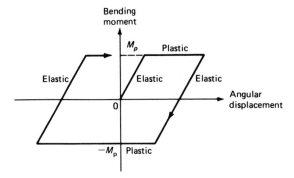

Fig. 19.5   Elastoplastic relation between bending moment and angular displacement at a section of a beam.

Fig. 19.5, the section that has been transformed into a hinge cannot support a moment higher than the plastic moment M_p but it may continue to deform plastically at a constant moment M_p. The relationship reverses to an elastic behavior when the angular displacement begins to decrease as shown in Fig. 19.5. We note the complete similarity for the behavior between an elastoplastic spring (Fig. 7.4) in a single degree-of-freedom system and an elastoplastic section of a beam (Fig. 19.5).

The stiffness matrix for a beam segment with a hinge at one end (Fig. 19.6) may be obtained by application of eq. (14.16) which is repeated here for convenience, namely

$$k_{ij} = \int_0^L EI\psi_i''(x)\,\psi_j''(x)\,dx, \tag{19.25}$$

where $\psi_i(x)$ and $\psi_j(x)$ are displacement functions. For a uniform beam in which the formation of the plastic hinge takes place at end $\textcircled{i}$ as shown in Fig. 19.6, the deflection functions corresponding to unit displacement at one of the nodal

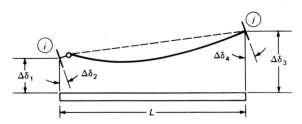

Fig. 19.6   Beam geometry with a plastic hinge at joint $\textcircled{i}$.

coordinates are

$$\psi_1(x) = 1 - \frac{3x}{2L} + \frac{x^3}{2L^3} \qquad (19.26a)$$

$$\psi_2(x) = 0 \qquad (19.26b)$$

$$\psi_3(x) = \frac{3x}{2L} - \frac{x^3}{2L^3} \qquad (19.26c)$$

$$\psi_4(x) = -\frac{x}{2} + \frac{x^3}{2L^2}. \qquad (19.26d)$$

For example, to calculate, k_{11}, we substitute the second derivative $\psi_1''(x)$ from eq. (19.26a) into eq. (19.25) and obtain

$$k_{11} = EI \int_0^L \left(\frac{3x}{L^3}\right)^2 dx = \frac{3EI}{L^3}. \qquad (19.27)$$

Similarly, all the other stiffness coefficients for the case in which the formation of the plastic hinge takes place at end ⓘ of a beam segment are determined using eq. (19.25) and the deflection functions given by eqs. (19.26). The resulting stiffness equation in increment form is

$$\begin{bmatrix} \Delta P_1 \\ \Delta P_2 \\ \Delta P_3 \\ \Delta P_4 \end{bmatrix} = \frac{EI}{L^3} \begin{bmatrix} 3 & 0 & -3 & 3L \\ 0 & 0 & 0 & 0 \\ -3 & 0 & 3 & -3L \\ 3L & 0 & -3L & 3L^2 \end{bmatrix} \begin{bmatrix} \Delta\delta_1 \\ \Delta\delta_2 \\ \Delta\delta_3 \\ \Delta\delta_4 \end{bmatrix}. \qquad (19.28)$$

It should be pointed out that $\Delta\delta_2$ is the incremental rotation of joint ⓘ at the frame and not the increase in rotation at end ⓘ of the beam under consideration. The incremental rotation of the plastic hinge is given by the difference between $\Delta\delta_2$ and the increase in rotation of the end ⓘ of the member. Hinge rotation may be calculated for the various cases with formulas developed in the next section. Analogous to eq. (19.28), the following equation gives the relationship between incremental forces and incremental displacements for a uniform beam with a hinge at end ⓙ, namely

$$\begin{bmatrix} \Delta P_1 \\ \Delta P_2 \\ \Delta P_3 \\ \Delta P_4 \end{bmatrix} = \frac{EI}{L^3} \begin{bmatrix} 3 & 3L & -3 & 0 \\ 3L & 3L^2 & -3L & 0 \\ -3 & -3L & 3 & 0 \\ 0 & 0 & 0 & 0 \end{bmatrix} \begin{bmatrix} \Delta\delta_1 \\ \Delta\delta_2 \\ \Delta\delta_3 \\ \Delta\delta_4 \end{bmatrix}. \qquad (19.29)$$

Finally, if hinges are formed at both ends of the beam, the stiffness matrix becomes null. Hence in this case the stiffness equation is

$$\begin{bmatrix} \Delta P_1 \\ \Delta P_2 \\ \Delta P_3 \\ \Delta P_4 \end{bmatrix} = \begin{bmatrix} 0 & 0 & 0 & 0 \\ 0 & 0 & 0 & 0 \\ 0 & 0 & 0 & 0 \\ 0 & 0 & 0 & 0 \end{bmatrix} \begin{bmatrix} \Delta \delta_1 \\ \Delta \delta_2 \\ \Delta \delta_3 \\ \Delta \delta_4 \end{bmatrix}. \tag{19.30}$$

19.8   ROTATION OF PLASTIC HINGES

In the solution process, at the end of each step interval it is necessary to calculate the end moments of every beam segment to check whether or not a plastic hinge has been formed. The calculation is done using the element incremental moment-displacement relationship. It is also necessary to check if the plastic deformation associated with a hinge is compatible with the sign of the moment. The plastic hinge is free to rotate in one direction only, and in the other direction the section returns to an elastic behavior. The assumed moment rotation characteristics of the member are of the type illustrated in Fig. 19.5. The conditions implied by this model are: (1) the moment cannot exceed the plastic moment; (2) if the moment is less than the plastic moment, the hinge cannot rotate; (3) if the moment is equal to the plastic moment, then the hinge may rotate in the direction consistent with the sign of the moment; and (4) if the hinge starts to rotate in a direction inconsistent with the sign of the moment, the hinge is removed.

The incremental rotation of a plastic hinge as already mentioned, is given, by the difference between the incremental joint rotation of the frame and the increase in rotation of the end of the member at that joint. For example, with a hinge at end ⓘ only (Fig. 19.6), the incremental joint rotation is $\Delta \delta_2$ and the increase in rotation of this end due to rotation $\Delta \delta_4$ is $-\Delta \delta_4/2$ and due to the displacements $\Delta \delta_1$ and $\Delta \delta_3$ is $1.5(\Delta \delta_3 - \Delta \delta_1)/L$. Hence the increment in rotation $\Delta \rho_i$ of a hinge formed at end ⓘ is given by

$$\Delta \rho_i = \Delta \delta_2 + \frac{1}{2} \Delta \delta_4 - 1.5 \frac{\Delta \delta_3 - \Delta \delta_1}{L}. \tag{19.31}$$

Similarly, with a hinge formed at end ⓙ only (Fig. 19.7), the increment in rotation of this hinge is given by

$$\Delta \rho_j = \Delta \delta_4 + \frac{1}{2} \Delta \delta_2 - 1.5 \frac{\Delta \delta_3 - \Delta \delta_1}{L}. \tag{19.32}$$

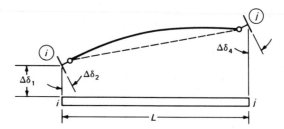

Fig. 19.7   Beam geometry with a plastic hinge at joint ⓙ.

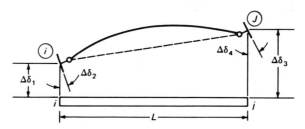

Fig. 19.8   Beam geometry with plastic hinges at both ends.

Finally, with hinges formed at both ends of a beam segment (Fig. 19.8), the rotations of the hinges are given by

$$\Delta\rho_i = \Delta\delta_1 - \frac{\Delta\delta_3 - \Delta\delta_1}{L}, \tag{19.33}$$

$$\Delta\rho_j = \Delta\delta_2 - \frac{\Delta\delta_3 - \Delta\delta_1}{L}. \tag{19.34}$$

19.9   CALCULATION OF MEMBER DUCTILITY RATIO

Nonlinear beam deformations are expressed in terms of the member ductility ratio, which is defined as the ratio of the maximum total end rotation of the member to the end rotation at the elastic limit. The elastic limit rotation is the angle developed when the member is subjected to antisymmetric yield moments M_y as shown in Fig. 19.9. In this case the relation between the end rotation and end moment is given by

$$\phi_y = \frac{M_y L}{6EI}. \tag{19.35}$$

The member ductility ratio μ is then defined as

$$\mu = \frac{\phi_y + \rho_{\max}}{\phi_y} \tag{19.36}$$

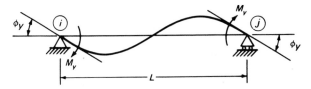

Fig. 19.9 Definition of yield rotation for beam segment.

which from eq. (19.35) becomes

$$\mu = 1 + \frac{6EI}{M_y L} \rho_{max} \qquad (19.37)$$

where ρ_{max} is the maximum rotation of the plastic hinge.

19.10 SUMMARY

The determination of the nonlinear response of multidegree-of-freedom structures requires the numerical integration of the governing equations of motion. There are many methods available for the solution of these equations. The step-by-step linear acceleration method with a modification known as the Wilson-θ method was presented in this chapter. This method is unconditionally stable, that is, numerical errors do not tend to accumulate during the integration process regardless of the magnitude selected for the time step. The basic assumption of the Wilson-θ method is that the acceleration varies linearly over the extended interval $\tau = \theta \Delta t$ in which $\theta \geqslant 1.38$ for unconditional stability.

Subroutine STEPM developed for numerical integration of linear systems is described and it is listed in the Appendix as Program 15. A computer program (SRSB) for the seismic analysis of multidegree-of-freedom shear buildings with nonlinear characteristics modeled as elastoplastic systems is also presented in this chapter and is listed in the Appendix as Program 16.

In the final sections of this chapter, stiffness matrices for elastoplastic behavior of framed structures are presented. Formulas to determine the plastic rotation of hinges and to calculate the corresponding ductility ratios are also presented in this chapter.

PROBLEMS

19.1 The stiffness and the mass matrices for a certain structure modeled as a two-degree-of-freedom system are

$$[K] = \begin{bmatrix} 100 & -50 \\ -50 & 50 \end{bmatrix} (k/\text{in}); \qquad [M] = \begin{bmatrix} 2 & 0 \\ 0 & 1 \end{bmatrix} (k \cdot \text{sec}^2/\text{in}).$$

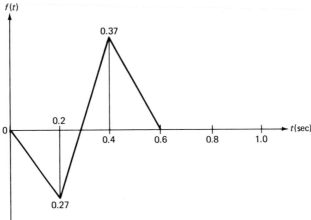

Fig. P19.1

Use Program 15 STEPM to determine the response when the structure is acted upon by the forces

$$\begin{bmatrix} F_1(t) \\ F_2(t) \end{bmatrix} = \begin{bmatrix} 772 \\ 386 \end{bmatrix} f(t) \text{ (kip)}$$

where $f(t)$ is given graphically in Fig. P19.1. Neglect damping in the system.

19.2 Solve Problem 19.1 considering that the damping present in the system results in the following damping matrix:

$$[C] = \begin{bmatrix} 10 & -5 \\ -5 & 5 \end{bmatrix} (k \cdot \text{sec/in}).$$

19.3 Use Program 15 STEPM to determine the response of the three-story

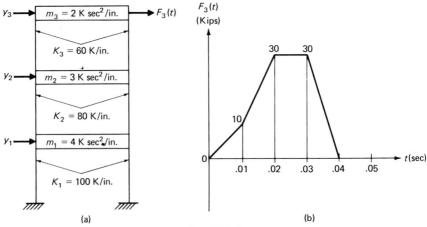

Fig. P19.3

shear building subjected to the force $F_3(t)$ as depicted in Fig. P19.3 applied at the level of the third floor. Neglect damping in the system.

19.4 Solve Problem 19.3 considering damping in the system of 10% in all the modes.

19.5 Use Program 16 SRSB to obtain the response in the elastic range for the structure of Problem 19.1 subjected to an acceleration at its foundation given by the function $f(t)$ shown in Fig. P19.1. (Note: To use this program in the elastic range it is necessary to give extremely large values to the plastic moments, say PM = 10^8 (K · in).)

19.6 Solve Problem 19.5 assigning values to the plastic moments of each story equal to half of the maximum elastic bending moments calculated from the elastic response. Neglect damping in the system.

19.7 Solve Problem 19.6 considering damping in the system as indicated in Problem 19.2.

19.8 Use Program 16 SRSB to obtain the response in the elastic range of the shear building shown in Fig. P19.3 when subjected to an acceleration of its foundation given by the function $f(t)$ depicted in Fig. P19.1. Neglect damping in the system.

19.9 Solve Problem 19.8 assuming values for the elastic moments of each story equal to half of the maximum elastic bending moments calculated from the elastic response.

Part **IV**

Structures modeled with distributed properties

20

Dynamics analysis of systems with distributed properties

The dynamic analysis of structures, modeled as lumped parameter systems with discrete coordinates was presented in Part I for single degree-of-freedom systems and in Parts II and III for multidegree-of-freedom systems. Modeling structures with discrete coordinates provides a practical approach for the analysis of structures subjected to dynamic loads. However, the results obtained from these discrete models can only give approximate solutions to the actual behavior of dynamic systems which have continuous distributed properties and, consequently, an infinite number of degrees of freedom.

The present chapter considers the dynamic theory of beams and rods having distributed mass and elasticity for which the governing equations of motion are partial differential equations. The integration of these equations is in general more complicated than the solution of ordinary differential equations governing discrete dynamic systems. Due to this mathematical complexity, the dynamic analysis of structures as continuous systems has limited use in practice. Nevertheless, the analysis, as continuous systems, of some simple structures provides, without much effort, results which are of great importance in assessing approximate methods based on discrete models.

20.1   FLEXURAL VIBRATION OF UNIFORM BEAMS

The treatment of beam flexure developed in this section is based on the simple bending theory as it is commonly used for engineering purposes. The method of analysis is known as the Bernoulli–Euler theory which assumes that a plane cross section of a beam remains plane during flexure.

Consider in Fig. 20.1 the free body diagram of a short segment of a beam. It is of length dx and is bounded by plane faces which are perpendicular to its axis. The forces and moments which act on the element are also shown in the figure: they are the shear forces V and $V + (\partial V/\partial x)$; the bending moments M and $M + (\partial M/\partial x)$; the lateral load pdx; and the inertia force $(mdx)\partial^2 y/dt^2$. In this notation m is the mass per unit length and $p = p(x, t)$ is the load per unit length. Partial derivatives are used to express acceleration and variations of shear and moment because these quantities are functions of two variables, position x along the beam and time t. If the deflection of the beam is small, as the theory presupposes, the inclination of the beam segment from the unloaded position is also small. Under these conditions, the equation of motion perpendicular to the x axis of the deflected beam obtained by equating to zero the sum of the forces in the free body diagram of Fig. 20.1(b) is

$$V - \left(V + \frac{\partial V}{\partial x}\, dx\right) + p(x, t)\, dx - mdx\, \frac{\partial^2 y}{\partial t^2} = 0$$

which, upon simplification, becomes

$$\frac{\partial V}{\partial x} + m\, \frac{\partial^2 y}{\partial t^2} = p(x, t). \qquad (20.1)$$

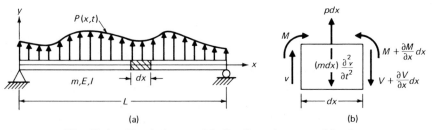

Fig. 20.1   Simple beam with distributed mass and load.

From simple bending theory, we have the relations

$$M = EI \frac{\partial^2 y}{\partial x^2} \tag{20.2}$$

and

$$V = \frac{\partial M}{\partial x} \tag{20.3}$$

where E is Young's modulus of elasticity and I is the moment of inertia of the cross-sectional area with respect to the neutral axis through the centroid. For a uniform beam, the combination of eqs. (20.1), (20.2), and (20.3) results in

$$V = EI \frac{\partial^3 y}{\partial x^3} \tag{20.4}$$

and

$$EI \frac{\partial^4 y}{\partial x^2} + m \frac{\partial^2 y}{\partial t^2} = p(x, t). \tag{20.5}$$

It is seen that eq. (20.5) is a partial differential equation of fourth order. It is an approximate equation. Only lateral flexural deflections were considered while the deflections due to shear forces and the inertial forces caused by the rotation of the cross section (rotary inertia) were neglected. The inclusion of shear deformations and rotary intertia in the differential equation of motion considerably increases its complexity. The equation taking into consideration shear deformation and rotary inertia is known as Timoshenko's equation. The differential equation, eq. (20.5), also does not include the flexural effects due to the presence of forces which may be applied axially to the beam. The axial effects will be discussed in Chapter 21.

20.2 SOLUTION OF THE EQUATION OF MOTION IN FREE VIBRATION

For free vibration ($p(x, t) = 0$), eq. (20.5) reduces to the homogeneous differential equation

$$EI \frac{\partial^4 y}{\partial x^4} + m \frac{\partial^2 y}{\partial t^2} = 0. \tag{20.6}$$

The solution of eq. (20.6) can be found by the method of separation of variables. In this method, it is assumed that the solution may be expressed as the product of a function of position $\Phi(x)$ and a function of time $f(t)$, that is,

$$y(x, t) = \Phi(x)f(t). \tag{20.7}$$

The substitution of eq. (20.7) in the differential equation, eq. (20.6), leads to

$$EIf(t) \frac{d^4\Phi(x)}{dx^4} + m\Phi(x) \frac{d^2f(t)}{dt^2} = 0. \qquad (20.8)$$

This last equation may be written as

$$\frac{EI}{m} \frac{\Phi^{IV}(x)}{\Phi(x)} = -\frac{\ddot{f}(t)}{f(t)}. \qquad (20.9)$$

In this notation Roman indices indicate derivatives with respect to x and over-dots indicate derivatives with respect to time. Since the left-hand side of eq. (20.9) is a function only of x while the right-hand side is a function only of t, each side of the equation must equal the same constant value; otherwise, the identity of eq. (20.9) cannot exist. We designate the constant by ω^2 which equated separately to each side of eq. (20.9) results in the two following differential equations:

$$\Phi^{IV}(x) - a^4\Phi(x) = 0 \qquad (20.10)$$

and

$$\ddot{f}(t) + \omega^2 f(t) = 0 \qquad (20.11)$$

where

$$a^4 = \frac{m\omega^2}{EI}. \qquad (20.12)$$

It is particularly convenient to solve eq. (20.12) for ω and to use the following notation, namely

$$\omega = C\sqrt{\frac{EI}{mL^4}} \qquad (20.13)$$

in which $C = (aL)^2$.

Equation (20.11) is the familiar free-vibration equation for the undamped single degree-of-freedom system and its solution from eq. (1.17) is

$$f(t) = A \cos \omega t + B \sin \omega t \qquad (20.14)$$

where A and B are constants of integration. Equation (20.10) can be solved by letting

$$\Phi(x) = Ce^{sx}. \qquad (20.15)$$

The substitution of eq. (20.15) into eq. (20.10) results in

$$(s^4 - a^4)Ce^{sx} = 0$$

which, for a nontrivial solution, requires that

$$s^4 - a^4 = 0. \qquad (20.16)$$

The roots of eq. (20.16) are

$$s_1 = a, \qquad s_3 = ai,$$
$$s_2 = -a, \qquad s_4 = -ai. \qquad (20.17)$$

The substitution of each of these roots into eq. (20.15) provides a solution of eq. (20.10). The general solution is then given by the superposition of these four possible solutions, namely

$$\Phi(x) = C_1 e^{ax} + C_2 e^{-ax} + C_3 e^{iax} + C_4 e^{-iax} \qquad (20.18)$$

where C_1, C_2, C_3, and C_4 are constants of integration. The exponential functions in eq. (20.18) may be expressed in terms of trigonometric and hyperbolic functions by means of the relations

$$e^{\pm ax} = \cosh ax \pm \sinh ax$$
$$e^{\pm iax} = \cos ax \pm i \sin ax. \qquad (20.19)$$

Substitution of these relations into eq. (20.18) yields

$$\Phi(x) = A \sin ax + B \cos ax + C \sinh ax + D \cosh ax \qquad (20.20)$$

where A, B, C, D are new constants of integration. These four constants of integration define the shape and the amplitude of the beam in free vibration; they are evaluated by considering the boundary conditions at the ends of the beam as illustrated in the examples presented in the following section.

20.3 NATURAL FREQUENCIES AND MODE SHAPES FOR UNIFORM BEAMS

20.3.1 Both Ends Simply Supported

In this case the displacements and bending moments must be zero at both ends of the beam; hence the boundary conditions for the simply supported beams are

$$y(0, t) = 0, \qquad M(0, t) = 0,$$
$$y(L, t) = 0, \qquad M(L, t) = 0.$$

In view of eqs. (20.2) and (20.7), these boundary conditions imply the following conditions on the shape function $\Phi(x)$.

At $x = 0$,

$$\Phi(0) = 0, \qquad \Phi''(0) = 0. \qquad (20.21)$$

At $x = L$,

$$\Phi(L) = 0, \qquad \Phi''(L) = 0. \qquad (20.22)$$

The substitution of the first two of these boundary conditions into eq. (20.20) yields

$$\Phi(0) = A0 + B1 + C0 + D1 = 0,$$

$$\Phi''(0) = a^2(-A0 - B1 + C0 + D1) = 0$$

which reduce to

$$B + D = 0$$

$$-B + D = 0.$$

Hence

$$B = D = 0.$$

Similarly, substituting the last two boundary conditions into eq. (20.20) and setting $B = D = 0$ leads to

$$\Phi(L) = A \sin aL + C \sinh aL = 0$$

$$\Phi''(L) = a^2(-A \sin aL + C \sinh aL) = 0 \qquad (20.23)$$

which, when added, give

$$2C \sinh aL = 0.$$

From this last relation, $C = 0$ since the hyperbolic sine function cannot vanish except for a zero argument. Thus eqs. (20.23) reduce to

$$A \sin aL = 0. \qquad (20.24)$$

Excluding the trivial solution ($A = 0$), we obtain the frequency equation

$$\sin aL = 0 \qquad (20.25)$$

which will be satisfied for

$$a_n L = n\pi, \qquad n = 0, 1, 2 \dots. \qquad (20.26)$$

Substitution of the roots, eq. (20.26), into eq. (20.13) yields

$$\omega_n = n^2 \pi^2 \sqrt{\frac{EI}{mL^4}} \qquad (20.27)$$

where the subscript n serves to indicate the order of the natural frequencies.
Since $B = C = D = 0$, it follows that eq. (20.20) reduces to

$$\Phi_n(x) = A \sin \frac{n\pi x}{L}$$

or simply

$$\Phi_n(x) = \sin\frac{n\pi x}{L} \tag{20.28}$$

We note that in eq. (20.28) the constant A is absorbed by the other constants in the modal response given below by eq. (20.29).

From eq. (20.7) a modal shape or normal mode of vibration is given by

$$y_n(x, t) = \Phi_n(x)f_n(t)$$

or from eqs. (20.14) and (20.28) by

$$y_n(x, t) = \sin\frac{n\pi x}{L}\left[A_n\cos\omega_n t + B_n\sin\omega_n t\right]. \tag{20.29}$$

The general solution of the equation of motion in free vibration which satisfies the boundary conditions, eqs. (20.21) and (20.22), is the sum of all the normal modes of vibration, eq. (20.29), namely

$$y(x, t) = \sum_{n=1}^{\infty}\sin\frac{n\pi x}{L}\left[A_n\cos\omega_n t + B_n\sin\omega_n t\right]. \tag{20.30}$$

The constants A_n and B_n are determined, as usual, from the initial conditions. If at $t = 0$, the shape of the beam is given by

$$y(x, 0) = \rho(x)$$

and the velocity by

$$\frac{\partial y(x, 0)}{\partial t} = \psi(x)$$

for $0 \leqslant x \leqslant L$, it follows from eq. (20.30) that

$$\sum_{n=1}^{\infty}A_n\sin\frac{n\pi x}{L} = \rho(x)$$

and

$$\sum_{n=1}^{\infty}B_n\omega_n\sin\frac{n\pi x}{L} = \psi(x).$$

Therefore, as shown in Chapter 5, Fourier coefficients are expressed as

$$A_n = \frac{2}{L}\int_0^L\rho(x)\sin\frac{n\pi x}{L}\,dx$$

$$B_n = \frac{2}{L}\int_0^L\psi(x)\sin\frac{n\pi x}{L}\,dx. \tag{20.31}$$

TABLE 20.1   Natural Frequencies and Normal Modes for Simply Supported Beams.

Natural Frequencies			Normal Modes
$\omega_n = C_n \sqrt{\dfrac{EI}{mL^4}}$			$\Phi_n = \sin \dfrac{nx}{L}$
n	C_n	I_n^*	Shapes
1	π^2	$4/\pi$	
2	$4\pi^2$	0	
3	$9\pi^2$	$4/3\pi$	
4	$16\pi^2$	0	
5	$25\pi^2$	$4/5\pi$	

$*I_n = \int_0^L \Phi_n(x)\,dx / \int_0^L \Phi^2(x)\,dx.$

The first five values for the natural frequencies and normal modes for the simply supported beam are presented in Table 20.1.

20.3.2   Both Ends Free (Free Beam)

The boundary conditions for a beam with both ends free are as follows.
At $x = 0$,

$$M(0, t) = 0 \qquad \text{or} \qquad \Phi''(0) = 0,$$

$$V(0, t) = 0 \qquad \text{or} \qquad \Phi'''(0) = 0.$$

At $x = L$,

$$M(L, t) = 0 \quad \text{or} \quad \Phi''(L) = 0,$$
$$V(L, t) = 0 \quad \text{or} \quad \Phi'''(L) = 0. \tag{20.33}$$

The substitutions of these conditions in eq. (20.20) yield

$$\Phi''(0) = a^2(-B + D) = 0,$$
$$\Phi'''(0) = a^3(-A + C) = 0,$$

and

$$\Phi''(L) = a^2(-A \sin aL - B \cos aL + C \sinh aL + D \cosh aL) = 0$$
$$\Phi'''(L) = a^3(-A \cos aL + B \sin aL + C \cosh aL + D \sinh aL) = 0.$$

From the first two equations we obtain

$$D = B, \qquad C = A \tag{20.34}$$

which, substituted into the last two equations, result in

$$(\sinh aL - \sin aL)A + (\cosh aL - \cos aL)B = 0,$$
$$(\cosh aL - \cos aL)A + (\sinh aL + \sin aL)B = 0. \tag{20.35}$$

For nontrivial solution of eqs. (20.35), it is required that the determinant of the unknown coefficients A and B be equal to zero; hence

$$\begin{vmatrix} \sinh al - \sin aL & \cosh al - \cos aL \\ \cosh aL - \cos aL & \sinh aL - \sin aL \end{vmatrix} = 0. \tag{20.36}$$

The expansion of this determinant provides the frequency equation for the free beam, namely

$$\cos aL - \cosh aL - 1 = 0. \tag{20.37}$$

The first five natural frequencies which are obtained by substituting the roots of eq. (20.37) into eq. (20.13) are presented in Table 20.2. The corresponding normal modes are obtained by letting $A = 1$ (normal modes are determined only to a relative magnitude), substituting in eqs. (20.35) the roots of a_n of eq. (20.37), solving one of these equations for B, and finally introducing into eq. (20.20) the constants C, D from eq. (20.34) together with B. Performing these operations, we obtain

$$\Phi_n(x) = \cosh a_n x + \cos a_n x - \sigma_n(\sinh a_n x + \sin a_n x) \tag{20.38}$$

where

$$\sigma_n = \frac{\cosh a_n L - \cos a_n L}{\sinh a_n L - \sin a_n L}. \tag{20.39}$$

TABLE 20.2   Natural Frequencies and Normal Modes for Free Beams.

Natural Frequencies				Normal Modes

$$\Phi_n(x) = \cosh a_n x + \cos a_n x - \sigma_n(\sinh a_n x + \sin a_n L)$$

$$\omega_n = C_n \sqrt{\frac{EI}{mL^4}} \qquad \sigma_n = \frac{\cosh a_n x - \cos a_n L}{\sinh a_n x - \sin a_n L}$$

n	$C_n = (a_n L)^2$	σ_n	I_n^*	Shapes
1	22.3733	0.982502	0.8308	0.224L / 0.776L
2	61.6728	1.000777	0	0.132L / 0.868L / 0.500L
3	120.9034	0.999967	0.3640	0.094L / 0.644L / 0.356L / 0.906L
4	199.8594	1.000001	0	0.073L / 0.500L / 0.927L / 0.277L / 0.723L
5	298.5555	1.000000	0.2323	0.060L / 0.409L / 0.774L / 0.226L / 0.591L / 0.940L

$*I_n = \int_0^L \Phi_n(x)\, dx / \int_0^L \Phi_n^2(x)\, dx.$

20.3.3   Both Ends Fixed

The boundary conditions for a beam with both ends fixed are as follows.
    At $x = 0$,

$$y(0, t) = 0 \qquad \text{or} \qquad \Phi(0) = 0,$$
$$y'(0, t) = 0 \qquad \text{or} \qquad \Phi'(0) = 0. \qquad (20.40)$$

At $x = L$,

$$y(L, t) = 0 \qquad \text{or} \qquad \Phi(L) = 0,$$
$$y'(L, t) = 0 \qquad \text{or} \qquad \Phi'(L) = 0. \qquad (20.41)$$

The use of the boundary conditions, eqs. (20.40), into eq. (20.20) gives

$$A = C = 0$$

while conditions, eqs. (20.41), yield the homogeneous system

$$(\cos aL - \cosh aL)B + (\sin aL - \sinh aL)D = 0,$$

$$-(\sin aL + \sinh al)B + (\cos aL - \cosh aL)D = 0. \qquad (20.42)$$

Equating to zero the determinant of the coefficients of this system results in the frequency equation

$$\cos a_n L \cosh a_n L - 1 = 0. \qquad (20.43)$$

TABLE 20.3   Natural Frequencies and Normal Modes for Fixed Beams.

Natural Frequencies

Normal Modes

$$\Phi_n(x) = \cosh a_n x - \cos a_n x - \sigma_n(\sinh a_n x - \sin a_n x)$$

$$\omega_n = C_n \sqrt{\frac{EI}{mL^4}}$$

$$\sigma_n = \frac{\cos a_n L - \cosh a_n L}{\sin a_n L - \sinh a_n L}$$

n	$C_n = (a_n L)^2$	σ_n	I_n^*	Shape
1	22.3733	0.982502	0.8308	
2	61.6728	1.000777	0	
3	120.9034	0.999967	0.3640	
4	199.8594	1.000001	0	
5	298.5555	1.000000	0.2323	

$*I_n = \int_0^L \Phi_n(x)\, dx / \int_0^L \Phi_n^2(x)\, dx.$

From the first of eqs. (20.42), it follows that

$$D = -\frac{\cos aL - \cosh aL}{\sin aL - \sinh aL} B,$$

(20.44)

where B is arbitrary. To each value of the natural frequency

$$\omega_n = (a_n L)^2 \sqrt{\frac{EI}{mL^4}}$$

(20.45)

obtained by the substitution of the roots of eq. (20.43) into eq. (20.13), there corresponds a normal mode

$$\Phi_n(x) = \cosh a_n x - \cos a_n x - \sigma_n(\sinh a_n x - \sin a_n x),$$

(20.46)

$$\sigma_n = \frac{\cos a_n L - \cosh a_n L}{\sin a_n L - \sinh a_n L}.$$

(20.47)

The first five natural frequencies calculated from eqs. (20.43) and (20.45) and the corresponding normal modes obtained from eq. (20.46) are presented in Table 20.3

20.3.4   One End Fixed and the Other End Free (Cantilever Beam)

At the fixed end $(x = 0)$ of the cantilever beam, the deflection and the slope must be zero and at the free end $(x = L)$ the bending moment and the shear force must be zero. Hence the boundary conditions for this beam are as follows.
At $x = 0$,

$$y(0, t) = 0 \qquad \text{or} \qquad \Phi(0) = 0,$$

$$y'(0, t) = 0 \qquad \text{or} \qquad \Phi'(0) = 0.$$

(20.48)

At $x = L$,

$$M(L, t) = 0 \qquad \text{or} \qquad \Phi''(L) = 0,$$

$$V(L, t) = 0 \qquad \text{or} \qquad \Phi'''(L) = 0.$$

(20.49)

These boundary conditions when substituted into the shape equation, eq. (20.20), lead to the frequency equation

$$\cos a_n L \cosh a_n + 1 = 0.$$

(20.50)

To each root of eq. (20.50) corresponds a natural frequency

$$\omega_n = (a_n L)^2 \sqrt{\frac{EI}{mL^4}}$$

(20.51)

and a normal shape

$$\Phi_n(x) = (\cosh a_n x - \cos a_n x) - \sigma_n(\sinh a_n x - \sin a_n x), \qquad (20.52)$$

where

$$\sigma_n = \frac{\cos a_n L + \cosh a_n L}{\sin a_n L + \sinh a_n L} \qquad (20.53)$$

The first five natural frequencies and the corresponding mode shapes for cantilever beams are presented in Table 20.4.

TABLE 20.4   Natural Frequencies and Normal Modes for Cantilever Beams.

Natural Frequencies		Normal Modes			
		$\Phi_n = (\cosh a_n x - \cos a_n x) - \sigma_n(\sinh a_n x - \sin a_n x)$			
$\omega_n = C_n \sqrt{\dfrac{EI}{mL^4}}$		$\sigma = \dfrac{\cos a_n L + \cosh a_n L}{\sin a_n L + \sinh a_n L}$			
n	$C_n = (a_n L)^2$	σ_n	I_n^*	Shape	
1	15.4118	1.000777	0.8600		
2	49.9648	1.000001	0.0826	0.774 L	
3	104.2477	1.000000	0.03345	0.501L 0.868L	
4	178.2697	1.000000	0.0434	0.356L 0.644L 0.906L	
5	272.0309	1.000000	0.2076	0.279L 0.500L 0.723L 0.926L	

$^*I_n = \int_0^L \Phi_n(x)\, dx / \int_0^L \Phi_n^2(x)\, dx.$

20.3.5   One End Fixed and the Other Simply Supported

The boundary conditions for a beam with one end fixed and the other simply supported are as follows.

At $x = 0$,

$$y(0, t) = 0 \quad \text{or} \quad \Phi(0) = 0,$$

$$y'(0, t) = 0 \quad \text{or} \quad \Phi'(0) = 0. \tag{20.54}$$

At $x = L$,

$$y(L, t) = 0 \quad \text{or} \quad \Phi(L) = 0,$$

$$M(L, t) = 0 \quad \text{or} \quad \Phi''(L) = 0. \tag{20.55}$$

TABLE 20.5   Natural Frequencies and Normal Modes for Fixed-Simply Supported Beams.

Natural Frequencies				Normal Modes	
$\omega_n = C_n \sqrt{\dfrac{EI}{mL^4}}$			$\Phi(x) = \cosh a_n x - \cos a_n x + \sigma_n(\sinh a_n x - \sin a_n x)$ $\sigma_n = \dfrac{\cos a_n L - \cosh a_n L}{\sinh a_n L - \sinh a_n L}$		
n	$C_n = (a_n L)^2$	σ_n	I_n^*	Shape	
1	15.4118	1.000777	0.8600		
2	49.9648	1.000001	0.0826		
3	104.2477	1.000000	0.3345		
4	178.2697	1.000000	0.0434		
5	272.0309	1.000000	0.2076		

$*I_n = \int_0^L \Phi_n(x)\, dx / \int_0^L \Phi_n^2(x)\, dx.$

The substitution of these boundary conditions into the shape equation, eq. (20.20), results in the frequency equation

$$\tan a_n L - \tanh a_n L = 0. \tag{20.56}$$

To each root of this last equation corresponds a natural frequency

$$\omega_n = (a_n L)^2 \sqrt{\frac{EI}{mL^4}} \tag{20.57}$$

and a normal mode

$$\Phi_n(x) = (\cosh a_n x - \cos a_n x) + \sigma_n(\sinh a_n x - \sin a_n x) \tag{20.58}$$

where

$$\sigma_n = \frac{\cos a_n L - \cosh a_n L}{\sin a_n L - \sinh a_n L}. \tag{20.59}$$

The first five natural frequencies for the fixed simply supported beam and corresponding mode shapes are presented in Table 20.5.

20.4   ORTHOGONALITY CONDITION BETWEEN NORMAL MODES

The most important property of the normal modes is that of orthogonality. It is this property which makes possible the uncoupling of the equations of motion as it has previously been shown for discrete systems. The orthogonality property for continuous systems can be demonstrated in essentially the same way as for discrete parameter systems.

Consider in Fig. 20.2 a beam subjected to the inertial forces resulting from the

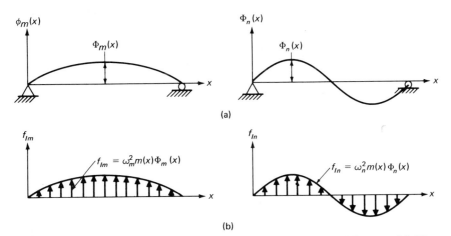

(a)

(b)

Fig. 20.2   Beam showing two modes of vibration and inertial forces. (a) Displacements. (b) Inertial forces.

vibrations of two different modes, $\Phi_m(x)$ and $\Phi_n(x)$. The deflection curves for these two modes and the corresponding inertial forces are depicted in the same figure. Betti's law is applied to these two deflection patterns. Accordingly, the work done by the inertial force, f_{In}, acting on the displacements of mode m is equal to the work of the inertial forces, f_{Im}, acting on the displacements of mode n, that is,

$$\int_0^L \Phi_m(x)\, f_{In}(x)\, dx = \int_0^L \Phi_n(x)\, f_{Im}(x)\, dx. \qquad (20.60)$$

The inertial force f_{In} per unit length along the beam is equal to the mass per unit length times the acceleration. Inasmuch as the vibratory motion in a normal mode is harmonic, the amplitude of the acceleration is given by $\omega_n^2 \Phi_n(x)$. Hence the inertial force per unit length along the beam for the nth mode is

$$f_{In} = \omega_n^2 m(x)\, \phi_n(x)$$

and for the mth mode

$$f_{Im} = \omega_m^2 m(x)\, \Phi_m(x). \qquad (20.61)$$

Substituting these expressions in eq. (20.60), we obtain

$$\omega_n^2 \int_0^L \Phi_m(x)\, m(x)\, \Phi_n(x)\, dx = \omega_m^2 \int_0^L \Phi_n(x)\, m(x)\, \Phi_m(x)\, dx$$

which may be written as

$$(\omega_n^2 - \omega_m^2) \int_0^L \Phi_m(x)\, \Phi_n(x)\, m(x)\, dx = 0. \qquad (20.62)$$

It follows that, for two different frequencies $\omega_n \neq \omega_m$, the normal modes must satisfy the relation

$$\int_0^L \Phi_m(x)\, \Phi_n(x)\, m(x)\, dx = 0 \qquad (20.63)$$

which is equivalent to the orthogonal condition between normal modes for discrete parameter systems, eq. (10.27).

20.5 FORCED VIBRATION OF BEAMS

For a uniform beam acted on by lateral forces $p(x, t)$, the equation of motion, eq. (20.5), may be written as

$$EI\, \frac{\partial^4 y}{\partial x^4} = p(x, t) - m\, \frac{\partial^2 y}{\partial t^2} \qquad (20.64)$$

in which $p(x, t)$ is the external load per unit length along the beam. We assume that the general solution of this equation may be expressed by the summation of the products of the normal modes $\Phi_n(x)$ multiplied by factors $z_n(t)$ which are to be determined. Hence

$$y(x, t) = \sum_{n=1}^{\infty} \Phi_n(x)\, z_n(t). \qquad (20.65)$$

The normal modes $\Phi_n(x)$ satisfy the differential equation, eq. (20.10), which by eq. (20.12) may be written as

$$EI\, \Phi_n^{IV}(x) = m\omega_n^2\, \Phi_n(x), \qquad n = 1, 2, 3, \ldots . \qquad (20.66)$$

The normal modes should also satisfy the specific force boundary conditions at the ends of the beam. Substitution of eq. (20.65) in eq. (20.64) gives

$$EI \sum_n \Phi_n^{IV}(x)\, z_n(t) = p(x, t) - m \sum_n \phi_n(x)\, \ddot{z}(t). \qquad (20.67)$$

In view of eq. (20.66), we can write eq. (20.67) as

$$\sum_n m\omega_n^2\, \Phi_n(x)\, z_n(t) = p(x, t) - m \sum_n \Phi_n(x)\, \ddot{z}(t). \qquad (20.68)$$

Multiplying both sides of eq. (20.68) by $\Phi_m(x)$ and integrating between 0 and L result in

$$\omega_m^2 z_m(t) \int_0^L m\Phi_m^2(x)\, dx = \int_0^L \phi_m(x)\, p(x, t) - \ddot{z}_m(t) \int_0^L m\Phi_m^2(x)\, dx.$$

$$(20.69)$$

We note that all the terms which contain products of different indices $(n \neq m)$ vanish from the summations in eq. (20.68) in view of the orthogonality conditions, eq. (20.63), between normal modes. Equation (20.69) may conveniently be written as

$$M_n \ddot{z}_n(t) + \omega_n^2 M_n z_n(t) = F_n(t), \qquad n = 1, 2, 3, \ldots, m, \ldots \qquad (20.70)$$

where

$$M_n = \int_0^L m\Phi_n^2(x)\, dx \qquad (20.71)$$

is the modal mass, and

$$F_n(t) = \int_0^L \Phi_n(x)\, p(x, t)\, dx \qquad (20.72)$$

is the modal force.

The equation of motion for the nth normal mode, eq. (20.70), is completely analogous to the modal equation, eq. (12.9), for discrete systems. Modal damping could certainly be introduced by simply adding the damping term in eq. (20.70); hence we would obtain

$$M_n \ddot{z}_n(t) + C_n \dot{z}_n(t) + K_n z_n(t) = F_n(t) \qquad (20.73)$$

which, upon dividing by M_n, gives

$$\ddot{z}_n(t) + 2\xi_n \omega_n \dot{z}_n(t) + \omega_n^2 z_n(t) = \frac{F_n(t)}{M_n} \qquad (20.74)$$

where $\xi_n = C_n/C_{n,\text{cr}}$ is the modal damping ratio and $K_n = M_n \omega_n^2$ is the modal stiffness. The total response is then obtained from eq. (20.65) as the superposition of the solution of the modal equation, eq. (20.74), for as many modes as desired. Though the summation in eq. (20.65) is over an infinite number of terms, in most structural problems only the first few modes have any significant contribution to the total response and in some cases the response is given essentially by the contribution of the first mode alone.

The modal equation, eq. (20.74), is completely general and applies to beams with any type of load distribution. If the loads are concentrated rather than distributed, the integral in eq. (20.72) merely becomes a summation having one term for each load. The computation of the integral in eqs. (20.71) and (20.72) becomes tedious except for the simply supported beam because the normal shapes are rather complicated functions. Values of the ratios of these integrals needed for problems with uniform distributed load are presented in the last columns of Tables 20.1 through 20.5 for some common types of beams.

Illustrative Example 20.1.   Consider in Fig. 20.3 a simply supported uniform beam subjected to a concentrated constant force suddenly applied at a section x_1 units from the left support. Determine the response using modal analysis.

The modal shapes of a simply supported beam by eq. (20.28) are

$$\Phi_n = \sin \frac{n\pi x}{L}, \qquad n = 1, 2, 3, \ldots \qquad (a)$$

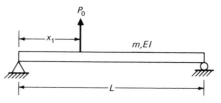

Fig. 20.3   Simply supported beam subjected to a suddenly applied force.

and the modal force by eq. (20.72) is

$$F_n(t) = \int_0^L \Phi_n(x)\, p(x, t)\, dx.$$

In this problem $p(x, t) = P_0$ at $x = x_1$; otherwise, $p(x, t) = 0$. Hence

$$F_n(t) = P_0 \Phi_n(x_1)$$

or using eq. (a), we obtain

$$F_n(t) = P_0 \sin \frac{n\pi x_1}{L}. \tag{b}$$

The modal mass by eq. (20.71) is

$$M_n = \int_0^L m\Phi^2(x)\, dx$$

$$= \int_0^L m \sin^2 \frac{n\pi x}{L}\, dx = \frac{mL}{2}. \tag{c}$$

Substituting the modal force, eq. (b), and the modal mass, eq. (c), into the modal equation, eq. (20.70), results in

$$\ddot{z}_n(t) + \omega_n^2 z_n(t) = \frac{P_0 \sin \dfrac{n\pi x_1}{L}}{mL/2}. \tag{d}$$

For initial conditions of zero displacement and zero velocity, the solution of eq. (d) from eqs. (4.5) is

$$z_n = (z_{st})_n (1 - \cos \omega_n t) \tag{e}$$

in which

$$(z_{st})_n = \frac{2P_0 \sin \dfrac{n\pi x_1}{L}}{\omega_n^2 mL} \tag{f}$$

so that

$$z_n = \frac{2P_0 \sin \dfrac{n\pi x_1}{L}}{\omega_n^2 mL}(1 - \cos \omega_n t). \tag{g}$$

The modal deflection at any section of the beam is

$$y_n(x, t) = \Phi_n(x)z_n(t) \tag{h}$$

which, upon substitution of eqs. (a) and (g), becomes

$$y_n(x, t) = \frac{2P_0 \sin \dfrac{n\pi x_1}{L}}{\omega_n^2 mL} (1 - \cos \omega_n t) \sin \frac{n\pi x}{L}. \tag{i}$$

By eq. (20.65), the total deflection is then

$$y(x, t) = \frac{2P_0}{mL} \sum_n \left[\frac{1}{\omega_n^2} \sin \frac{n\pi x_1}{L} (1 - \cos \omega_n t) \sin \frac{n\pi x}{L} \right]. \tag{j}$$

As a special case, let us consider the force applied at midspan, i.e., $x_1 = L/2$. Hence eq. (j) becomes in this case

$$y(x, t) = \frac{2P_0}{mL} \sum_n \left[\frac{1}{\omega_n^2} \sin \frac{n\pi}{2} (1 - \cos \omega_n t) \sin \frac{n\pi x}{L} \right]. \tag{20.75}$$

From the latter (due to the presence of the factor $\sin n\pi/2$) it is apparent that all the even modes do not contribute to the deflection at any point. This is true because such modes are antisymmetrical (shapes in Table 20.1) and are not excited by a symmetrical load.

It is also of interest to compare the contribution of the various modes to the deflection at midspan. This comparison will be done on the basis of maximum modal displacement disregarding the manner in which these displacements combine. The amplitudes will indicate the relative importance of the modes. The dynamic load factor $(1 - \cos \omega_n t)$ in eq. (20.75) has a maximum value of 2 for all the modes. Furthermore, since all sines are unity for odd modes and zero for even modes, the modal contributions are simply in proportion to $1/\omega_n^2$. Hence the maximum modal deflections are in proportion to 1, 1/81, and 1/625 for the first, third, and fifth modes, respectively. It is apparent, in this example, that the higher modes contribute very little to the midspan deflection.

Illustrative Example 20.2.   Determine the maximum deflection at the midpoint of the fixed beam shown in Fig. 20.4 subjected to a harmonic load $p(x, t) = p_0 \sin 300t$ lb/in uniformly distributed along the span. Consider in the analysis the first three modes contributing to the response.

The natural frequencies for uniform beams are given by eq. (20.13) as

$$\omega_n = C_n \sqrt{\frac{EI}{mL^4}}$$

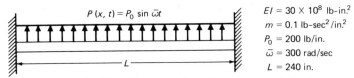

$$P(x, t) = P_0 \sin \bar{\omega} t$$

$EI = 30 \times 10^8$ lb-in.^{2}
$m = 0.1$ lb-sec^2/in.^{2}
$P_0 = 200$ lb/in.
$\bar{\omega} = 300$ rad/sec
$L = 240$ in.

Fig. 20.4   Fixed beam with uniform harmonic load.

or, substituting numerical values for this example, we get

$$\omega_n = C_n \sqrt{\frac{30 \times 10^8}{0.1(240)^4}} \tag{a}$$

where the values of C_n are given for the first five modes in Table 20.3. The deflection of the beam is given by eq. (20.65) as

$$y(x, t) = \sum_{n=1}^{\infty} \Phi_n(x) z_n(t) \tag{b}$$

in which $\Phi_n(x)$ is the modal shape defined for a fixed beam by eq. (20.46) and $z_n(t)$ is the modal response.

The modal equation by eq. (20.70) (neglecting damping) may be written as

$$\ddot{z}_n(t) + \omega_n^2 z(t) = \frac{\displaystyle\int_0^L p(x, t) \phi_n(x)\, dx}{\displaystyle\int_0^L m\phi_n^2(x)\, dx}$$

Then, substituting numerical values to this example, we obtain

$$\ddot{z}_n(t) + \omega_n^2 z_n(t) = \frac{200 \displaystyle\int_0^L \phi_n(x)\, dx}{0.1 \displaystyle\int_0^L \phi_n^2(x)\, dx} \sin 300t$$

or

$$\ddot{z}_n(t) + \omega_n^2 z_n(t) = 2000 I_n \sin 300t \tag{c}$$

in which

$$I_n = \frac{\displaystyle\int_0^L \phi_n(x)\, dx}{\displaystyle\int_0^L \phi^2(x)\, dx}$$

TABLE 20.6 Modal Response at Midspan for the Beam in Fig. 20.4.

Mode	$\omega_n \left(\dfrac{rad}{sec}\right)$	$a_n L$	I_n	$z_n = \dfrac{2000\,I_n}{\omega_n^2 - \bar{\omega}^2}$ (in)	$\Phi_n \left(x = \dfrac{L}{2}\right)$
1	67.28	4.730	0.8380	-0.0194	1.588
2	185.45	7.853	0	0	0
3	363.56	10.996	0.3640	0.0173	-1.410
4	500.98	14.137	0	0	0
5	897.76	17.279	0.2323	0.00065	1.414

is given for the first five modes in Table 20.3. The modal steady-state response is

$$z_n(t) = \frac{2000 I_n}{\omega_n^2 - (300)^2} \sin 300t. \tag{d}$$

The numerical calculations are conveniently presented in Table 20.6.

The deflections at midspan of the beam are then calculated from eq. (b) and values in Table 20.6 as

$$y\left(\frac{L}{2}, t\right) = [(1.588)(-0.0194) + (-1.410)(0.0173) + (1.414)(0.00065)] \sin 300t$$

$$y\left(\frac{L}{2}, t\right) = -0.0266 \sin 300t \text{ (in).}$$

20.6 DYNAMIC STRESSES IN BEAMS

To determine stresses in beams, we apply the following well-known relationships for bending moment M and shear force V, namely

$$M = EI \frac{\partial^2 y}{\partial x^2},$$

$$V = \frac{\partial M}{\partial x} = EI \frac{\partial^3 y}{\partial x^2}.$$

Therefore, the calculation of the bending moment or the shear force requires only differentiation of the deflection function $y = y(x, t)$ with respect to x. For example, in the case of the simple supported beam with a concentrated load suddenly applied at its center, differentiation of the deflection function, eq. (20.75), gives

$$M = -\frac{2\pi^2 P_0 EI}{mL^3} \sum_n \left[\frac{n^2}{\omega_n^2} \sin \frac{n\pi}{2} (1 - \cos \omega_n t) \sin \frac{n\pi x}{L} \right] \tag{20.76}$$

$$V = -\frac{2\pi^3 P_0 EI}{mL^4} \sum_n \left[\frac{n^3}{\omega_n^2} \sin \frac{n\pi}{2} (1 - \cos \omega_n t) \cos \frac{n\pi x}{L} \right]. \qquad (20.77)$$

We note that the higher modes are increasingly more important for moments than for deflections and even more so for shear force, as indicated by the factors 1, n^2, and n^3, respectively in eqs. (20.75), (20.76), and (20.77).

To illustrate, we compare the amplitudes for the first and third modes at their maximum values. Noting that ω_n^2 is proportional to n^4, we obtain from eqs. (20.75), (20.76), and (20.77) the following ratios:

$$\frac{y_1}{y_3} = 3^4 = 81$$

$$\frac{M_1}{M_3} = 3^2 = 9$$

$$\frac{V_1}{V_3} = 3.$$

This tendency in which higher modes have increasing importance in moment and shear calculation is generally true of beam response.

In those cases in which the first mode dominates the response, it is possible to obtain approximate deflections and stresses from static values of these quantities amplified by the dynamic load factor. For example, the maximum deflection of a simple supported beam with a concentrated force at midspan may be closely approximated by

$$y\left(x = \frac{L}{2}\right) = \frac{P_0 L^3}{48 EI} (1 - \cos \omega_1 t).$$

If we consider only the first mode, the corresponding value given by eq. (20.75) is

$$y\left(x = \frac{L}{2}\right) = \frac{2P_0}{mL\omega_1^2} (1 - \cos \omega_1 t).$$

Since $\omega_1^2 = \pi^4 EI/mL^4$, it follows that

$$y\left(x = \frac{L}{2}\right) = \frac{2P_0 L^3}{\pi^4 EI} (1 - \cos \omega_1 t)$$

$$= \frac{P_0 L^3}{48.7 EI} (1 - \cos \omega_1 t).$$

The close agreement between these two computations is due to the fact that static deflections can also be expressed in terms of modal components, and for a beam supporting a concentrated load at midspan the first mode dominates both static and dynamic response.

20.7 SUMMARY

The dynamic analysis of single span beams with distributed properties (mass and elasticity) and subjected to flexural loading was presented in this chapter. The extension of this analysis to multispan or continuous beams and other structures, though possible, becomes increasingly complex and impractical. The results obtained from these single span beams are particularly important in evaluating approximate methods based on discrete models as those presented in preceding chapters. From such evaluation, it has been found that the stiffness method of dynamic analysis in conjunction with the consistent mass formulation provides in general satisfactory results even with a rather coarse discretization of the structure.

The natural frequencies and corresponding normal modes of single-span beams with different supports are determined by solving the differential equation of motion and imposing the corresponding boundary conditions. The normal modes satisfy the orthogonality condition between any two modes m and n, namely,

$$\int_0^L \phi_m(x)\,\phi_n(x)\,m\,dx = 0. \qquad (m \neq n).$$

The response of a continuous system may be determined as the superposition of nodal contributions, that is

$$y(x, t) = \sum_n \phi_n(x)\,z_n(t)$$

where $z_n(t)$ is the solution of n modal equation

$$\ddot{z}(t) + 2\xi_n\omega_n\dot{z}(t) + \omega_n^2 z(t) = F_n(t)/M_n$$

in which

$$F_n(t) = \int_0^L \phi_n(x)\,p(x, t)\,dx$$

and

$$M_n = \int_0^L m\phi_n^2(x)\,dx.$$

The bending moment M and the shear for V at any section of a beam are calculated from the well-known relations

$$M = EI \frac{\partial^2 y}{\partial x^2},$$

$$V = EI \frac{\partial^3 y}{\partial_x^3}.$$

PROBLEMS

20.1 Determine the first three natural frequencies and corresponding modal shapes of a simply supported reinforced concrete beam having a cross section 10 in wide by 24 in deep with a span of 36 ft. Assume the flexural stiffness of the beam, $EI = 3.5 \times 10^9$ lb · in^2 and weight per unit volume $W = 150$ lb/ft^3. (Neglect shear distortion and rotary inertia.)

20.2 Solve Problem 20.1 for the beam with its two ends fixed.

20.3 Solve Problem 20.1 for the beam with one end fixed and the other simply supported.

20.4 Determine the maximum deflection at the center of the simply supported beam of Problem 20.1 when a constant force of 2000 lb is suddenly applied at 9 ft from the left support.

20.5 A simply supported beam is prismatic and has the following properties: $m = 0.3$ lb · sec^2/in per inch of span, $EI = 10^6$ lb · in^2, and $L = 150$ in. The beam is subjected to a uniform distributed static load p_0 which is suddenly removed. Write the series expression for the resulting free vibration and determine the amplitude of the first mode in terms of p_0.

20.6 The beam of Problem 20.5 is acted upon by a concentrated force given by $P(t) = 1000 \sin 500t$ lb applied at its midspan. Determine the amplitude of the steady-state motion at a quarter point from the left support in each of the first two modes. Neglect damping.

20.7 Solve Problem 20.6 assuming 10% of critical damping in each mode. Also determine the steady-state motion at the quarter point considering the first two modes.

20.8 The cantilever beam shown in Fig. P20.8 is prismatic and has the following properties: $m = 0.5$ lb · sec^2/in per inch of span, $E = 30 \times 10^6$ psi,

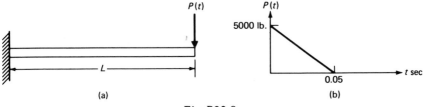

(a)   (b)

Fig. P20.8

$L = 100$ in, and $I = 120$ in^4. Considering only the first mode, compute the maximum deflection and the maximum dynamic bending moment in the beam due to the load time function of Fig. P20.8. (Chart in Fig. 4.5 may be used.)

20.9 A prismatic simply supported beam of the following properties: $L = 120$ in, $EI = 10^7$ lb $\cdot$ in^2, and $m = 0.5$ lb $\cdot$ sec^2/in per inch of span is loaded as shown in Fig. P20.9. Write the series expression for the deflection at the midsection of the beam.

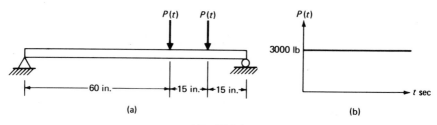

(a)   (b)

Fig. P20.9

20.10 Assuming that the forces on the beam of Problem 20.9 are applied for only a time duration $t_d = 0.1$ sec, and considering only the first mode, determine the maximum deflection at each of the load points of the beam. (Chart in Fig. 4.4 may be used.)

20.11 A prismatic beam with its two ends fixed has the following properties: $L = 180$ in, $EI = 30 \times 10^8$ lb in^2, $m = 1$ lb $\cdot$ sec^2/in per inch of span. The beam is acted upon by a uniformly distributed impulsive force $p(x, t) = 2000 \sin 400t$ lb during a time interval equal to half of the period of the sinusoidal load function ($t_d = \pi/400$ sec). Determine the maximum deflection at the midsection. Considering only the first mode, determine the maximum deflection at the midsection. (Chart in Fig. 8.3 may be used.)

20.12 Solve Problem 20.11 considering the first two modes.

21

Discretization of continuous systems

The modal superposition method of analysis was applied in the preceding chapter to some simple structures having distributed properties. The determination of the response by this method requires the evaluation of several natural frequencies and corresponding mode shapes. The calculation of these dynamic properties is rather laborious, as we have seen, even for simple structures as one-span uniform beams. The problem becomes increasingly more complicated and unmanageable as this method of solution is applied to more complex structures. However, the analysis of such structures becomes relatively simple if for each segment or element of the structure the properties are expressed in terms of dynamic coefficients much in the same manner as done previously when static deflection functions were used as an approximation to dynamic deflections in determining stiffness, mass, and other coefficients.

In this chapter the dynamic coefficients relating harmonic forces and displacements at the nodal coordinates of a beam segment are obtained from dynamic deflection functions.

These coefficients can then be used to assemble the dynamic matrix for the whole structure by the direct method as shown in the preceding chapters for assembling the system stiffness and mass matrices. Also, in the present chapter, the mathematical relationship between the dynamic coefficients based on dynamic displacement functions and the coefficients of the stiffness and consistent mass matrices derived from static displacement functions is established.

21.1   DYNAMIC MATRIX FOR FLEXURAL EFFECTS

As in the case of static influence coefficients (stiffness coefficients for example), the dynamic influence coefficients also relate forces and displacements at the nodal coordinates of a beam element. The difference between the dynamic and static coefficients is that the dynamic coefficients refer to nodal forces and displacements that vary harmonically while the static coefficients relate static forces and displacements at the nodal coordinates. The dynamic influence coefficient S_{ij} is then defined as the harmonic force of frequency $\bar{\omega}$ at nodal coordinates i, due to a harmonic displacement of a unit amplitude and of the same frequency at nodal coordinate j.

To determine the expressions for the various dynamic coefficients for a uniform beam segment as shown in Fig. 21.1, we refer to the differential equation of motion, eq. (20.5), which in the absence of external loads in the span, that is $p(x, t) = 0$, is

$$EI\frac{\partial^4 y}{\partial x^4} + m\frac{\partial^2 y}{\partial t^2} = 0. \tag{21.1}$$

For harmonic boundary displacements of frequency $\bar{\omega}$, we introduce in eq. (21.1) the trial solution

$$y(x, t) = \Phi(x) \sin \bar{\omega} t. \tag{21.2}$$

Substitution of eq. (21.2) into eq. (21.1) yields

$$\Phi^{IV}(x) - \bar{a}^4 \Phi(x) = 0 \tag{21.3}$$

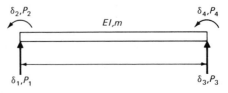

Fig. 21.1   Nodal coordinates of a flexural beam segment.

where

$$\bar{a}^4 = \frac{m\bar{\omega}^2}{EI}.$$ (21.4)

We note that eq. (21.3) is equivalent to eq. (20.10) which is the differential equation for the shape function of a beam segment in free vibration. The difference between these two equations is that eq. (21.3) is a function of the parameter $\bar{a}$ which, in turn, is a function of the forcing frequency $\bar{\omega}$, while "a" in (20.10) depends on the natural frequency ω. The solution of eq. (21.3) is of the same form as the solution of eq. (20.10). Thus by analogy with eq. (20.20), we can write

$$\Phi(x) = C_1 \sin \bar{a}x + C_2 \cos \bar{a}x + C_3 \sinh \bar{a}x + C_4 \cosh \bar{a}x.$$ (21.5)

Now, to obtain the dynamic coefficient for the beam segment, boundary conditions indicated by eqs. (21.6) and (21.7) are imposed.

$$\Phi(0) = \delta_1, \qquad \Phi(L) = \delta_3,$$

$$\Phi'(0) = \delta_2, \qquad \Phi'(L) = \delta_4.$$ (21.6)

Also

$$\Phi'''(0) = \frac{P_1}{EI}, \qquad \Phi'''(L) = -\frac{P_3}{EI},$$

$$\Phi''(0) = -\frac{P_2}{EI}, \qquad \Phi''(L) = \frac{P_4}{EI}.$$ (21.7)

In eqs. (21.6), $\delta_1, \delta_2, \delta_3$, and δ_4 are amplitudes of linear and angular harmonic displacements at the nodal coordinates while in eqs. (21.7) P_1, P_2, P_3, and P_4 are the corresponding harmonic forces and moments as shown in Fig. 21.1. The substitution of the boundary conditions, eqs. (21.6) and (21.7), into eq. (21.5) results in

$$\begin{bmatrix} \delta_1 \\ \delta_2 \\ \delta_3 \\ \delta_4 \end{bmatrix} = \begin{bmatrix} 0 & 1 & 0 & 1 \\ \bar{a} & 0 & \bar{a} & 0 \\ s & c & S & C \\ \bar{a}c & -\bar{a}s & \bar{a}C & \bar{a}S \end{bmatrix} \begin{bmatrix} C_1 \\ C_2 \\ C_3 \\ C_4 \end{bmatrix}$$ (21.8)

and

$$\begin{bmatrix} P_1 \\ P_2 \\ P_3 \\ P_4 \end{bmatrix} = EI \begin{bmatrix} -\bar{a}^3 & 0 & \bar{a}^3 & 0 \\ 0 & \bar{a}^2 & 0 & -\bar{a}^2 \\ \bar{a}^3c & -\bar{a}^3s & -\bar{a}^3C & -\bar{a}^3S \\ -\bar{a}^2s & -\bar{a}^2c & \bar{a}^2S & \bar{a}^2C \end{bmatrix} \begin{bmatrix} C_1 \\ C_2 \\ C_3 \\ C_4 \end{bmatrix}$$ (21.9)

in which

$$s = \sin \bar{a}L, \qquad S = \sinh \bar{a}L,$$

$$c = \cos \bar{a}L, \qquad C = \cosh \bar{a}L. \qquad (21.10)$$

Next, eq. (21.8) is solved for the constants of integration C_1, C_2, C_3, C_4, which are subsequently substituted into eq. (21.9). We thus obtain the dynamic matrix relating harmonic displacements and harmonic forces at the nodal coordinate of the beam segment, namely

$$\begin{bmatrix} P_1 \\ P_2 \\ P_3 \\ P_4 \end{bmatrix} = B \begin{bmatrix} \bar{a}^2(cS + sC) & & \text{Symmetric} & \\ \bar{a}\,sS & sC - cS & & \\ -\bar{a}^2(s + S) & \bar{a}(c - C) & \bar{a}^2(cS + sC) & \\ \bar{a}(C - c) & S - s & -\bar{a}\,sS & sC - cS \end{bmatrix} \begin{bmatrix} \delta_1 \\ \delta_2 \\ \delta_3 \\ \delta_4 \end{bmatrix}$$

$$(21.11)$$

where

$$B = \frac{\bar{a}EI}{1 - cC}. \qquad (21.12)$$

We require the denominator to be different from zero, that is,

$$1 - \cos \bar{a}L \cosh \bar{a}L \neq 0. \qquad (21.13)$$

The element dynamic matrix in eq. (21.11) can then be used to assemble the system dynamic matrix for a continuous beam or a plane frame in a manner entirely analogous to the assemblage of the system stiffness matrix from element stiffness matrices.

21.2. DYNAMIC MATRIX FOR AXIAL EFFECTS

The governing equation for axial vibration of a beam element is obtained by establishing the dynamic equilibrium of a differential element dx of the beam, as shown in Fig. 21.2. Thus

$$\left(P + \frac{\partial P}{\partial x}\,dx\right) - P - (mdx)\,\frac{\partial^2 u}{\partial t^2} = 0,$$

$$\frac{\partial P}{\partial x} = m\,\frac{\partial^2 u}{\partial t^2}, \qquad (21.14)$$

where u is the displacement at x. The displacement at $x + dx$ will then be $u + (\partial u/dx)\,dx$. It is evident that the element dx in the new position has changed length by an amount $(\partial u/\partial x)\,dx$, and thus the strain is $\partial u/\partial x$. Since from Hooke's

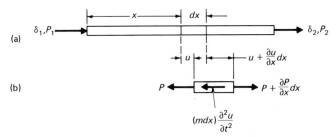

Fig. 21.2   Axial effects on a beam. (a) Nodal axial coordinates. (b) Forces acting on a differential element.

law the ratio of stress to strain is equal to the modulus of elasticity E, we can write

$$\frac{\partial u}{\partial x} = \frac{P}{AE},$$ (21.15)

where A is the cross-sectional area of the beam. Differentiating with respect to x results in

$$AE \frac{\partial^2 u}{\partial x^2} = \frac{\partial P}{\partial X}$$ (21.16)

and combining eqs. (21.14) and (21.16) yields the differential equation for axial vibration of a beam segment, namely

$$\frac{\partial^2 u}{\partial x^2} - \frac{m}{AE} \frac{\partial^2 u}{\partial t^2} = 0.$$ (21.17)

A solution of eq. (21.17) of the form

$$u(x, t) = U(x) \sin \overline{\omega} t$$ (21.18)

will result in a harmonic motion of amplitude

$$U(x) = C_1 \sin bx + C_2 \cos bx$$ (21.19)

where

$$b = \sqrt{\frac{m \overline{\omega}^2}{AE}}$$ (21.20)

and C_1, C_2 are constants of integration.

To obtain the dynamic matrix for the axially vibrating beam segment, boundary conditions indicated by eqs. (21.21) and (21.22) are imposed, namely

$$U(0) = \delta_1, \qquad U(L) = \delta_2,$$ (21.21)

$$U'(0) = -\frac{P_1}{AE}, \qquad U'(L) = \frac{P_2}{AE},$$ (21.22)

where δ_1 and δ_2 are the displacements and P_1 and P_2 are the forces at the nodal coordinates of the beam segment as shown in Fig. 21.2.

Substitution of the boundary conditions, eqs. (21.21) and (21.22), into eq. (21.19) results in

$$\begin{bmatrix} \delta_1 \\ \delta_2 \end{bmatrix} = \begin{bmatrix} 0 & 1 \\ \sin bL & \cos bL \end{bmatrix} \begin{bmatrix} C_1 \\ C_2 \end{bmatrix} \tag{21.23}$$

and

$$\begin{bmatrix} P_1 \\ P_2 \end{bmatrix} = AEb \begin{bmatrix} -1 & 0 \\ \cos bL & -\sin bL \end{bmatrix} \begin{bmatrix} C_1 \\ C_2 \end{bmatrix}. \tag{21.24}$$

Then solving eq. (21.23) for the constants of integration, we obtain

$$\begin{bmatrix} C_1 \\ C_2 \end{bmatrix} = \begin{bmatrix} -\cot bL & \csc bL \\ 1 & 0 \end{bmatrix} \begin{bmatrix} \delta_1 \\ \delta_2 \end{bmatrix} \tag{21.25}$$

subject to the condition

$$\sin bL \neq 0. \tag{21.26}$$

Finally, the substitution of eq. (21.25) into eq. (21.24) results in eq. (21.27) relating harmonic forces and displacement at the nodal coordinates through the dynamic matrix for an axially vibrating beam segment. Thus we have

$$\begin{bmatrix} P_1 \\ P_2 \end{bmatrix} = EAb \begin{bmatrix} \cot bL & -\csc bL \\ -\csc bL & \cot bL \end{bmatrix} \begin{bmatrix} \delta_1 \\ \delta_2 \end{bmatrix}. \tag{21.27}$$

21.3   DYNAMIC MATRIX FOR TORSIONAL EFFECTS

The equation of motion of a beam segment in torsional vibration is similar to that of the axial vibration of beams discussed in the preceding section. Let x (Fig. 21.3) be measured along the length of the beam. Then the angle of twist

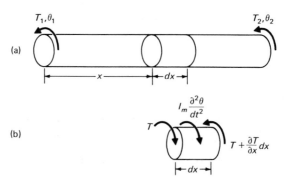

Fig. 21.3   Torsional effects on a beam. (a) Nodal torsional coordinates. (b) Moments acting on a differential element.

for any element of length dx of the beam due to a torque T is

$$d\theta = \frac{T dx}{J_T G} \qquad (21.28)$$

where $J_T G$ is the torsional stiffness given by the product of the torsional constant J_T (J_T is the polar moment of inertia for circular sections) and the shear modulus of elasticity G. The torque applied on the faces of the element are T and $T + (\partial T / \partial x) dx$ as shown in Fig. 21.3. From eq. (21.28), the net torque is then

$$\frac{\partial T}{\partial x} dx = J_T G \frac{\partial^2 \theta}{\partial x^2} dx. \qquad (21.29)$$

Equating this torque to the product of the mass moment of inertia $I_m\, dx$ of the element dx and the angular acceleration $\partial^2 \theta / \partial t^2$, we obtain the differential equation of motion

$$J_T G \frac{\partial^2 \theta}{\partial x^2} dx = I_m \frac{\partial^2 \theta}{\partial t^2} dx$$

or

$$\frac{\partial^2 \theta}{\partial x^2} - \frac{I_m}{J_T G} \frac{\partial^2 \theta}{\partial t^2} = 0, \qquad (20.30)$$

where I_m is the mass moment of inertia per unit length about the longitudinal axis x.

We seek a solution of eq. (21.30) in the form

$$\theta(x, t) = \theta(x) \sin \overline{\omega} t \qquad (21.31)$$

which, upon substitution into eq. (21.30), results in a harmonic torsional motion of amplitude

$$\theta(x) = C_1 \sin cx + C_2 \cos cx \qquad (21.32)$$

in which

$$c = \sqrt{\frac{I_m \overline{\omega}^2}{J_T G}}. \qquad (21.33)$$

For a circular section, the torsional constant J_T is equal to the polar moment of inertia J_0. Thus eq. (21.33) reduces to

$$c = \sqrt{\frac{m \overline{\omega}^2}{AG}} \qquad (21.34)$$

since $I_m = J_0 m / A$.

We note that eq. (21.30) for torsional vibration is analogous to eq. (21.17) for axial vibration of beam segments. It follows that by analogy to eq. (21.27) we can write the dynamic relation between torsional moments and rotations in a beam segment. Hence

$$\begin{bmatrix} T_1 \\ T_2 \end{bmatrix} = J_T Gc \begin{bmatrix} \cot cL & -\csc cL \\ -\csc cL & \cot cL \end{bmatrix} \begin{bmatrix} \theta_1 \\ \theta_2 \end{bmatrix}. \qquad (21.35)$$

21.4   BEAM FLEXURE INCLUDING AXIAL-FORCE EFFECT

When a beam is subjected to a force along its longitudinal axis in addition to lateral loading, the dynamic equilibrium equation for a differential element of the beam is affected by the presence of this force. Consider the beam shown in Fig. 21.4 in which the axial force is assumed to remain constant during flexure with respect to both magnitude and direction. The dynamic equilibrium for a differential element dx of the beam [Fig. 21.4(b)] is established by equating to zero both the sum of the forces and the sum of the moments.

Summing forces in the y direction, we obtain

$$V + p(x, t)\, dx - \left(V + \frac{\partial V}{\partial x}\, dx\right) - (m\, dx)\frac{\partial^2 y}{\partial t^2} = 0 \qquad (21.36)$$

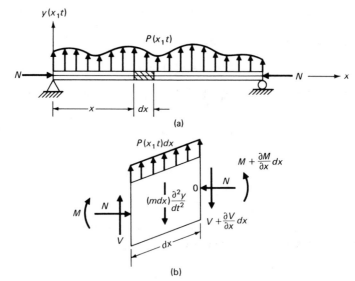

Fig. 21.4   Beam supporting constant axial force and lateral dynamic load. (a) Loaded beam. (b) Forces acting on a differential element.

which, upon reduction, yields

$$\frac{\partial V}{\partial x} + m \frac{\partial^2 y}{\partial t^2} = p(x, t). \tag{21.37}$$

The summation of moments about point 0 gives

$$M + V \, dx - \left(M + \frac{\partial M}{\partial x} \, dx\right) + \frac{1}{2}\left(p(x, t) - m \frac{\partial^2 y}{\partial t^2}\right) dx^2 - N \frac{\partial y}{\partial x} \, dx = 0. \tag{21.38}$$

Discarding higher order terms, we obtain for the shear force the expression

$$V = N \frac{\partial y}{\partial x} + \frac{\partial M}{\partial x}. \tag{21.39}$$

Then using the familiar relations from bending theory

$$M = EI \frac{\partial^2 y}{\partial x^2} \tag{21.40}$$

and combining eqs. (21.37), (21.39), and (21.40) we obtain the equation of motion of a beam segment including the effect of the axial forces, that is,

$$EI \frac{\partial^4 y}{\partial x^4} + N \frac{\partial^2 y}{\partial x^2} + m \frac{\partial^2 y}{\partial t^2} = p(x, t). \tag{21.41}$$

A comparison of eqs. (21.41) and (20.5) reveals that the presence of the axial force gives rise to an additional transverse force acting on the beam. As indicated previously in Section 20.1, in the derivation of eq. (21.41) it has been assumed that the deflections are small and that the deflections due to shear forces or rotary inertia are negligible.

In the absence of external loads applied to the span of the beam, eq. (21.41) reduces to

$$EI \frac{\partial^4 y}{\partial x^4} + N \frac{\partial^2 y}{\partial x^2} + m \frac{\partial^2 y}{\partial t^2} = 0. \tag{21.42}$$

The solution of eq. (21.42) is found as before by substituting

$$y(x, t) = \Phi(x) \sin \omega t. \tag{21.43}$$

We thereby obtain the ordinary differential equation

$$\frac{d^4 \Phi}{dx^4} + \frac{N}{EI} \frac{d^2 \Phi}{dx^2} - \frac{m \omega^2}{EI} \Phi = 0. \tag{21.44}$$

The solution of eq. (21.44) is

$$\Phi(x) = A \sin p_2 x + B \cos p_2 x + C \sinh p_1 x + D \cosh p_1 x, \tag{21.45}$$

where A, B, C, D are constants of integration and

$$p_1 = \sqrt{\frac{-\alpha}{2} + \sqrt{\left(\frac{\alpha}{2}\right)^2 + \beta}},$$

$$p_2 = \sqrt{\frac{\alpha}{2} + \sqrt{\left(\frac{\alpha}{2}\right)^2 + \beta}}, \tag{21.46}$$

$$\alpha = \frac{N}{EI}, \tag{21.47}$$

$$\beta = \frac{m\omega^2}{EI}. \tag{21.48}$$

To obtain the dynamic matrix (which in this case includes the effect of axial forces) for the transverse vibration of the beam element, the boundary conditions, eqs. (21.49), are imposed, namely

$$\Phi(0) = \delta_1, \qquad \Phi(L) = \delta_3,$$

$$\frac{d\Phi(0)}{dx} = \delta_2, \qquad \frac{d\Phi(L)}{dx} = \delta_4,$$

$$\frac{d^3\Phi(0)}{dx^3} = \frac{P_1}{EI} - \frac{N}{EI}\delta_2, \qquad \frac{d^3\Phi(L)}{dx^3} = -\frac{P_3}{EI} - \frac{N}{EI}\delta_4,$$

$$\frac{d^2\Phi(0)}{dx^2} = -\frac{P_2}{EI}, \qquad \frac{d^2\Phi(L)}{dx^2} = \frac{P_4}{EI} \tag{21.49}$$

In eqs. (21.49) δ_1, δ_3, and δ_2, δ_4, are, respectively, the transverse and angular displacements at the ends of the beam, while P_1, P_3, and P_2, P_4 are corresponding forces and moments at these nodal coordinates. The substitution into eq. (21.45) of the boundary conditions given by eqs. (21.49) results in a system of eight algebraic equations which upon elimination of the four constants of integration A, B, C, D yields the dynamic matrix (including the effect of axial forces) relating harmonic forces and displacements at the nodal coordinates of a beam segment. The final result is

$$\begin{bmatrix} P_1 \\ P_2 \\ P_3 \\ P_4 \end{bmatrix} = \begin{bmatrix} S_{11} & & \text{Symmetric} & \\ S_{21} & S_{22} & & \\ S_{31} & S_{32} & S_{33} & \\ S_{41} & S_{42} & S_{43} & S_{43} \end{bmatrix} \begin{bmatrix} \delta_1 \\ \delta_2 \\ \delta_3 \\ \delta_4 \end{bmatrix} \tag{21.50}$$

where

$$S_{11} = S_{33} = B\left[(p_1^2 p_2^3 + p_1^4 p_2)cS + (p_1 p_2^4 + p_1^3 p_2^2)sC\right],$$

$$S_{21} = -S_{43} = B\left[(p_1 p_2^3 - p_1^3 p_2) + (p_1^3 p_2 - p_1 p_2^3)cC + 2p_1^2 p_2^2 sS\right],$$

$$S_{22} = S_{44} = B\left[(p_2^2 p_1 + p_1^3)sC - (p_2^3 + p_1^2 p_2)cS\right],$$

$$S_{41} = -S_{32} = B\left[(p_1 p_2^3 + p_1^3 p_2)(C - c)\right],$$

$$S_{31} = B\left[(-p_1^2 p_2^3 - p_1^4 p_2)S - (p_1^3 p_2^2 + p_1 p_2^4)s\right],$$

$$S_{42} = B\left[(p_1^2 p_2 + p_2^3)S - (p_1 p_2^2 + p_1^3)s\right]. \tag{21.51}$$

In the above, the letters $s, c, S,$ and C denote

$$s = \sin p_2 L, \qquad S = \sinh p_1 L,$$

$$c = \cos p_2 L, \qquad C = \cosh p_1 L,$$

and the letter B denotes

$$B = \frac{EI}{2p_1 p_2 - 2p_1 p_2\, cC + (p_1^2 - p_2^2)sS}. \tag{21.52}$$

Furthermore, eq. (21.50) is subject to the condition

$$2p_1 p_2 - 2p_1 p_2\, cC + (p_1^2 - p_2^2)sS \neq 0. \tag{21.53}$$

21.5   POWER SERIES EXPANSION OF THE DYNAMIC MATRIX FOR FLEXURAL EFFECTS

It is of interest to demonstrate that the influence coefficients of the stiffness matrix, eq. (14.20), and of the consistent mass matrix, eq. (14.34), may be obtained by expanding the influence coefficients of the dynamic matrix in a Taylor's series.[1] For the sake of the discussion, we consider the dynamic coefficient from the second row and first column of the dynamic matrix, eq. (21.11)

$$S_{21} = \frac{\bar{a}^2 EI \sin \bar{a}L \, \sinh \bar{a}L}{1 - \cos \bar{a}L \, \cosh \bar{a}L}. \tag{21.54}$$

In the following derivation, operations with power series, including addition, subtraction, multiplication, and division, are employed. The validity of these operations and convergence of the resulting series is proved in Knopp.[2] In general,

[1]Paz, M., "Mathematical observations in structural dynamics," *Int. J. Computers and Structures*, Vol. 3, pp. 385–396, 1973.
[2]Knopp, K., *Theory and Application of Infinite Series*, Blackie, London, 1963.

convergent power series may be added, subtracted, or multiplied and the resulting series will converge at least in the common interval of convergence of the two original series. The operation of division of two power series may be carried out formally; however, the determination of the radius of convergence of the resulting series is more complicated. It requires the use of theorems in the field of complex variables and it is related to analytical continuation. Very briefly, it can be said that the power series obtained by division of two convergent power series about a complex point Z_0 will be convergent in a circle with center Z_0 and of radius given by the closest singularity to Z_0 of the functions represented by the series in the numerator and denominator.

The known expansions in power series about the origin of trigonometric and hyperbolic functions are used in the intermediate steps in expanding the function in eq. (21.54), namely

$$\cos x \cosh x = 1 - \frac{x^4}{6} + \frac{x^8}{2520} - \frac{x^{12}}{7,484,400} + \cdots,$$

$$(1 - \cos x \cosh x)^{-1} = \frac{6}{x^4} + \frac{1}{70} + \frac{85x^4}{2,910,600} + \cdots,$$

$$\sin x \sinh x = x^2 - \frac{x^6}{90} + \frac{x^{10}}{113,400} - \cdots,$$

where $x = \bar{a}L$. Substitution of these series equations in the dynamic coefficient, eq. (21.54), yields

$$S_{21} = \frac{\bar{a}^2 EI \sin \bar{a}L \sinh \bar{a}L}{1 - \cos \bar{a}L \cosh \bar{a}L} = \frac{6EI}{L^2} - \frac{11mL^2\bar{\omega}^2}{210} - \frac{223m^2L^6\bar{\omega}^4}{2,910,600EI} - \cdots.$$

(21.55)

The first term on the right-hand side of eq. (21.55) is the stiffness coefficient k_{21} in the stiffness matrix, eq. (14.20), and the second term, the consistent mass coefficient m_{21} in the mass matrix, eq. (14.34). The series expansion, eq. (21.55), is convergent in the positive real field for

$$0 < \bar{a}L < 4.7$$

(21.56)

or from eq. (21.4)

$$0 < \bar{\omega} < (4.7)^2 \sqrt{\frac{EI}{mL^4}}.$$

(21.57)

In eq. (21.56) the numerical value 4.7 is an approximation of the closest singularity to the origin of the functions in the quotient expanded in eq. (21.54).

The series expansions for all the coefficients in the dynamic matrix, eq. (21.11), are obtained by the method explained in obtaining the expansion of the coefficient S_{21}. These series expansions are:

$$S_{33} = S_{11} = \frac{12EI}{L^3} - \frac{13Lm\bar{\omega}^2}{35} - \frac{59L^5 m^2 \bar{\omega}^4}{161,700EI} - \cdots ,$$

$$S_{21} = -S_{43} = \frac{6EI}{L^2} - \frac{11L^2 m\bar{\omega}^2}{210} - \frac{223L^6 m^2 \bar{\omega}^4}{2,910,600EI} - \cdots ,$$

$$S_{41} = -S_{32} = \frac{6EI}{L^2} + \frac{13L^2 m\bar{\omega}^2}{420} + \frac{1681L^6 m^2 \bar{\omega}^4}{23,284,800EI} - \cdots ,$$

$$S_{22} = S_{44} = \frac{4EI}{L} - \frac{L^3 m\bar{\omega}^2}{105} - \frac{71L^7 m^2 \bar{\omega}^4}{4,365,900EI} - \cdots ,$$

$$S_{31} = -\frac{12EI}{L^3} - \frac{9Lm\bar{\omega}^2}{70} - \frac{1279L^5 m^2 \bar{\omega}^4}{3,880,800EI} - \cdots ,$$

$$S_{42} = \frac{2EI}{L} + \frac{L^3 m\bar{\omega}^2}{140} + \frac{1097L^7 m^2 \bar{\omega}^4}{69,854,400EI} - \cdots . \tag{21.58}$$

21.6 POWER SERIES EXPANSION OF THE DYNAMIC MATRIX FOR AXIAL AND FOR TORSIONAL EFFECTS

Proceeding in a manner entirely analogous to expansion of the dynamic coefficients for flexural effects, we can also expand the dynamic coefficients for axial and for torsional effects. The Taylor's series expansions, up to three terms, of the coefficients of the dynamic matrix in eq. (21.27) (axial effects) are

$$AEb \cos bL = \frac{AE}{L} - \frac{m\bar{\omega}^2 L}{3} - \frac{L^3 m^2 \bar{\omega}^4}{45AE} - \cdots ,$$

$$-AEb \csc bL = -\frac{AE}{L} - \frac{m\bar{\omega}^2 L}{6} - \frac{7L^3 m^2 \bar{\omega}^4}{300AE} - \cdots . \tag{21.59}$$

It may be seen that the first term in each series of eq. (21.59) is equal to the corresponding stiffness coefficient of the matrix in eq. (15.3) and the second term, to the consistent mass coefficient of the matrix in (15.26). Similarly, the Taylor's series expansions of the coefficients of the dynamic matrix for torsional effects, eq. (21.35), are

$$J_T Gc \cot cL = \frac{J_T G}{L} - \frac{J_0 m \overline{\omega}^2}{A} - \frac{L^3 m^2 \overline{\omega}^4 J_0}{A^2 G J_T} - \cdots ,$$

$$-J_T Gc \operatorname{cosec} cL = -\frac{J_T G}{L} - \frac{J_0 m \overline{\omega}^2}{GA} - \frac{7L^3 m^3 \overline{\omega}^4}{360 A^2 G J_T} - \cdots . \qquad (21.60)$$

Comparing the first two terms of the above series with the stiffness and mass influence coefficients of the matrices in eqs. (16.16) and (16.17), we find that for torsional effects the first term is also equal to the stiffness coefficient and the second term, to the consistent mass coefficient.

21.7   POWER SERIES EXPANSION OF THE DYNAMIC MATRIX INCLUDING THE EFFECT OF AXIAL FORCES

The series expansions of the coefficients of the dynamic matrix, eq. (21.50), (with axial effects) are obtained by the method described in the last two sections. Detailed derivation of these expansions are given by Paz and Dung.[3] The series expansion of the dynamic matrix, eq. (21.50), is

$$[S] = [K] - [G_0] N - [M_0] \overline{\omega}^2 - [A_1] N \overline{\omega}^2 - [G_1] N^2 - [M_1] \overline{\omega}^4 - \cdots$$

$$(21.61)$$

where the first three matrices in this expansion $[K]$, $[G_0]$, and $[M_0]$ are, respectively, the stiffness, geometric, and mass matrices which were obtained in previous chapters on the basis of static displacement functions. These matrices are given, respectively, by eqs. (14.20), (14.45), and (14.34). The other matrices in eq. (21.61) corresponding to higher order terms are represented as follows. The second-order mass-geometrical matrix:

$$[A_1] = \frac{mL^3}{EI} \begin{bmatrix} \dfrac{1}{3150} & & \text{Symmetric} & \\ \dfrac{L}{1260} & \dfrac{L^2}{3150} & & \\ -\dfrac{1}{3150} & \dfrac{L}{1680} & \dfrac{1}{3150} & \\ -\dfrac{L}{1680} & \dfrac{L^2}{3600} & -\dfrac{L}{1260} & \dfrac{L^2}{3150} \end{bmatrix}$$

[3] Paz, M., and Dung, L., "Power series expansion of the general stiffness matrix for beam elements," *Int. J. Numerical Methods in Engineering*, Vol. 9, pp. 449–459, 1975.

The second-order geometrical matrix:

$$[G_1] = \frac{1}{EI} \begin{bmatrix} \dfrac{L}{700} & & \text{Symmetric} & \\[2ex] \dfrac{L^2}{1400} & \dfrac{11}{6300}L^3 & & \\[2ex] -\dfrac{L}{700} & -\dfrac{1}{1400}L^2 & \dfrac{L}{700} & \\[2ex] \dfrac{L^2}{1400} & -\dfrac{13L^3}{12600} & -\dfrac{1}{1400}L^2 & \dfrac{11L^3}{6300} \end{bmatrix}$$

The second-order mass matrix:

$$[M_1] = \frac{m^2 L^3}{1000EI} \begin{bmatrix} \dfrac{59}{161.7} & & \text{Symmetric} & \\[2ex] \dfrac{223L}{2910.6} & \dfrac{71L^2}{4365.9} & & \\[2ex] \dfrac{1279}{3880.8} & \dfrac{1681L}{23284.8} & \dfrac{59}{161.7} & \\[2ex] \dfrac{1681L}{23284.8} & \dfrac{1097L^2}{69854.4} & \dfrac{223L}{2910.6} & \dfrac{71L^2}{4365.9} \end{bmatrix}$$

21.8 SUMMARY

The dynamic coefficients relating harmonic forces and displacements at the nodal coordinates of a beam segment were obtained from dynamic deflection equations. This coefficient can then be used in assembling the dynamic matrix for the entire structure by the same procedure (direct method) employed in assembling the stiffness and mass matrices for discrete systems.

In this chapter it has been demonstrated that the stiffness, consistent mass, and other influence coefficients may be obtained by expanding the dynamic influence coefficients in Taylor's series. This mathematical approach also provides higher order influence coefficients and the determination of the radius of convergence of the series expansion.

Appendix

Computer programs

The computer programs listed in the following pages are written in the Fortran language. These programs are discussed in the appropriate chapters of the text. They are intended for instructional purposes and are not highly sophisticated programs which may compete with commercially available ones.

Six of the programs listed are short computer programs designed to test corresponding subroutines. A list of these subroutines, together with testing program numbers, chapter presented, and brief descriptions of their purposes, is given in Table A.1. The subroutines listed in this table are required to perform specific calculations in the following structural programs: Program 11, BEAM; Program 12, FRAME; Program 13, GRID; and Program 14, TRUSS. Consequently, subroutines listed in Table A.1 should be available to these programs when required to perform any of the specific calculations indicated. For these structural programs to call any of the subroutines, the letter "C" (Comment) should be removed from the corresponding calling statement.

TABLE A.1 **Subroutines Required for Specific Calculation by Programs 11, 12, 13, and 14.**

Subroutine	Program	Chapter Presented	Purpose
JACOBI	5	10	Natural frequencies and normal modes
MODAL	6	11	Response by modal superposition
HARMO	8	11	Response for harmonic excitation
DAMP	9	12	Damping matrix from damping ratios
CONDE	10	13	Condense stiffness and mass matrices
STEPM	15	19	Response by Wilson-θ method

To facilitate the use of the programs listed in the Appendix, Table A.2 provides a brief description of the purpose of each program and indicates the corresponding chapter where each program is developed.

TABLE A.2 **List of Computer Programs.**

	Program	Chapter	Purpose
1	DUHAMEL	4	Response SDFS* by Duhamel's integral
2	FOURIER	5	Response SDFS to periodic excitation
3	FREQRESP	5	Response SDFS in the frequency domain
4	STEPS	7	Response SDFS with elastoplastic behavior
5	JACOBI	10	Natural frequencies and normal modes
6	MODAL	11	Response by modal superposition
7	SRESB	11	Response shear building to seismic excitation
8	HARMO	11	Response for harmonic excitation
9	DAMP	12	Damping matrix from damping ratios
10	CONDE	13	Condense stiffness and mass matrices
11	BEAM	14	Dynamic analysis of beams
12	FRAME	15	Dynamic analysis of plane frames
13	GRID	16	Dynamic analysis of grids
14	TRUSS	18	Dynamic analysis of plane trusses
15	STEPM	19	Response by Wilson-θ method
16	SRSB	19	Seismic response shear building with elastoplastic behavior

*SDFS = single degree-of-freedom system.

PROGRAM 1: DUHAMEL

Objective: To determine the response of a damped single degree-of-freedom system excited by a time-dependent force applied to the mass or by an acceleration acting at the support.

Method: Duhamel's integral.

Description: Chapter 4, Tables 4.2 and 4.3.

```
C
C       RESPONSE SINGLE DEGREE OF FREEDOM SYSTEM
C       USING DUHAMEL INTEGRAL
C
        IMPLICIT REAL*8(A-H,O-Z)
        REAL*8INT1,INT2,INT3,INT4,K,M
        DIMENSION T(200),F(200)
C
C STEMENT FUNCTIONS
C
        INT1(TAU)=DEXP(XIWD*TAU)*(XIWD*DCOS(WD*TAU)+WD*DSIN(WD*TAU))/DWSQ
        INT2(TAU)=DEXP(XIWD*TAU)*(XIWD*DSIN(WD*TAU)-WD*DCOS(WD*TAU))/DWSQ
        INT3(TAU)=TAU*INT2(TAU)-XIWD*INT2(TAU)/DWSQ+WD*INT1(TAU)/DWSQ
        INT4(TAU)=TAU*INT1(TAU)-XIWD*INT1(TAU)/DWSQ-WD*INT2(TAU)/DWSQ
C
C
C READ DATA
C
        WRITE(6,105)
  105 FORMAT(1H1,10X,'INPUT DATA'//)
        READ(5,100) NTYPE,N,M,K,XI,TMAX,DT,INT,GR
        WRITE(6,100)NTYPE,N,M,K,XI,TMAX,DT,INT,GR
  100 FORMAT(2I5,5F10.3,I5,F10.0)
        READ(5,200) (T(I),F(I),I=1,N)
        WRITE(6,200)(T(I),F(I),I=1,N)
  200 FORMAT(8E10.2)
        IF(INT.EQ.0) GO TO 21
        CALL INTER(N,T,F,DT,TMAX,INT)
C SET INITIAL VALUES
C
   21 IF(NTYPE.EQ.0) GO TO 6
        DO 5 I=1,N
    5 F(I)=F(I)*GR
    6 FIM1=F(1)
        TIM1=T(1)
        ATI=0.0
        BTI=0.0
        DAT=0.0
        DBT=0.0
        Y=0.0
        V=0.0
        P=0.0
        YMAX=0.0
        VMAX=0.0
        AMAX = 0.0
        PMAX=0.0
        OMEGA=DSQRT(K/M)
        CRIT=2*DSQRT(K*M)
        C=XI*CRIT
        WD=OMEGA*DSQRT(1.-(XI**2))
        XIWD=XI*OMEGA
        DWSQ=XIWD**2+WD**2
        ACC=FIM1/M
C
C WRITE THE HEADINGS
C
        WRITE (6,700) M,K,OMEGA,WD,C,XI
  700 FORMAT (5X,'RESPONSE OF SINGLE DEGREE SYSTEM USING DUHAMEL INTEGRA
```

```
      1L',//,5X,'MASS',15X,'=',F10.4,/,5X,'SPRING CONSTANT',4X,'=',F10.2,
     2/,5X,'NATURAL FREQUENCY  =',F10.3,'  RADIANS/SEC',/,5X,'DAMPED ','
     3FREQUENCY',3X,'=',F10.3,'  RADIANS/SEC',/,5X,'DAMPING CONSTANT',3X
     4,'=',F10.3,/,5X,'RELATIVE DAMPING',3X,'=',F10.4,//)
      WRITE (6,900)
  900 FORMAT (9X,'TIME',7X,'FORCE',6X,'DISPL.',4X,'VELOCITY',8X,'ACC.',
     12X,'SUP. REAC.',/)
      IF ( NTYPE .EQ. 0 ) GO TO 12
      ACC=-FIM1
      FIM1=-FIM1*M
   12 WRITE (6,400) TIM1,FIM1,Y,V,ACC,P
C
C LOOP OVER TIME INTERVALS
C
      NM1=N-1
      DO 1 I=1,NM1
C
C SOLVE FOR THE DISPLACEMENT
C
      FI=F(I+1)
      TI=T(I+1)
      IF ( NTYPE .NE. 0 ) FI=-FI*M
      DFTI=FI-FIM1
      DTI=TI-TIM1
      FT=DFTI/DTI
      G=FIM1-TIM1*FT
      AI=INT1(TI)-INT1(TIM1)
      BI=INT2(TI)-INT2(TIM1)
      VS=INT3(TI)-INT3(TIM1)
      VC=INT4(TI)-INT4(TIM1)
      AI=AI*G
      BI=BI*G
      BI=BI+FT*VS
      AI=AI+FT*VC
      ATI=ATI+AI
      BTI=BTI+BI
      Y=DEXP(-XIWD*TI)*(ATI*DSIN(WD*TI)-BTI*DCOS(WD*TI))/(M*WD)
C
C SOLVE FOR THE VELOCITY
C
      DA=(WD*BTI-XIWD*ATI)*DSIN(WD*TI)
      DB=(WD*ATI+XIWD*BTI)*DCOS(WD*TI)
      V=DEXP(-XIWD*TI)*(DA+DB)/(M*WD)
C
C SOLVE FOR THE ACCELERATION
C
      ACC=(FI-C*V-K*Y)/M
C
C SOLVE FOR  SPRING AND DAMPING FORCES
C
      FS=Y*K
      FD=V*C
      P=DSQRT(FS**2+FD**2)
C
C SAVE THE MAXIMUM VALUES
C
      IF (DABS(Y) .GT. DABS(YMAX)) YMAX=Y
      IF (DABS(V) .GT. DABS(VMAX)) VMAX=V
```

```
      IF (DABS(ACC) .GT. DABS(AMAX)) AMAX=ACC
      IF (DABS(P) .GT. DABS(PMAX)) PMAX=P
C
C INCREMENT VARIABLES
C
      TIM1=TI
      FIM1=FI
C
C PRINT THE RESPONSE
C
      WRITE (6,400) TI,FI,Y,V,ACC,P
  400 FORMAT (1X,F12.4,F12.2,2F12.3,2F12.2)
    1 CONTINUE
C
C PRINT THE MAXIMUM VALUES
C
      WRITE (6,300) YMAX,VMAX,AMAX,PMAX
  300 FORMAT (//5X,'MAX DISPL.          =',F10.2,/,
     15X,'MAX VELOC.         =',F10.2,/,
     25X,'MAX ACC.           =',F10.2,/,
     35X,'MAX SUP. FORCE     =',F10.2,//)
      RETURN
      END




C
C     INTERPOLATION BETWEEN DATA POINTS
C
      SUBROUTINE INTER(N,TC,X,DT,TMAX,INT)
      IMPLICIT REAL*8(A-H,O-Z)
      DIMENSION TC(200),X(200),F(200)
      NT=TMAX/DT
      NT1=NT+1
      DO 5 I=1,NT1
    5 F(I)=0.0
      F(1)=X(1)
      ANN=0.0
      II=1
      DO 10 I=2,NT1
      AI=I-1
      T=AI*DT
      IF(T.GT.TC(N)) GO TO 12
      IF(T.LE.TC(II+1)) GO TO 9
      ANN=-TC(II+1)+T-DT
      II=II+1
    9 ANN=ANN+DT
      F(I)=X(II)+(X(II+1)-X(II))*ANN/(TC(II+1)-TC(II))
   10 CONTINUE
   12 TC1=TC(1)
      N=NT1
      DO 20 I=1,NT1
      X(I)=F(I)
      AL=I
   20 TC(I)=TC1+DT*(AL-1.)
      RETURN
      END
```

PROGRAM 2: FOURIER

Objective: To calculate the response of a damped single degree-of-freedom system excited by a periodic time-dependent force applied to the mass or by a periodic time-dependent acceleration applied to the support.

Method: Superposition of the steady-state response to the terms of the Fourier series expansion of the excitation function.

Description: Chapter 5, Tables 5.1 and 5.2.

```
C
C         RESPONSE OF A SINGLE DEGREE SYSTEM USING FOURIER SERIES
C
          IMPLICIT REAL*8(A-H,O-Z)
          DIMENSION A(200),B(200),F(200),T(200)
C
C   STATEMENT FUNCTIONS
C
          AI(T1,T2,F1,F2,WB)=(F1-T1*(F2-F1)/(T2-T1))*(DSIN(WB*T2)-DSIN(WB*
         1T1))/WB+(F2-F1)*((DCOS(WB*T2)-DCOS(WB*T1))+WB*(T2*DSIN(WB*T2)
         2-T1*DSIN(WB*T1)))/(WB*WB*(T2-T1))
C
          BI(T1,T2,F1,F2,WB)=(F1-T1*(F2-F1)/(T2-T1))*(DCOS(WB*T1)-DCOS(
         1WB*T2))/WB+(F2-F1)*((DSIN(WB*T2)-DSIN(WB*T1))-WB*(T2*DCOS(WB*T2)
         2-T1*DCOS(WB*T1)))/((T2-T1)*WB*WB)
C
          S(AN,BN,RN,XSI)=(AN*2.*RN*XSI+BN*(1.-RN*RN))/((1.-RN*RN)**2
         1+(2.*RN*XSI)**2)
C
          C(AN,BN,RN,XSI)=(AN*(1.-RN*RN)-BN*2.*RN*XSI)/((1.-RN*RN)**2+(
         12.*RN*XSI)**2)
C
          WRITE(6,100)
      100 FORMAT(' RESPONCE OF A SINGLE DEGREE SYSTEM USING FOURIER SERIES'/
         *)
C
C   READ INPUT DATA
C
          WRITE(6,130)
      130 FORMAT(1H1,10X,'INPUT DATA',/)
          READ(5,102)NTYPE,N,NT,DT,TP,TT,AK,AM,XSI,INT,NPRT
          WRITE(6,102)NTYPE,N,NT,DT,TP,TT,AK,AM,XSI,INT,NPRT
      102 FORMAT (3I5,6F10.2,2I2)
          READ(5,103)(T(I),F(I),I=1,N)
      103 FORMAT(8F10.3)
          DO 10 I=1,N
       10 WRITE(6,103)T(I),F(I)
C
C         INTERPOLATE BETWEEN DATA POINTS
C
          IF(INT.EQ.0) GO TO 17
          CALL INTER(N,T,F,DT,TP)
       17 DO 300 L=1,N
          IF(NTYPE.EQ.0) GO TO 300
          F(L)=-F(L)*AM
      300 WRITE (6,103) T(L),F(L)
          WRITE(6,150)
```

```
  150 FORMAT(1H1,9X,'OUTPUT RESULTS'//,14X,'FREQUENCY',6X,
     1 'FOURIER FORCE COEFF.   FOURIER REPS. CCEFF.'/5X,'N',
     13X,'RAD/SEC',7X,'CPS',8X,'A(N)',8X,'B(N)',8X,'A(N)',8X,'B(N)'/)
C
C   INITIALIZE VARIABLES
C
      W=6.283185307/TP
      AZ=0.
      DO 50 I=2,N
      AZ=AZ+(T(I)-T(I-1))*(F(I)+F(I-1))/2.
   50 CONTINUE
      AZ=AZ/TP
      CO=AZ/AK
      WRITE(6,104) AZ,CO
  104 FORMAT(5X,'0',9X,'0',9X,'0',E12.4,12X,E12.4,/)
C
C   CALCULATE FOURIER COEFFICIENTS
C
      DO 30 J=1,NT
      A(J)=0.
      B(J)=0.
      AJ=J
      WB=W*AJ
      DO 20 I=2,N
      T1=T(I-1)
      T2=T(I)
      F1=F(I-1)
      F2=F(I)
      A(J)=A(J)+AI(T1,T2,F1,F2,WB)
   20 B(J)=B(J)+BI(T1,T2,F1,F2,WB)
      A(J)=A(J)*2./TP
      B(J)=B(J)*2./TP
      CPS=WB/6.283185
      RN=WB/DSQRT(AK/AM)
      CC=C(A(J),B(J),RN,XSI)/AK
      CS=S(A(J),B(J),RN,XSI)/AK
      WRITE(6,140) J,WB,CPS,A(J),B(J),CC,CS
  140 FORMAT (1X,I5,2F10.2,4E12.4)
   30 CONTINUE
      IF(NPRT.EQ.0) GO TO 99
C
C     CALCULATE STEADY STATE RESPONSE
C
      NI=TT/DT+2
      CD=2.0*DSQRT(AK*AM)*XSI
      WRITE(6,160)
  160 FORMAT(1H1,15X,'STEADY-STATE RESPONSE',//,7X,'TIME',7X,
     1 'FORCE',6X,'DISPL.',8X,'VELOC.',8X,'ACC.',3X,'FOUND.FORCE'/)
      DO 70 J=1,NI
      AJ=J-1
      TAU=AJ*DT
      FT=AZ
      X=AZ/AK
      V=0.0
      ACC=0.0
      DO 65 I=1,NT
      AA=I
      WB=AA*W
      RN=WB/DSQRT(AK/AM)
      FT=FT+A(I)*DCOS(WB*TAU)+B(I)*DSIN(WB*TAU)
```

```
    X=X+(S(A(I),B(I),RN,XSI)*DSIN(WB*TAU)+C(A(I),B(I),RN,XSI)*
   1DCOS(WB*TAU))/AK
    V=V+(S(A(I),B(I),RN,XSI)*WB*DCOS(WB*TAU)-C(A(I),B(I),RN,XSI)*
   1WB*DSIN(WB*TAU))/AK
    ACC=ACC+(-S(A(I),B(I),RN,XSI)*WB*WB*DSIN(WB*TAU)-C(A(I),B(I),RN
   1,XSI)*WB*WB*DCOS(WB*TAU))/AK
    FTR=DSQRT((CD*V)**2+(AK*X)**2)
 65 CONTINUE
    WRITE(6,200) TAU,FT,X,V,ACC,FTR
200 FORMAT (F11.4,5E13.4)
 70 CONTINUE
 99 STOP
    END


C
C    INTERPOLATION BETWEEN DATA POINTS
C
    SUBROUTINE INTER(N,TC,X,DT,TMAX)
    IMPLICIT REAL*8(A-H,O-Z)
    DIMENSION TC(200),X(200),F(200)
    NT=TMAX/DT
    NT1=NT+2
    DO 5 I=1,NT1
  5 F(I)=0.0
    F(1)=X(1)
    ANN=0.0
    II=1
    DO 10 I=2,NT1
    AI=I-1
    T=AI*DT
    IF(T.GT.TC(N)) GO TO 12
    IF(T.LE.TC(II+1)) GO TO 9
    ANN=-TC(II+1)+T-DT
    II=II+1
  9 ANN=ANN+DT
    F(I)=X(II)+(X(II+1)-X(II))*ANN/(TC(II+1)-TC(II))
 10 CONTINUE
 12 TC1=TC(1)
    N=NT1
    DO 20 I=1,NT1
    X(I)=F(I)
    AL=I
 20 TC(I)=TC1+DT*(AL-1.)
    RETURN
    END
```

PROGRAM 3: FREQRESP

Objective: To calculate the response in the frequency domain for a damped
single degree-of-freedom system.

Method: Fast Fourier transform (FFT[1]).

Method: Chapter 5, Tables 5.9 and 5.10.

---

[1]Cooley, J. W., Lewis, P. A. W., and Welch, P. D., *IEEE Trans. Education*, Vol. 12, No. 1,
pp. 27-34, March 1969.

```
C
C       FREQUENCY RESPONSE USING FFT
C
        COMPLEX A(1024),CC,F(1024)
        DIMENSION P(200),TC(200)
C
C       READ INPUT DATA AND INITIALIZE
C
        WRITE (6,110)
        READ (5,105) M,T,AK,C,AM,INT,NEQ
  105 FORMAT(I10,4F10.2,2I5)
        WRITE(6,105) M,T,AK,C,AM,INT,NEQ
  110 FORMAT(1H1,11X,'INPUT DATA',//)
        N = 2**M
        AN = N
        DT = T/AN
        READ (5,120) (TC(L),P(L),L=1,NEQ)
  120 FORMAT (8F10.2)
        WRITE(6,120) (TC(L),P(L),L=1,NEQ)
    3 IF(INT.EQ.0) GO TO 13
        CALL INTER (NEQ,TC,P,DT,T)
C
C       CALCULATION OF FORCE FOURIER COEFFICIENTS
C
   13 WRITE (6,130)
        DO 6 I=1,N
    6 A(I)=CMPLX(P(I),0.)*DT
        FAC=-1.0
        CALL FFT(A,FAC,M)
  130 FORMAT(//,10X,'OUTPUT RESULTS'///,10X,'FORCE FOURIER ',
     1'COEFFICIENTS'//,9X,'N',11X,'REAL',11X,'IMAG'/)
  140 FORMAT (I10,2F15.2)
        DO 17 I=1,N
        NI=I-1
   17 WRITE(6,140)NI,A(I)
C
C       CALCULATION OF RESPONSE AND FORCE FOR CHECKING
C
        Z = C/(2.* SQRT(AK*AM))
        WB = 2.*3.14159265358979/T
        WF= SQRT(AK/AM)
        R1 = WB/WF
        WRITE(6,150)
  150 FORMAT(/,6X,'TIME',4X,'DISPL. REAL',3X,'DISPL. IMAG.',
     1 6X,'FORCE REAL',4X,'FORCE IMAG.'/)
        DO 10 I = 1,N
        AI = I-1
        CC = CMPLX(1.-(AI*R1)**2, 2.*AI*R1*Z)
        F(I)=A(I)/T
   10 A(I)=A(I)*WB/(2.*3.14159265358979*AK*CC)
        FAC = 1.0
        CALL FFT(A,FAC,M)
        CALL FFT(F,FAC,M)
        DO 20 I= 1,N
        AI=I-1
        TIME=AI*DT
   20 WRITE(6,160) TIME,A(I),F(I)
  160 FORMAT (F10.3,2F15.4,2F15.2)

        STOP
        END
```

```
      SUBROUTINE FFT(A,SIGN,M)
      COMPLEX A(1024),U,W,H
      N=2**M
      NV2=N/2
      NM1=N-1
      J=1
      DO 7 I=1,NM1
      IF(I.GE.J) GO TO 5
      H=A(J)
      A(J)=A(I)
      A(I)=H
    5 K=NV2
    6 IF(K.GE.J) GO TO 7
      J=J-K
      K=K/2
      GO TO 6
    7 J=J+K
      PI=3.14159265358979
      DO 20 L=1,M
      LE=2**L
      LE1=LE/2
      U=(1.0,0.)
      W=CMPLX(COS(PI/LE1),+SIGN*SIN(PI/LE1))
      DO 20 J=1,LE1
      DO 10 I=J,N,LE
      IP=I+LE1
      H=A(IP)*U
      A(IP)=A(I)-H
      A(I)=A(I)+H
   10 CONTINUE
      U=U*W
   20 CONTINUE
      RETURN
      END
C
C
C     INTERPOLATION BETWEEN DATA POINTS
C
      SUBROUTINE INTER(N,TC,X,DT,TMAX)
      DIMENSION TC(200),X(200),F(200)
      NT=TMAX/DT
      NT1=NT+1
      DO 5 I=1,NT1
    5 F(I)=0.0
      F(1)=X(1)
      ANN=0.0
      II=1
      DO 10 I=2,NT1
      AI=I-1
      T=AI*DT
      IF(T.GT.TC(N)) GO TO 12
      IF(T.LE.TC(II+1)) GO TO 9
      ANN=-TC(II+1)+T-DT
      II=II+1
    9 ANN=ANN+DT
      F(I)=X(II)+(X(II+1)-X(II))*ANN/(TC(II+1)-TC(II))
   10 CONTINUE
   12 TC1=TC(1)
      N=NT1
      DO 20 I=1,NT1
      X(I)=F(I)
      AL=I
```

```
   20  TC(I)=TC1+DT*(AL-1.)
       WRITE(6,100) (X(I),I=1,NT1)
       WRITE(6,100) (TC(I),I=1,NT1)
  100  FORMAT(5E12.4)
       RETURN
       END
```

PROGRAM 4: STEPS

Objective: To determine the response of a damped single degree-of-freedom system with elastoplastic behavior excited by a force applied to the mass or an acceleration at the support.

Method: Step-by-step linear acceleration method.

Description: Chapter 7, Tables 7.2 and 7.3.

```
C
C        ELASTIC-PERFECTLY PLASTIC SINGLE DEGREE SYSTEM
C
         DIMENSION X(100),F(100),TC(100)
C
C        READ INPUT DATA AND INITIALIZE
C
         WRITE(6,105)
         READ(5,100) NTYPE,NEQ,SK,SM,SC,DT,RT,RC
         WRITE(6,100)NTYPE,NEQ,SK,SM,SC,DT,RT,RC
  100  FORMAT (2I5,3F10.0,F10.2,2F10.1)
         READ(5,110) (TC(I),X(I),I=1,NEQ)
         WRITE(6,110)(TC(I),X(I),I=1,NEQ)
  110  FORMAT (8F10.2)
         IF (NTYPE.EQ.0) GO TO 5
         DO 7 I=1,NEQ
    7  X(I)=X(I)*SM
    5  UD=0.0
         UV=0.0
         UA=X(1)/SM
         F(1)=X(1)
         NT=TC(NEQ)/DT
         NT1=NT+1
         ANN=0.0
         NTM1=NT-1
         A1=3.0/DT
         A2=6.0/DT
         A3 = DT/2.0
         A4 = 6./DT**2
         YT=RT/SK
         YC=RC/SK
         KEY=0
         SKP=SK
         II=1
C
C        PRINT HEADINGS
C
  105  FORMAT (1H1,10X,'INPUT DATA'//)
C
C        INTERPOLATION BETWEEN LOAD DATA POINTS
```

```
C
      DO 10 I=2,NT1
      AI=I-1
      T=AI*DT
      IF(T.GT.TC(NEQ)) GO TO 99
      IF(T.LE.TC(II+1)) GO TO 9
      ANN=-TC(II+1)+T-DT
      II=II+1
    9 ANN=ANN+DT
      F(I)=X(II)+(X(II+1)-X(II))*ANN/(TC(II+1)-TC(II))
   10 CONTINUE
   99 CONTINUE
C
C     STEP BY STEP LOOP TO CALCULATE RESPONSE
C
      WRITE (6,115)
  115 FORMAT (//,10X,'RESPONSE ELASTIC-PERFECTLY PLASTIC'/,10X,
     1 'SINGLE DEGREE FREEDOM SYSTEM'//,6X,'TIME',6X,'DISPL.',7X
     1 ,'VELOCITY',8X,'ACC.',5X,'KEY'/)
      DO 90 L=1,NTM1
      AL=L
      T=DT*AL
      SKB=SKP+A4*SM+A1*SC
      DFB=F(L+1)-F(L)+(A2*SM+3*SC)*UV+(3.*SM+A3*SC)*UA
      DUD=DFB/SKB
      DUV=3.*DUD/DT -3.0*UV-UA*DT/2.0
      UD=UD+DUD
      UV=UV+DUV
      IF(KEY) 11,12,13
   12 R=RT-(YT-UD)*SK
      SKP=SK
      IF(UD.GT.YC.AND.UD.LT.YT) GO TO 20
      IF (UD.LT.YC) GO TO 15
      KEY=1
      SKP=0.
      R=RT
      GO TO 20
   13 IF (UV.GT.0.) GO TO 20
      KEY=0
      SKP=SK
      YT=UD
      YC=UD-(RT-RC)/SK
      R=RT-(YT-UD)*SK
      GO TO 20
   11 IF (UV.LT.0.) GO TO 20
      KEY=0
      SKP=SK
      YT=UD+(RT-RC)/SK
      YC=UD
      R=RT-(YT-UD)*SK
      GO TO 20
   15 KEY=-1
      R=RC
      SKP=0.
   20 UA=(F(L+1)-SC*UV-R)/SM
   90 WRITE(6,120) T,UD,UV,UA,KEY
  120 FORMAT(F10.3,3F12.4,I8)
      STOP
      END
```

SENTRY

PROGRAM 5: JACOBI

(Testing Program and Subroutine)

Objective: To solve for the eigenvalues (square of the natural frequencies) and eigenvectors (modal shapes) of a multidegree-of-freedom structural system.

Method: Generalized Jacobi method.[2]

Description: Chapter 10, Tables 10.1 and 10.2.

```
L
C       MAIN PROGRAM TO TEST SUBROUTINE JACOBI
C
        IMPLICIT REAL*8(A-H,O-Z)
        DIMENSION A(30,30),B(30,30),X(30,30),EIGV(30),D(30)
        RTCL = 1.E-12
        NSMAX = 15
C
C       INPUT DATA
C
        WRITE (6,105)
    105 FORMAT (1H1,10X,'INPUT DATA',/)
        READ(5,100)N,IFPR
        WRITE(6,100) N,IFPR
        DO 10 I=1,N
        READ(5,110)(A(I,J),J=1,N)
        WRITE(6,110)(A(I,J),J=1,N)
     10 CONTINUE
        DO 20 I=1,N
        READ(5,110)(B(I,J),J=1,N)
        WRITE(6,110)(B(I,J),J=1,N)
     20 CONTINUE
    100 FORMAT(2I10)
    110 FORMAT(9F10.4)
        CALL JACOBI (A,B,X,EIGV,D,N,IFPR)
        STOP
        END

        SUBROUTINE JACOBI (A,B,X,EIGV,D,N,IFPR)
        IMPLICIT REAL*8(A-H,O-Z)
        DIMENSION A(30,30),B(30,30),X(30,30),EIGV(30),D(30)
C
C       INITIALIZE EIGENVALUE AND EIGENVECTOR MATRICES
C
        NSMAX = 15
        WRITE (6,1980)
        RTCL = 1.D-12
        IOUT=6
        DO 10 I=1,N
        IF(A(I,I).GT.0. .AND. B(I,I).GT.0.)GO TO 4
        WRITE(IOUT,2020)
        STOP
      4 D(I)=A(I,I)/B(I,I)
     10 EIGV(I)=D(I)
        DO 30 I=1,N
        DO 20 J=1,N
```

[2] Bathe, K. J., and Wilson, E. L., *Numerical Methods in Finite Element Analysis*, Prentice Hall, Englewood Cliffs, N.J., 1976.

```
   20 X(I,J)=0.
   30 X(I,I)=1.
      IF(N.EQ.1) RETURN
C
C     INITIALIZE SWEEP COUNTER AND BEGIN ITERATION
C
      NSWEEP=0
      NR=N-1
   40 NSWEEP=NSWEEP+1
      IF(IFPR.EQ.1)WRITE(IOUT,2000)NSWEEP
C
C     CHECK IF PRESENT OFF-DIAGONAL ELEMENT IS LARGE
C
      EPS=(.01**NSWEEP)**2
      BK=B(I,K)
      A(J,I)=AJ+CG*AK
      B(J,I)=BJ+CG*BK
      A(I,K)=AK+CA*AJ
  180 B(I,K)=BK+CA*BJ
  190 AK=A(K,K)
      BK=B(K,K)
      A(K,K)=AK+2.*CA*A(J,K)+CA*CA*A(J,J)
      B(K,K)=BK+2.*CA*B(J,K)+CA*CA*B(J,J)
      A(J,J)=A(J,J)+2.*CG*A(J,K)+CG*CG*AK
      B(J,J)=B(J,J)+2.*CG*B(J,K)+CG*CG*BK
      A(J,K)=0.
      B(J,K)=0.
C
C     UPDATE THE EIGENVECTOR MATRIX AFTER EACH ROTATION
C
      DO 200 I=1,N
      XJ=X(I,J)
      XK=X(I,K)
      X(I,J)=XJ+CG*XK
  200 X(I,K)=XK+CA*XJ
  210 CONTINUE
C
C     UPDATE THE EIGENVALUES AFTER EACH SWEEP
C
      DO 220 I=1,N
      IF (A(I,I).GT.0. .AND. B(I,I).GT.0.) GO TO 220
      WRITE(IOUT,2020)
      STOP
  220 EIGV(I)=A(I,I)/B(I,I)
      IF(IFPR.EQ.0)GO TO 230
      WRITE(IOUT,2010) (EIGV(I),I=1,N)
C
C     CHECK FOR CONVERGENCE
C
  230 DO 240 I=1,N
      TOL=RTOL*D(I)
      DIF=DABS(EIGV(I)-D(I))
      IF(DIF.GT.TOL)GO TO 280
  240 CONTINUE
C
C     CHECK ALL OFF-DIAGONAL ELEMENTS  TO SEE IF ANOTHER SWEEP IS
C     REQUIRED
C
      EPS=RTOL**2
      DO 250 J=1,NR
```

```
      JJ=J+1
       DO 250 K=JJ,N
      EPSA=(A(J,K)*A(J,K))/(A(J,J)*A(K,K))
      EPSB=(B(J,K)*B(J,K))/(B(J,J)*B(K,K))
      IF((EPSA.LT.EPS).AND.(EPSB.LT.EPS))GO TO 250
       GO TO 280
  250 CONTINUE
C
C     FILL OUT BOTTOM TRIANGLE OF RESULTANT MATRICES
C     AND SCALE EIGENVECTORS
C
  255 DO 260 I=1,N
      DO 260 J=1,N
      A(J,I)=A(I,J)
  260 B(J,I)=B(I,J)
      DO 210 J=1,NR
      JJ=J+1
      DO 210 K=JJ,N
      EPTOLA=(A(J,K)*A(J,K))/(A(J,J)*A(K,K))
      EPTOLB=(B(J,K)*B(J,K))/(B(J,J)*B(K,K))
      IF((EPTOLA.LT.EPS).AND.(EPTOLB.LT.EPS))GO TO 210
C
C     IF ZEROING IS REQUIRED,CALCULATE THE ROTATION MATRIX ELEMENT CA,CG
C
      AKK=A(K,K)*B(J,K)-B(K,K)*A(J,K)
      AJJ=A(J,J)*B(J,K)-B(J,J)*A(J,K)
      AB=A(J,J)*B(K,K)-A(K,K)*B(J,J)
      CHECK=(AB*AB+4.*AKK*AJJ)/4.
      IF(CHECK)50,60,60
   50 WRITE(IOUT,2020)
      STOP
   60 SQCH=DSQRT(CHECK)
      D1=AB/2.+SQCH
      D2=AB/2.-SQCH
      DEN=D1
      IF(DABS(D2).GT.DABS(D1))DEN=D2
      IF(DEN)80,70,80
   70 CA=0.
      CG=-A(J,K)/A(K,K)
      CG=-A(J,K)/A(K,K)
      GO TO 90
   80 CA=AKK/DEN
      CG=-AJJ/DEN
C
C     GENERALIZED ROTATION TO ZERO THE PRESENT OFF-DIAGONAL ELEMENT
C
   90 IF(N-2)100,190,100
  100 JP1=J+1
      JM1=J-1
      KP1=K+1
      KM1=K-1
      IF(JM1-1)130,110,110
  110 DO 120 I=1,JM1
      AJ=A(I,J)
      BJ=B(I,J)
      AK=A(I,K)
      BK=B(I,K)
      A(I,J)=AJ+CG*AK
      B(I,J)=BJ+CG*BK
      A(I,K)=AK+CA*AJ
```

```
 120  B(I,K)=BK+CA*BJ
 130  IF (KP1-N)140,140,160
 140  DO 150 I=KP1,N
      AJ=A(J,I)
      BJ=B(J,I)
      AK=A(K,I)
      BK=B(K,I)
      A(J,I)=AJ+CG*AK
      B(J,I)=BJ+CG*BK
      A(K,I)=AK+CA*AJ
 150  B(K,I)=BK+CA*BJ
 160  IF(JP1-KM1)170,170,190
 170  DO 180 I=JP1,KM1
      AJ=A(J,I)
      BJ=B(J,I)
      AK=A(I,K)
      DO 270 J=1,N
      BB=DSQRT(B(J,J))
      DO 270 K=1,N
 270  X(K,J)=X(K,J)/BB
C
C     UPDATE MATRIX AND START NEW SWEEP,IF ALLOWED
C
      WRITE(6,1990)
      DO 1991 LI=1,N
 1991 WRITE(6,2010) (X(LI,LJ),LJ=1,N)
 1990 FORMAT(/10X,'EIGENVECTORS',/)
      RETURN
  280 DO 290 I=1,N
  290 D(I)=EIGV(I)
      IF(NSWEEP.LT.NSMAX)GO TO 40
      GO TO 255
 1980 FORMAT (//,10X,'EIGENVALUES',/)
 2000 FORMAT (/, ' SWEEP=',I2)
 2010 FORMAT(1H0,5X,6E14.5/)
 2020 FORMAT (25H0*** ERROR  SOLUTION STOP  /
     1        30H MATRICES NOT POSITVE DEFINITE)
      END
```

PROGRAM 6: MODAL

(Subroutine and Testing Program)

Objective: Subroutine MODAL is designed to calculate the response of a multi-degree-of-freedom system.

Method: Modal superpostion method.

Description: Chapter 11, Tables 11.1 and 11.2.

```
C
C     MAIN PROGRAM TO TEST MODAL
C
      IMPLICIT REAL*8(A-H,O-Z)
      DIMENSION EIGEN(40),X(40,40),SM(40,40),F(40,40)
C
C     READ INPUT DATA AND INITIALIZE
C
```

```
      READ (5,100) ND,GR
      WRITE (6,100) ND ,GR
      IF (GR.EQ.0.) GO TO 5
      DO 4 I=1,ND
      READ (5,110) (SM(I,J),J=1,ND)
    4 WRITE (6,110)(SM(I,J),J=1,ND)
  100 FORMAT(I10,F10.0)
    5 READ(5,110) (EIGEN(I),I=1,ND)
      WRITE(6,110)(EIGEN(I),I=1,ND)
  110 FORMAT(8F10.0)
  112 FORMAT(8F10.5)
      DO 10 I=1,ND
      READ(5,112) (X(I,J),J=1,ND)
   10 WRITE(6,112)(X(I,J),J=1,ND)
C
      CALL MODAL (ND,EIGEN,X,F,GR,SM)
C
      STOP
      END

      SUBROUTINE MODAL (ND,EIGEN,X,F,GR,SM)
      IMPLICIT REAL*8(A-H,O-Z)
      REAL*8INT1,INT2,INT3,INT4,K,M
      DIMENSION EIGEN(40),X(40,40),XIS(40),F(40,40),P(40),T(40),Y(40,40)
    1 ,UD(40),FF(40),NG(40),SM(40,40)
C
C STEMENT FUNCTIONS
C
      INT1(TAU)=DEXP(XIWD*TAU)*(XIWD*DCOS(WD*TAU)+WD*DSIN(WD*TAU))/DWSQ
      INT2(TAU)=DEXP(XIWD*TAU)*(XIWD*DSIN(WD*TAU)-WD*DCOS(WD*TAU))/DWSQ
      INT3(TAU)=TAU*INT2(TAU)-XIWD*INT2(TAU)/DWSQ+WD*INT1(TAU)/DWSQ
      INT4(TAU)=TAU*INT1(TAU)-XIWD*INT1(TAU)/DWSQ-WD*INT2(TAU)/DWSQ
C
C     READ FORCING FUNTIONS AND INTERPOLATE
C
      NG=ND
      IF(GR.NE.0.) NG=1
      NNN=40
      READ(5,110) DT,TMAX,(NQ(L),L=1,NG)
      WRITE(6,110)DT,TMAX,(NQ(L),L=1,NG)
  110 FORMAT(2F10.4,12I5)
      DO 76 I=1,NNN
      FF(I)=0.0
      DO 76 J=1,NNN
   76 F(I,J)=0.0
      DO 77 ID=1,NG
      NEQ=NQ(ID)
      IF(NEQ.EQ.0) GO TO 77
      READ(5,120) (T(L),P(L),L=1,NEQ)
      WRITE(6,120)( T(L),P(L),L=1,NEQ)
  120 FORMAT(4F10.2)
      NT= T(NEQ)/DT
      IF (NT.GT.TMAX/DT) NT=TMAX/DT
      NT1=NT+1
      FF(1)=P(1)
      ANN=0.0
      II=1
      DO 19 I=2,NT1
      AI=I-1
      TA=AI*DT
      IF(TA.GT.T (NEQ)) GO TO 160
```

```
      IF(TA.LE.T (II+1)) GO TO 9
      ANN= -T(II+1)+TA-DT
      II=II+1
    9 ANN=ANN+DT
      FF(I)=P(II)+(P(II+1)-P(II))*ANN/( T(II+1)- T(II))
      IF (GR.NE.0.) FF(I)=FF(I)*GR
       F(ID,I)=FF(I)
   19 CONTINUE
  160 CONTINUE
   77 CONTINUE
C
C     DETERMINE TIME AND EQUIVALENT FORCES
C
      NT=TMAX/DT
      DO 17 L=1,NNN
      AL=L-1
       T(L)= T(1)+AL*DT
      IF(GR.EQ.0.) GO TO 17
      DO 18 ID=1,ND
   18 F(ID,L)=-FF(L)*SM(ID,ID)
   17 CONTINUE
C
C     READ DAMPING RATIOS AND SET INITIAL VALUES
C
      READ(5,100) (XIS(L),L=1,ND)
      WRITE(6,100)(XIS(L),L=1,ND)
  100 FORMAT(8F10.3)
C
C     WRITE HEADINGS
C
      WRITE (6,700)
  700 FORMAT(1H1,8X,'RESPONSE FOR ELASTIC SYSTEM',//,
     1 6X,'TIME',5X,'DISPLACEMENTS AT NODAL COORDINATES',/)
      NT1=NT+1
      DO 50 ID=1,ND
      DO 10 IT=1,NT1
      P(IT)=0.0
      DO 10 I=1,ND
   10 P(IT)=P(IT)+F(I,IT) *X(I,ID)
      M=1.0
      K=EIGEN(ID)
      XI=XIS(ID)
    6 FIM1=P(I)
      TIM1=T(1)
      ATI=0.0
      BTI=0.0
      DAT=0.0
      DBT=0.0
      Y(ID,1)=0.0
      OMEGA=DSQRT(K/M)
      CRIT=2*DSQRT(K*M)
      C=XI*CRIT
      WD=OMEGA*DSQRT(1.-(XI**2))
      XIWD=XI*OMEGA
      DWSQ=XIWD**2+WD**2
C
C     LOOP OVER TIME AND SOLVE FOR MODAL DISPLACEMENTS
C
      NM1=NT-1
      DO 1 I=1,NM1
      FI=P(I+1)
```

```
      TI=T(I+1)
      DFTI=FI-FIM1
      DTI=TI-TIM1
      FT=DFTI/DTI
      G=FIM1-TIM1*FT
      AI=INT1(TI)-INT1(TIM1)
      BI=INT2(TI)-INT2(TIM1)
      VS=INT3(TI)-INT3(TIM1)
      VC=INT4(TI)-INT4(TIM1)
      AI=AI*G
      AI=AI+FT*VC
      ATI=ATI+AI
      BI=BI*G
      BI=BI+FT*VS
      BTI=BTI+BI
      Y(ID,I+1) =DEXP(-XIWD*TI)*(ATI*DSIN(WD*TI)-BTI*DCOS(WD*TI))/(M*WD)
      TIM1=TI
      FIM1=FI
    1 CONTINUE
   50 CONTINUE
      DO 53 IT=1,NT
      DO 52 I=1,ND
      UD(I)=0.0
      DO 52 J=1,ND
   52 UD(I)=UD(I)+X(I,J)*Y (J,IT)
   53 WRITE(6,301) T(IT),(UD(L),L=1,ND)
  301 FORMAT(F10.3,6E14.4)
      RETURN
```

PROGRAM 7: SRESB

(Seismic Response of an Elastic Shear Building)

Objective: To determine the response in the linear range of structures modeled as shear buildings and subjected to excitation at the foundation.

Method: Modal superposition method of analysis.

Description: Chapter 11, Tables 11.5 and 11.6.

Subroutines Required: JACOBI (listed in Program 5) and MODAL (listed in Program 6).

```
C
C       SEISMIC RESPONSE ELASTIC SHEAR BUILDING
C
      IMPLICIT REAL*8 (A-H,O-Z)
      DIMENSION SK(40,40),SM(40,40),SC(40,40),X(40,40),
     1 DUA(40),UD(40),UV(40),UA(40),S(40),EIGEN(40)
C
C       READ INPUT DATA AND INITIALIZE
C
      READ (5,100) E,GR,ND,IFPR
      WRITE (6,100)E,GR,ND,IFPR
  100 FORMAT (2F10.0,2I5)
      DO 2 I=1,ND
      DO 2 J=1,ND
      SM(I,J)=0.0
```

```
      SC(I,J)=0.0
      X(I,J)=0.0
    2 SK(I,J)=0.0
      ND1=ND+1
      DO 7 I=1,ND
      READ(5,110) SI,SL,SM(I,I)
      WRITE(6,110)SI,SL,SM(I,I)
  110 FORMAT(3F10.2,F10.0)
      S(I)=12.0*E*SI/SL**3
      SC(I,I)=SM(I,I)
      UD(I)=0.0
    7 UV(I)=0.0
C
C     ASSEMBLE STIFFNESS MATRIX
C
      S(ND+1)=0.0
      DO 19 I=1,ND
      IF(I.EQ.1) GO TO 19
      SK(I,I-1)=-S(I)
      SK(I-1,I)=-S(I)
   19 SK(I,I)=S(I)+S(I+1)
C
C     DETERMINE NATURAL FREQUENCIES AND MODE SHAPES
C
      CALL JACOBI (SK,SC,X,EIGEN,S,ND,IFPR)
C
C     RESPONSE USING MODAL SUPERPOSITION
C
      CALL MODAL(ND,EIGEN,X,SC,GR,SM)
C
      STOP
      END
```

PROGRAM 8: HARMO

(Subroutine and Testing Program)

Objective: To determine the steady-state response of a damped multidegree-of-freedom system subjected to harmonic forces.

Method: Gauss elimination to solve the complex system of algebraic equations for the amplitudes of the steady-state motion.

Description: Chapter 11, Tables 11.9 and 11.10.

```
$JOB              ,PAGES=5,TIME=5,LINES=400
C
C     MAIN PROGRAM TEST HARMO
C
      IMPLICIT REAL * 8 (A-H,O-Z)
      COMPLEX*16 A(30,10),DET
      DIMENSION SK(30,30),SM(30,30)
C
C     INPUT DATA
C
      WRITE (6,120)
```

```
120 FORMAT (1H1,5X,'INPUT DATA',//)
    READ(5,100) ND
    WRITE(6,100)ND
100 FORMAT(3I5)
    DO 10I = 1,ND
    READ(5,110) (SK(I,J),J=1,ND)
 10 WRITE(6,110)(SK(I,J),J=1,ND)
    DO20 I=1,ND
    READ (5,110) (SM(I,J),J=1,ND)
 20 WRITE(6,110) (SM(I,J),J=1,ND)
110 FORMAT(8F10.0)
    CALL HARMO (ND,SK,SM)
    STOP
    END

    SUBROUTINE HARMO (ND,SK,SM)
    IMPLICIT REAL * 8 (A-H,O-Z)
    COMPLEX * 16 A(30,30),DET
    DIMENSION SK(30,30),SM(30,30)
    READ(5,110) FACK,FACM,W
    WRITE(6,110)FACK,FACM,W
    DO 10 I=1,ND
    DO 10 J = 1,ND
    AR = SK(I,J)-W*W*SM(I,J)
    AI = SK (I,J)*W*FACK+SM(I,J)*W*FACM
 10 A(I,J)=DCMPLX(AR,AI)
110 FORMAT (8F10.2)
    READ (5,110) (A(L,ND+1),L=1,ND)
    WRITE(6,110) (A(L,ND+1),L=1,ND)
    DO 7 L=1,ND
  7 A(L,ND+1)=DCONJG(A(L,ND+1))
    EPS=1.0E-10
    NPLUSM=ND+1
    DET=DCMPLX(1.D0,0.D0)
    DO 9 K=1,ND
    DET=DET*A(K,K)
    IF(CDABS(A(K,K)).GT.EPS) GO TO 5
    WRITE(6,202)
    GO TO99
  5 KP1=K+1
    DO 6 J=KP1,  NPLUSM
  6 A(K,J)=A(K,J)/A(K,K)
    A(K,K)=DCMPLX(1.D0,0.D0)
    DO 9 I=1,ND
    IF (I.EQ.K.OR.CDABS(A(I,K)).EQ.0.) GO TO 9
    DO 8 J=KP1,NPLUSM
  8 A(I,J)=A(I,J)-A(I,K)*A(K,J)
    A(I,K)=DCMPLX(0.D0,0.D0)
  9 CONTINUE
202 FORMAT(37HOSMALL PIVOT -MATRIX MAY BE SINGULAR )
    WRITE (6,170)
170 FORMAT(//,5X,'THE STEADY-STATE RESPONSE IS'//, 4X,'COORD.',5X,
   1' COS COMP.',4X,'  SIN COMP.',/)
    DO 87 I=1,ND
    A(I,ND+1)=DCONJG(A(I,ND+1))
 87 WRITE(6,122)I, A(I,ND+1)
122 FORMAT(I10,2D15.4)
 99 RETURN
    END
```

PROGRAM 9: DAMP

(Subroutine and Testing Program)

Objective: To calculate for a multidegree-of-freedom system the damping matrix from specified modal damping ratios.

Method: Definition of damping coefficients to satisfy the orthoganality condition between normal modes, eq. (12.22).

Description: Chapter 12, Tables 12.1 and 12.2.

```
C
C       MAIN PROGRAM TO TEST SUBROUTINE DAMP
C
        IMPLICIT REAL*8(A-H,O-Z)
        DIMENSICN EIGEN(30),X(30,30),SM(30,30),SC(30,30)
C
C       READ INPUT DATA AND INITIALIZE
C
        WRITE (6,115)
  115 FORMAT (1H1,10X,'INPUT DATA',//)
        READ (5,100)NL
        WRITE(6,100)NL
  100 FORMAT(I10)
        READ(5,110) (EIGEN(I),I=1,NL)
        WRITE(6,110)(EIGEN(I),I=1,NL)
  110 FORMAT(8F10.4)
        DO 10 I=1,NL
        READ (5,110) (X(I,J),J=1,NL)
   10 WRITE(6,110) (X(I,J),J=1,NL)
        DO 20 I=1,NL
        READ(5,110) (SM(I,J),J=1,NL)
   20 WRITE(6,110)(SM(I,J),J=1,NL)
        CALL DAMP (NL,X,SM, SC,EIGEN)
        STOP
        END

        SUBROUTINE DAMP(NL,X,SM,SC,EIGEN)
        IMPLICIT REAL*8(A-H,C-Z)
        DIMENSION X(30,30),T(30,30),SM(30,30),SC(30,30),EIGEN(30),XIS(30)
        READ (5,110) (XIS(L),L=1,NL)
        WRITE(6,110) (XIS(L),L=1,NL)
        WRITE (6,125)
  125 FORMAT (//,5X,'THE DAMPING MATRIX IS',/)
        DO 10 I=1,NL
        EIGEN(I)=DSQRT(EIGEN(I))
        DO 10 J=1,NL
   10 SC(I,J) =0.0
        DO 20 II=1,NL
        DA = 2.*XIS(II)*EIGEN(II)
        DO 20 I=1,NL
        DO20 J=1,NL
   20 SC(I,J)=SC(I,J)+X(I,II)*X(J,II)*DA
        DO 30 I=1,NL
        DO 30 J=1,NL
        T(I,J)=0.0
        DO 30 K = 1,NL
```

```
 30 T(I,J) = T(I,J)+SM(I,K)*SC(K,J)
    DO 40 I=1,NL
    DO 40 J=1,NL
    SC(I,J)=0.0
    DO 40 K=1,NL
 40 SC(I,J) = SC(I,J)+T(I,K)*SM(K,J)
    DO 50 I=1,NL
 50 WRITE(6,120) (SC(I,J),J=1,NL)
110 FORMAT(3F10.2)
120 FORMAT (6D14.4)
    RETURN
    END
$ENTRY
```

PROGRAM 10: CONDE

(Subroutine and Testing Program)

Objective: To reduce the dimensions of the stiffness and mass matrices.

Method: Static condensation.

Description: Chapter 13, Tables 13.1 and 13.2.

```
C
C       MAIN PROGRAM TO TEST CONDE
C
        IMPLICIT REAL*8(A-H,O-Z)
        DIMENSION SK(30,30),SM(30,30),SC(30,30),T(30,30)
C
C       READ INPUT DATA AND INITIALIZE
C
        WRITE (6,125)
    125 FORMAT (1H1,5X,'INPUT DATA',/)
C
        READ (5,100) ND,NCR
        WRITE(6,100) ND,NCR
    100 FORMAT(3I10)
        DO 10 I=1,ND
        READ(5,110) (SK(I,J),J=1,ND)
     10 WRITE(6,110)(SK(I,J),J=1,ND)
    110 FORMAT(8F10.0)
        DO 20 I=1,ND
        READ(5,110)(SM(I,J),J=1,ND)
     20 WRITE(6,110)(SM(I,J),J=1,ND)
        CALL CONDE (ND,NCR,SK,SM,SC,T)
        STOP
        END

        SUBROUTINE CONDE (ND,NCR,SK,SM,SC,T)
        IMPLICIT REAL*8(A-H,O-Z)
        DIMENSION SK(30,30),SM(30,30),T(30,30),TT(30),SC(30,30)
C
C       CALCULATE THE REDUCED STIFFNESS MATRIX AND THE TRANSFORMATION MATRIX
C
        NL=ND-NCR
        DO 9 K=1,NCR
```

```
      IF (DABS(SK(K,K)).GT.1.D-10) GO TO 5
      WRITE (6,202) K
  202 FORMAT ('                      PIVOT TOO SMALL',I10)
      GO TO 99
    5 KP1 = K+1
      DO 6 J=KP1,ND
    6 SK(K,J) = SK(K,J)/SK(K,K)
      SK(K,K) =1.
      DO 9 I = 1,ND
      IF (I.EQ.K.OR. SK(I,K) .EQ.0) GO TO 9
      DO 8 J=KP1,ND
    8 SK(I,J) = SK(I,J) - SK(I,K)* SK(K,J)
      SK(I,K) = 0.0
    9 CONTINUE
      DO 30 I = 1,NCR
      DO 30 J = 1,NL
      JJ = J+NCR
   30 T(I,J) = -SK(I,JJ)
      DO 40 I=1,NL
      II = I + NCR
      DO 50 J = 1,NL
   50 T(II,J) = 0.0
      T(II,I) = 1.0
   40 CONTINUE
      DO 20 I= 1,NL
      DO 20 J = 1,NL
      II = I + NCR
      JJ = J+NCR
   20 SK(I,J) = SK(II,JJ)
      WRITE (6,169)
  169 FORMAT(1H0,5X,'THE REDUCED STIFFNESS MATRIX IS'/)
      DO 80 I=1,NL
   80 WRITE (6,190) (SK(I,J),J=1,NL)
      WRITE(6,170)
  170 FORMAT(/6X,'THE TRANSFORMATION MATRIX IS'/)
      DO 81 I = 1,ND
   81 WRITE(6,190) (T(I,J),J = 1,NL)
  190 FORMAT (6E14.4)
C
C     CALCULATE THE REDUCED MASS MATRIX
C
      DO 68 J = 1,ND
      DO 60 I=1,NL
      TT(I) = 0.0
      DO 60 K=1,ND
   60 TT(I) = TT(I) + T(K,I) *SM(K,J)
      DO 65 K = 1,NL
   65 SM(K,J) = TT(K)
   68 CONTINUE
      DO 78 I = 1,NL
      DO 70 J=1,NL
      TT(J)=0.0
      DO 70 K=1,ND
   70 TT(J)=TT(J)+SM(I,K)*T(K,J)
      DO 75 K=1,NL
   75 SM(I,K)=TT(K)
   78 CONTINUE
      DO 83 I=1,NL
      DO 83 J=1,NL
   83 SC(I,J)=SM(I,J)
      WRITE(6,172)
```

```
172 FORMAT(/,6X,'THE REDUCED MASS MATRIX IS'/)
  7 DO 82 I = 1,NL
 82 WRITE (6,190) (SM(I,J),J=1,NL)
 99 RETURN
    END
```

PROGRAM 11: BEAM

Objective: Dynamic analysis of beams.

Method: Stiffness matrix method.

Description: Chapter 14, Tables 14.3 and 14.4.

Subroutines: Table A.1.

```
$JOB              ,PAGES=5,TIME=5,LINES=400
C
C       DYNAMIC ANALYSIS OF UNIFORM BEAMS
C
        IMPLICIT REAL*8(A-H,O-Z)
        DIMENSION SK(30,30),SM(30,30),T(30,30),TT(30),EIGEN(30),NC(4),
       1 BK(4,4),BM(4,4),JC(30),CM(30),SC(30,30)
C
C       READ GENERAL DATA AND INITIALIZE VARIABLES
C
        WRITE (6,120)
  120 FORMAT (1H1,5X,'INPUT DATA',//)
    2 READ (5,100) NE,ND,NCR,NCM,LOC,IFPR,E
        IF (NE.EQ.0) GO TO 99
        WRITE(6,100) NE,ND,NCR,NCM,LOC,IFPR,E
  100 FORMAT(6I5, F20.0)
        NL=ND-NCR
        DO 5 I=1,ND
        DO 5 J=1,ND
        SK(I,J) = 0.0
    5 SM(I,J) = 0.0
C
C       ELEMENT STIFFNESS AND MASS MATRICES
C
C
C       LOOP OVER BEAM SEGMENTS
C
        DO 30 IF =1,NE
        READ (5,101) LE,SL,SI,SMA,(NC(L),L=1,4)
        WRITE(6,101) LE,SL,SI,SMA,(NC(L),L=1,4)
  101 FORMAT(I10,F10.2,F10.4,F10.2,4I5)
        A1=E*SI/SL**3
        A2 = SMA*SL/420.
        BK(1,1) = 12.*A1
        BK(1,2) = 6.*A1*SL
        BK(1,3) = -12.*A1
        BK(1,4) = 6.*A1*SL
        BK(2,2) = 4.* A1*SL**2
        BK(2,3) = -6.*A1*SL
        BK(2,4) = 2.*A1*SL**2
        BK(3,3) = 12.*A1
        BK(3,4) = -6.*A1*SL
        BK(4,4) = 4.*A1*SL**2
```

```
      IF(LOC.EQ.1) GO TO 11
      DO 6 I=1,4
      DO 6 J=1,4
    6 BM(I,J) = 0.0
      BM(1,1) = SL*SMA/2.0
      BM(3,3) = SL*SMA/2.0
      GO TO 12
   11 BM(1,1) = 156.*A2
      BM(1,2)= 22.*A2*SL
      BM(1,3) = 54.*A2
      BM(1,4) = -13.*A2*SL
      BM(2,2) = 4.*A2*SL**2
      BM(2,3) = 13.*A2*SL
      BM(2,4) = -3.*A2*SL**2
      BM(3,3) =156.*A2
      BM(3,4) = -22.*A2*SL
      BM(4,4) = 4.*A2*SL**2
   12 DO 15 I=2,4
      L=I-1
      DO 15 J=1,L
      BK(I,J)=BK(J,I)
   15 BM(I,J)=BM(J,I)
C
C     ASSEMBLE SYSTEM STIFFNESS AND MASS MATRICES
C
      DO 13 II=1,4
      I=NC(II)
      IF(I.GT.ND) GO TO 13
      DO 10 JJ=1,4
      J=NC(JJ)
      IF(J.GT.ND) GO TO 10
   16 SK (I,J) = SK(I,J) + BK(II,JJ)
      SM(I,J) = SM(I,J) + BM(II,JJ)
   10 CONTINUE
   13 CONTINUE
   30 CONTINUE
C
C     CONCENTRATED MASSES
C
      IF (NCM.EQ.0) GO TO 41
      READ(5,103) (JC(L),CM(L), L= 1,NCM)
      WRITE(6,103)(JC(L),CM(L), L= 1,NCM)
  103 FORMAT(8(I2,F8.2))
      DO 40 L= NCM
      I=JC(L)
   40 SM(I,I)= SM(I,I)+ CM(L)
   41 CONTINUE
      DO 81 I=1,ND
      DO 81 J=1,ND
   81 SC(I,J)=SM(I,J)
C
C     FOR STATIC CONDENSATION CALL CONDE
C
      CALL CONDE(ND,NCR,LOC,SK,SM,SC,T)
C
C     FOR NATURAL FREQUENCIES  AND NORMAL MODES CALL JACOBI
C
      CALL JACOBI (SK,SC,T,EIGEN,TT,NL,IFPR)
C
C     FOR DAMPING MATRIX CALL DAMP
```

```
C
C       CALL DAMP (NL,T,SM,SC,EIGEN)
C
C       FOR HARMONIC FORCES CALL HARMO
C
C       CALL HARMO(NL,SK,SM)
C
C        FOR STEP BY STEP SOLUTION CALL STEP
C
C       CALL STEP(SK,SM,SC,NL)
C
        CALL MODAL (ND,EIGEN,T,SC,GR,SM)
C
        GO TO 2
     99 CONTINUE
        STOP
        END
```

PROGRAM 12: FRAME

Objective: Dynamic analysis of plane frames subjected to loads in the plane of
the frame.

Method: Stiffness matrix method.

Description: Chapter 15, Tables 15.1 and 15.2.

Subroutines: Table A.1.

```
C       DYNAMIC ANALYSIS OF PLANE FRAMES
        IMPLICIT REAL * 8 (A-H,O-Z)
        DIMENSION SK(30,30),SM(30,30),T(30,30),TT(30),EIGEN(30),NC(6),
       1 BK(6,6),BM(6,6),JC(30),CM(30),TK(30),TM(30),SC(30,30)
C
C       READ GENERAL DATA AND INITIALIZE VARIABLES
C
        READ (5,100) NE,ND,NCR,NCM,NBW,LOC,IFPR,E ,GR
        WRITE(6,100) NE,ND,NCR,NCM,NBW,LOC,IFPR,E ,GR
    100 FORMAT(7I5,2F10.0)
        NL=ND
        DO 5 I=1,ND
        DO 5 J=1,NBW
        SK(I,J) = 0.0
      5 SM(I,J) = 0.0
C
C       ELEMENT STIFFNESS AND MASS MATRICES
C
C       LOOP OVER BEAM SEGMENTS
C
        DO 90 IF =1,NE
        READ (5,101) LE,SL,AR,SI,SMA,TH,(NC(L),L=1,6)
        WRITE(6,101) LE,SL,AR,SI,SMA,TH,(NC(L),L=1,6)
    101 FORMAT(I10,5F10.0,6I2)
        DO 7 I=1,6
        DO 7 J=1,6
        BK(I,J)=0.0
        BM(I,J)=0.0
```

```
     7 T(I,J)=0.0
       ST=DSIN(TH/57.295779)
       CT=DCOS(TH/57.295779)
       A1=E*SI/SL**3
       A2 = SMA*SL/420.
       BK(1,1) = AR*E/SL
       BK(1,4)=-BK(1,1)
       BK(4,4)=BK(1,1)
       BK(2,2)=12.*A1
       BK(2,3) = 6.*A1*SL
       BK(2,5) = -12.*A1
       BK(2,6) = 6.*A1*SL
       BK(3,3) = 4.* A1*SL**2
       BK(3,5) = -6.*A1*SL
       BK(3,6) = 2.*A1*SL**2
       BK(5,5) = 12.*A1
       BK(5,6) = -6.*A1*SL
       BK(6,6) = 4.*A1*SL**2
       IF(LOC.EQ.1) GO TO 11
       BM(1,1) = SL*SMA/2.0
       BM(2,2) = SL*SMA/2.0
       BM(4,4)= SL*SMA/2.0
       BM(5,5)=SL*SMA/2.0
       GO TO 12
    11 BM(1,1)= 140.0*A2
       BM(1,4) =BM(1,1)/2.0
       BM(2,2) = 156.*A2
       BM(2,3)= 22.*A2*SL
       BM(2,5) = 54.*A2
       BM(2,6) = -13.*A2*SL
       BM(3,3) = 4.*A2*SL**2
       BM(3,5) = 13.*A2*SL
       BM(3,6) = -3.*A2*SL**2
       BM(4,4)=BM(1,1)
       BM(5,5) =156.*A2
       BM(5,6) = -22.*A2*SL
       BM(6,6) = 4.*A2*SL**2
    12 DO 15 I=2,6
       L=I-1
       DO 15 J=1,L
       BK(I,J)=BK(J,I)
    15 BM(I,J)=BM(J,I)
C
C      TRANSFORMATION MATRIX
C
       T(1,1) = CT
       T(1,2) = ST
       T(2,1)= -ST
       T(2,2) = CT
       T(3,3) = 1.0
       T(4,4) = CT
       T(4,5) = ST
       T(5,4) = -ST
       T(5,5) = CT
       T(6,6) = 1.0
C
C      ELEMENT  MATRICES IN GLOBAL COORDINATES
C
       DO 30 J=1,6
       DO 25 I=1,6
       TK(I) = 0.0
```

```
      TM(I) = 0.0
      DO 25 K=1,6
      TK(I)=TK(I)+T(K,I)*BK(K,J)
   25 TM(I)=TM(I)+T(K,I)*BM(K,J)
      DO 28 K=1,6
      BM(K,J)=TM(K)
   28 BK(K,J)=TK(K)
   30 CONTINUE
      DO 40 I=1,6
      DO 35 J=1,6
      TK(J)=0.0
      TM(J)=0.0
      DO 35 K=1,6
      TK(J) = TK(J) + BK(I,K)*T(K,J)
   35 TM(J) = TM(J)+BM(I,K)*T(K,J)
      DO 38 K=1,6
      BK(I,K) = TK(K)
   38 BM(I,K) = TM(K)
   40 CONTINUE
C
C     ASSEMBLE SYSTEM STIFFNESS AND MASS MATRICES
C
      DO 13 II=1,6
      I=NC(II)
      IF(I.GT.ND) GO TO 13
      DO 10 JJ=1,6
      J=NC(JJ)
      IF(J.GT.ND) GO TO 10
      IF (NBW.EQ.ND) GO TO16
      J=J-I+1
   16 SK (I,J) = SK(I,J) + BK(II,JJ)
      SM(I,J) = SM(I,J) + BM(II,JJ)
   10 CONTINUE
   13 CONTINUE
   90 CONTINUE
C
C     CONCENTRATED MASSES
C
      IF (NCM.EQ.0) GO TO 41
      READ(5,103) (JC(L),CM(L), L= 1,NCM)
      WRITE(6,103)(JC(L),CM(L), L= 1,NCM)
  103 FORMAT(8(I2,F8.2))
      DO 43 L= NCM
      I=JC(L)
   43 SM(I,I)= SM(I,I)+ CM(L)
   41 CONTINUE
      DO 79 I=1,ND
      DO 79 J=1,NBW
   79 SC(I,J)=SM(I,J)
C
C     FOR STATIC CONDENSATION CALL CONDE
C
C     CALL CONDE(ND,NCR,LOC,SK,SM,SC,T)
C
C     FOR NATURAL  FREQUENCIES  AND NORMAL MODES CALL JACOBI
C
C     CALL JACOBI (SK,SC,T,EIGEN,TT,NL,IFPR)
C
C     CALL MODAL (ND,EIGEN,T,SC,GR,SM)
C
C     FOR DAMPING MATRIX CALL DAMP
C
```

```
          CALL DAMP (NL,T,SM,SC,EIGEN)
C
C
C         FOR HARMONIC FORCES CALL HARMO
C
C         CALL HARMO(ND,NBW,SK,SM)
C
C          FOR STEP BY STEP SOLUTION CALL STEP
C
C
          CALL STEP(SK,SM,SC,NL)
C
C
          STOP
          END
```

PROGRAM 13: GRID

Objective: Dynamic analysis of grids (plane frames with normal loads).

Method: Stiffness matrix method.

Description: Chapter 16, Tables 16.1 and 16.2.

Subroutines: Table A.1.

```
C
C         DYNAMIC ANALYSIS OF PLANE GRIDS
C
          IMPLICIT REAL*8(A-H,O-Z)
          DIMENSION SK(40,40),SM(40,40),T(40,40),TT(40),EIGEN(40),NC(6),
         1 BK(6,6),BM(6,6),JC(40),CM(40),TK(40),TM(40),SC(40,40),
         2 NQ(40),F(40),P(40),TC(40)
C
C         READ GENERAL DATA AND INITIALIZE VARIABLES
C
          READ (5,100) NE,ND,NCR,NCM,LOC,IFPR,E,G
          WRITE(6,100) NE,ND,NCR,NCM,LOC,IFPR,E,G
      100 FORMAT (6I5,3F10.0)
          NNN=40
          NL=ND
          DO 5 I=1,ND
          DO 5 J=1,ND
          SK(I,J)=0.0
          SC(I,J)=0.0
        5 SM(I,J)=0.0
C
C         ELEMENT STIFFNESS AND ELEMENT MATRICES IN LOCAL COORD.
C
C         LOOP OVER BEAM SEGMENTS
C
          DO 90 IE=1,NE
          READ(5,101) LE,AR,SL,SI,SJ,SMA,TH,(NC(L),L=1,6)
          WRITE(6,101)LE,AR,SL,SI,SJ,SMA,TH,(NC(L),L=1,6)
      101 FORMAT(I2,2F8.2,4F10.2,6I2)
          DO 7 I=1,6
          DO 7 J=1,6
```

```
      BK(I,J)=0.D0
      T(I,J)=0.0
    7 BM(I,J) = 0.0
      ST=DSIN(TH/57.29578)
      CT=DCOS(TH/57.29578)
      A1 = SJ*G/SL
      A2=6.*E*SI/SL**2
      A3=12.*E*SI/SL**3
      A4=2.*E*SI/SL
      A5=SJ*SMA*SL/(6.*AR)
      BK(1,1)=A1
      BK(1,4)=-A1
      BK(2,2)=2.*A4
      BK(2,3)=A2
      BK(2,5)=A4
      BK(2,6)=-A2
      BK(3,3)=A3
      BK(3,5)=A2
      BK(3,6)=-A3
      BK(4,4)=A1
      BK(5,5)=2.*A4
      BK(5,6)=-A2
      BK(6,6)=A3
      IF(LOC.EQ.1) GO TO 11
      BM(3,3)=SL*SMA/2.0
      BM(6,6)=SL*SMA/2.0
      GO TO 12
   11 BM(1,1) = 2.*A5
      BM(1,4)=-A5
      BM(2,2)=4.*SMA*SL**3/420.
      BM(2,3)=22.*SMA*SL**2/420.
      BM(2,5)=-3.*SMA*SL**3/420.
      BM(2,6) = 13.*SMA*SL**2/420.
      BM(3,3)=156.*SMA*SL
      BM(3,5)=-13.*SMA*SL**2/420.
      BM(3,6)=54.*SMA*SL/420.
      BM(4,4)=BM(1,1)
      BM(5,5)=4.*SMA*SL**3/420.
      BM(5,6)=-22.*SMA*SL**2/420.
      BM(6,6) = 156.*SMA*SL/420.
   12 DO 15 I=2,6
      L=I-1
      DO 15 J=1,L
      BK(I,J)=BK(J,I)
   15 BM(I,J)=BM(J,I)
C
C     TRANSFORMATION MATRIX
C
      T(1,1) = CT
      T(1,2) = ST
      T(2,1)= -ST
      T(2,2) = CT
      T(3,3) = 1.0
      T(4,4) = CT
      T(4,5) = ST
      T(5,4) = -ST
      T(5,5) = CT
      T(6,6) = 1.0
C
C     ELEMENT STIFFNESS AND MASS MATRICES IN SYSTEM COORD.
```

```
C
      DO 30 J=1,6
      DO 25 I=1,6
      TK(I) = 0.0
      TM(I) = 0.0
      DO 25 K=1,6
      TK(I)=TK(I)+T(K,I)*BK(K,J)
   25 TM(I)=TM(I)+T(K,I)*BM(K,J)
      DO 28 K=1,6
      BM(K,J)=TM(K)
   28 BK(K,J)=TK(K)
   30 CONTINUE
      DO 40 I=1,6
      DO 35 J=1,6
      TK(J)=0.0
      TM(J)=0.0
      DO 35 K=1,6
      TK(J) = TK(J) + BK(I,K)*T(K,J)
   35 TM(J) = TM(J)+BM(I,K)*T(K,J)
      DO 38 K=1,6
      BK(I,K) = TK(K)
   38 BM(I,K) = TM(K)
   40 CONTINUE
C
C     ASSEMBLE SYSTEM STIFFNESS AND MASS MATRICES
C
      DO 13 II=1,6
      I=NC(II)
      IF(I.GT.ND) GO TO 13
      DO 10 JJ=1,6
      J=NC(JJ)
      IF(J.GT.ND) GO TO 10
   16 SK (I,J) = SK(I,J) + BK(II,JJ)
      SM(I,J) = SM(I,J) + BM(II,JJ)
   10 CONTINUE
   13 CONTINUE
   90 CONTINUE
C
C     CONCENTRATED MASSES
C
      IF (NCM.EQ.0) GO TO 41
      READ(5,103) (JC(L),CM(L), L= 1,NCM)
      WRITE(6,103)(JC(L),CM(L), L= 1,NCM)
  103 FORMAT(8(I2,F8.2))
      DO 43 L= NCM
      I=JC(L)
   43 SM(I,I)= SM(I,I)+ CM(L)
   41 CONTINUE
      DO 87 I=1,ND
      DO 87 J=1,ND
      SC(I,J)=SM(I,J)
   87 CONTINUE
C
C     FOR STATIC CONDENSATION CALL CONDE
C
C     CALL CONDE(ND,NCR,SK,SM,SC,T)
C
C     FOR NATURAL  FREQUENCIES  AND NORMAL MODES CALL JACOBI
C
      CALL JACOBI (SK,SC,T,EIGEN,TT,NL,IFPR)
```

```
C
C     CALL MODAL (ND,EIGEN,T,SC,GR,SM)
C
C     FOR DAMPING MATRIX CALL DAMP
C
C     CALL DAMP (NL,T,SM,SC,EIGEN)
C
C     FOR HARMONIC FORCES CALL HARMO
C
C     CALL HARMO(ND,SK,SM)
C
C      FOR STEP BY STEP SOLUTION CALL STEP
C
C     CALL STEP(SK,SM,SC,NL)
C
      STOP
      END
```

PROGRAM 14: TRUSS

Objective: Dynamic analysis of plane trusses.

Method: Stiffness matrix method.

Description: Chapter 18, Tables 18.1 and 18.2.

Subroutines: Table A.1.

```
C
C     DYNAMIC ANALYSIS OF PLANE TRUSSES
C
      IMPLICIT REAL*8(A-H,O-Z)
      DIMENSION SK(30,30),SM(30,30),T(30,30),SC(30,30),NC(4)
     1 ,BK(4,4),BM(4,4),JC(30),TK(30),TM(30),CM(30),EIGEN(30)
C
C     GENERAL DATA AND INITIALIZE
C
      READ (5,100) NE,ND,NCR,NCM,LOC,IFPR,E
      WRITE(6,100) NE,ND,NCR,NCM,LOC,IFPR,E
100   FORMAT (6I5,2F10.0)
      NL=ND
      DO 5 I=1,ND
      DO 5 J=1,ND
      SK(I,J)=0.0
5     SM(I,J)=0.0
C
C     ELEMENT STIFFNESS AND ELEMENT MATRICES
C
C     LOOP OVER TRUSS MEMBERS
C
      DO 90 IE=1,NE
      READ(5,101) LE,SL,AR,SMA,TH,(NC(L),L=1,4)
      WRITE(6,101) LE,SL,AR,SMA,TH,(NC(L),L=1,4)
101   FORMAT (I10,4F10.2,4I2)
      DO 7 I=1,4
      DO 7 J=1,4
      BK(I,J)=0.0
      BM(I,J)=0.0
7     T(I,J)=0.0
      ST=DSIN(TH/57.2958)
      CT=DCOS(TH/57.2958)
      A1=AR*E/SL
      A2=SMA*SL/6.0
      BK(1,1)=A1
      BK(1,3)=-A1
      BK(3,3)=A1
      IF(LOC.EQ.1) GO TO 11
```

```
      BM(1,1)=SMA*SL/2.0
      BM(2,2)=SMA*SL/2.0
      BM(3,3)=SMA*SL/2.0
      BM(4,4)=SMA*SL/2.0
      GO TO 12
   11 BM(1,1)=2.*A2
      BM(2,2)=2.*A2
      BM(3,3)=2.*A2
      BM(4,4)=2.*A2
      BM(1,3)=A2
      BM(2,4)=A2
   12 DO 15 I=2,4
      L=I-1
      DO 15 J=1,L
      BK(I,J)=BK(J,I)
   15 BM(I,J)=BM(J,I)
C
C     TRANSFORMATION MATRIX
C
      DO 14 I=1,4
      DO 14 J=1,4
   14 T(I,J)=0.0
      T(1,1)=CT
      T(1,2)=ST
      T(2,1)=-ST
      T(2,2)=CT
      T(3,3)=CT
      T(3,4)=ST
      T(4,3)=-ST
      T(4,4)=CT
C
C     MEMBER MATRIX IN GLOBAL COORDINATES
C
      DO 30 J=1,4
      DO 25 I=1,4
      TK(I)=0.0
      TM(I)=0.0
      DO 25 K=1,4
      TK(I)=TK(I)+T(K,I)*BK(K,J)
   25 TM(I)=TM(I)+T(K,I)*BM(K,J)
      DO 28 K=1,4
      BM(K,J)=TM(K)
   28 BK(K,J)=TK(K)
   30 CONTINUE
      DO 40 I=1,4
      DO 35 J=1,4
      TK(J)=0.0
      TM(J)=0.0
      DO 35 K=1,4
      TK(J)=TK(J)+BK(I,K)*T(K,J)
   35 TM(J)=TM(J)+BM(I,K)*T(K,J)
      DO 38 K=1,4
      BK(I,K)=TK(K)
   38 BM(I,K)=TM(K)
   40 CONTINUE
C
C     ASSEMBLE SYSTEM STIFFNESS AND MASS MATRICES
C
      DO 401 I=1,4
  401 WRITE (6,402) (BM(I,J),J=1,4)
  402 FORMAT(4E20.4)
      DO 13 II=1,4
      I=NC(II)
      IF (I.GT.ND) GO TO 13
      DO 10 JJ=1,4
      J=NC(JJ)
      IF (J.GT.ND) GO TO 10
   16 SK(I,J)=SK(I,J)+BK(II,JJ)
      SM(I,J)=SM(I,J)+BM(II,JJ)
   10 CONTINUE
   13 CONTINUE
   90 CONTINUE
C
C     CONCENTRATED MASSES
C
      IF (NCM.EQ.0) GO TO 41
      READ (5,103) (JC(L),CM(L),L=1,NCM)
      WRITE(6,103) (JC(L),CM(L),L=1,NCM)
  103 FORMAT (8(I2,F8.2))
      DO 43 L=1,NCM
      I=JC(L)
   43 SM(I,I)=SM(I,I)+CM(L)
```

```
 41 CONTINUE
    DO 79 I=1,ND
    DO 79 J=1,ND
 79 SC(I,J)=SM(I,J)
    DO 80 I=1,ND
 80 WRITE(6,190) (SK(I,J),J=1,ND)
    DO 81 I=1,ND
 81 WRITE(6,190) (SM(I,J),J=1,ND)
190 FORMAT(6E14.4)
C
C     FOR STATIC CONDENSATION CALL CONDE
C
C     CALL CONDE(ND,NCR,SK,SM,SC,T)
C
C     FOR NATURAL  FREQUENCIES  AND NORMAL MODES CALL JACOBI
C
C     CALL JACOBI (SK,SC,T,EIGEN,TK,ND,IFPR)
C
C     CALL MODAL (ND,EIGEN,T,SC,GK,SM)
C
C     FOR DAMPING MATRIX CALL DAMP
C
C     CALL DAMP (NL,T,SM,SC,EIGEN)
C
C     FOR HARMONIC FORCES CALL HARMO
C
C     CALL HARMO(ND,NBW,SK,SM)
C
C      FOR STEP BY STEP SOLUTION CALL STEP
C
C     CALL STEP(SK,SM,SC,NL)
C
      STOP
      END
```

PROGRAM 15: STEPM

(Subroutine and Testing Program)

Objective: Dynamic response of a linear multidegree-of-freedom system.

Method: Step-by-step linear acceleration method with Wilson-θ modification.

Description: Chapter 19, Tables 19.1 and 19.2.

```
C     MAIN PROGRAM STEP BY STEP
C
      IMPLICIT REAL * 8 (A-H,O-Z)
      DIMENSION SK(30,30),SM(30,30),SC(30,30)
      READ (5,100) ND
      WRITE(6,100) ND
100 FORMAT (I10)
      DO 10 I=1,ND
      READ(5,110) (SK(I,J),J=1,ND)
 10 WRITE(6,110)(SK(I,J),J=1,ND)
110 FORMAT(8F10.0)
      DO 20 I = 1,ND
      READ (5,110) (SM(I,J),J=1,ND)
 20 WRITE(6,110) (SM(I,J),J=1,ND)
      DO 30 I=1,ND
      READ(5,110)(SC(I,J),J=1,ND)
 30 WRITE(6,110)(SC(I,J),J=1,ND)
      CALL STEP (SK,SM,SC,ND)
      STOP
      END

      SUBROUTINE STEP (SK,SM,SC,ND)
```

```
C
      IMPLICIT REAL *P(A-H,O-Z)
      DIMENSION SK(30,30),SM(30,30),SC(30,30),F(30,30),X(30,30),
     1DUA(30),UD(30),UV(30),UA(30),TC(30),P(30),NEQ(30)
C      INITIALIZE AND READ INPUT DATA
C
C
      ND1=ND+1
      DO 1 I=1,30
      UD(I)=0.0
      UV(I)=0.0
      DO 1 J=1,30
    1 F(I,J)=0.0
      READ(5,100) THETA,DT,TMAX,(NEQ(L),L=1,ND)
      WRITE(6,100)THETA,DT,TMAX,(NEQ(L),L=1,ND)
  100 FORMAT(3F10.3,3I5)
C
C     INTERPOLATE BETWEEN EXCITATION DATA POINTS
C
  110 FORMAT(8F10.2)
      ANN=0.0
      II=1
      DO 12 ID=1,ND
      NE=NEQ(ID)
      IF(NE.EQ.0) GO TO 12
      IF(NE.GT.TMAX/DT) NE=TMAX/DT
      READ(5,110) (TC(J),P(J),J=1,NE)
      WRITE(6,110)(TC(J),P(J),J=1,NE)
      NT=TC(NE)/DT
      NT1=NT+1
      NT2=NT+2
      F(ID,1) = P(1)
      ANN=0.0
      II=1
      DO 10 I=2,NT2
      AI=I-1
      T=AI*DT
      IF (T.GT.TC(NE)) GO TO 12
      IF (T.LE.TC(II+1)) GO TO 9
      ANN=-TC(II+1)+T-DT
      II=II+1
    9 ANN=ANN+DT
      F(ID,I)=P(II)+(P(II+1)-P(II))*ANN/(TC(II+1)-TC(II))
C     WRITE(6,110) T,F(ID,I)
   10 CONTINUE
   12 CONTINUE
C
C     DETERMINE INITIAL ACCELERATION
C
      NT=TMAX/DT
      NT1=NT+1
      DO 13 I=1,ND
      X(I,ND1) = F(I,1)
      DO 13 J=1,ND
   13 X(I,J) = SM(I,J)
      DO 15 I=1,ND
      DO 15 J=1,ND
   15 X(I,ND1) = X(I,ND1)-SC(I,J)*UV(J)-SK(I,J)*UD(J)
C     DO 7 IJ=1,ND
C   7 WRITE (6,111) (X(IJ,JJ),JJ=1,ND1)
      CALL SOLVE (ND,X)
```

```
C     WRITE(6,111) (X(IJ,ND1),IJ=1,ND)
C 111 FORMAT(6E14.4)
      DO 23 I=1,ND
   23 UA(I)=X(I,ND1)
C
C     INITIALIZE CONSTANTS AND LOOP OVERTIME STEPS
C
      TU = THETA*DT
      A1=3./TU
      A2 = 6./TU
      A3 = TU/2.0
      A4 = A2/TU
      WRITE(6,110) TU,A1,A2,A3,A4
      DO 90 L=1,NT1
      DO 25 I=1,ND
      DC 25 J=1,ND
   25 X(I,J) = SK(I,J)+A4*SM(I,J)+A1*SC(I,J)
      AL = L
      T=AL*DT
      DO 35 I=1,ND
      X(I,ND1) = F(I,L+1) + (F(I,L+2)-F(I,L+1))*(THETA-1.0)-F(I,L)
      DO 30 J=1,ND
   30 X(I,ND1) = X(I,ND1) + (SM(I,J)*A2+SC(I,J)*3.0)*UV(J)
     1 + (SM(I,J)*3.0+A3*SC(I,J))*UA(J)
   35 CONTINUE
      CALL SOLVE (ND,X)
      DO 40 I=1,ND
      DUA(I)=A4*X(I,ND1)-A2*UV(I)-3.0*UA(I)
      DUA(I) = DUA(I)/THETA
      DUV  =          DT*UA(I)+DT*DUA(I)/2.0
      UD(I) = UD(I)+DT*UV(I)+DT*DT*UA(I)/2.0+DT*DT*DUA(I)/6.0
      UV(I)=UV(I)+DUV
   40 CONTINUE
      DO 50 I=1,ND
      X(I,ND1) = F(I,L+1)
      DO 45 J=1,ND
      X(I,ND1) = X(I,ND1) - SK(I,J)*UD(J)-SC(I,J)*UV(J)
   45 X(I,J) = SM(I,J)
   50 CONTINUE
      CALL SOLVE (ND,X)
      DO 60 I=1,ND
      UA(I) = X(I,ND1)
   60 WRITE(6,140)I,T,UD(I),UV(I),UA(I)
  140 FORMAT(I10,F10.3,3F20.3)
   90 CONTINUE
      RETURN
      END

      SUBROUTINE SOLVE (N,A)
      IMPLICIT REAL * 8 (A-H,O-Z)
      DIMENSION           A(30,30)
      M=1
      EPS=1.0E-10
      NPLUSM=N+M
      DET=1.0
      DO 9 K=1,N
      DET=DET*A(K,K)
      IF(DABS(A(K,K)).GT.EPS) GO TO 5
      WRITE(6,202)
      GO TO99
    5 KP1=K+1
```

```
      DO 6 J=KP1,  NPLUSM
    6 A(K,J)=A(K,J)/A(K,K)
      A(K,K)=1.
      DO 9 I=1,N
      IF (I.EQ.K.OR.A(I,K).EQ.0.) GO TO 9
      DO 8 J=KP1,NPLUSM
    8 A(I,J)=A(I,J)-A(I,K)*A(K,J)
      A(I,K)=0.D00
    9 CONTINUE
  202 FORMAT(37HOSMALL PIVOT -MATRIX MAY BE SINGULAR )
   99 RETURN
      END
```

PROGRAM 16: SRSB

(Seismic Response Nonlinear Shear Building)

Objective: Seismic response for inelastic behavior of structures modeled as shear buildings.

Method: Step-by-step linear acceleration with Wilson-θ modification for the integration of the nonlinear equations of motion (elastoplastic behavior).

Description: Chapter 19, Tables 19.5 and 19.6.

```
C
C       SEISMIC RESPONSE ELASTOPLASTIC SHEAR BUILDING
C
      IMPLICIT REAL*8(A-H,O-Z)
      DIMENSION SK(30,30),SM(30,30),SC(30,30),F(30),X(30,30),DD(30),
     1 DUA(30),UD(30),UV(30),UA(30),TC(30),P(30),SKP(30),RT(30),
     1 R(30),YT(30),YC(30),S(30),SP(30),KEY(30),EIGEN(30)
C
C       READ INPUT DATA AND INITIALIZE
C
      READ(5,100) THETA,DT,E,GR,TMAX,NEQ,ND,IFPR
      WRITE(6,100) THETA,DT,E,GR,TMAX,NEQ,ND,IFPR
  100 FORMAT(2F10.2,3F10.0,3I5)
      NX=TMAX/DT+5
      DO 1 I=1,NX
    1 F(I)=0.0
      DO 2 I=1,ND
      DO 2 J=1,ND
      SM(I,J)=0.0
      SC(I,J)=0.0
      X(I,J)=0.0
    2 SK(I,J)=0.0
      ND1=ND+1
      TU=THETA*DT
      A1=3./TU
      A2=6./TU
      A3=TU/2
      A4=A2/TU
      DO 7 I=1,ND
      READ(5,110) SI,SL,SM(I,I),PM
      WRITE(6,110) SI,SL,SM(I,I),PM
  110 FORMAT(3F10.2,F10.0)
      S(I)=12.0*E*SI/SL**3
      SP(I)=S(I)
      RT(I)=2*PM/SL
      SC(I,I)=SM(I,I)
      UD(I)=0.0
      UV(I)=0.0
      YT(I)=RT(I)/S(I)
      YC(I)=-RT(I)/S(I)
      KEY(I)=0
    7 SP(I)=S(I)
```

```
C
C     ASSEMBLE STIFFNESS MATRIX
C
      S(ND+1)=0.0
      DO 19 I=1,ND
      IF(I.EQ.1) GO TO 19
      SK(I,I-1)=-S(I)
      SK(I-1,I)=-S(I)
   19 SK(I,I)=S(I)+S(I+1)
C
C     DETERMINE NATURAL FREQUENCIES AND MODE SHAPES
C
      CALL JACOBI(SK,SC,X,EIGEN,TC,ND,IFPR)
C
C     DETERMINE DAMPING MATRIX
C
      CALL DAMP(ND,X,SM,SC,EIGEN)
      READ(5,120) (TC(L),P(L),L=1,NEQ)
      WRITE(6,120)(TC(L),P(L),L=1,NEQ)
  120 FORMAT(4F10.2)
      DO 4 I=1,NEQ
    4 P(I)=P(I)*GR
C
C     INTERPOLATION BETWEEN DATA POINTS
C
      NT=TC(NEQ)/DT
      NT1=NT+1
      F(1)=P(1)
      ANN=0.0
      II=1
      DO 10 I=2,NT1
      AI=I-1
      T=AI*DT
      IF(T.GT.TC(NEQ)) GO TO 16
      IF(T.LE.TC(II+1)) GO TO 9
      ANN=TC(II+1)+T-DT
      II=II+1
    9 ANN=ANN+DT
      F(I)=P(II)+(P(II+1)-P(II))*ANN/(TC(II+1)-TC(II))
   10 CONTINUE
   16 CONTINUE
C
C     INITIALIZE AND DETERMINE INITIAL ACCELERATION
C
      DO 22 I=1,ND
      X(I,ND)=-F(1)*SM(I,I)
      DO 22 J=1,ND
   22 X(I,J)=SM(I,J)
      CALL SOLVE(ND,X)
      DO 23 I=1,ND
   23 UA(I)=X(I,ND)
      SP(ND+1)=0.0
      R(ND+1)=0.0
C
C     LOOP OVER TIME CALCULATING RESPONSE
C
      WRITE (6,170)
      DO 90 L=1,NT
      AL = L
      T=DT*AL
      DO 20 I=1,ND
      IF(I.EQ.1) GO TO 20
      SK(I,I-1) = -SP(I)
      SK(I-1,I)=-SP(I)
   20 SK(I,I)=SP(I)+SP(I+1)
      DO 25 I=1,ND
      DO 25 J=1,ND
   25 X(I,J)=SK(I,J)+A4*SM(I,J)+A1*SC(I,J)
      DO 35 I=1,ND
      X(I,ND)=(F(L+1)+(F(L+2)-F(L+1))*(THETA-1.0)-F(L))*(-SM(I,I))
      DO 30 J=1,ND
   30 X(I,ND)=X(I,ND)+(SM(I,J)*A2+SC(I,J)*3.0)*UV(J)
    1 +(SM(I,J)*2.0+A3*SC(I,J))*UA(J)
   35 CONTINUE
      CALL SOLVE(ND,X)
      DO 38 I=1,ND
      DUA(I)=A4*X(I,ND)-A2*UV(I)-3.0*UA(I)
      DUA(I)=DUA(I)/THETA
      DUV=DT*UA(I)+DT*DUA(I)/2.0
```

```
         UD(I)=     UD(I)+DT*UV(I)+DT*DT*UA(I)/2.0+DT*DT*DUA(I)/6.0
 38      UV(I)=UV(I)+DUV
         DD(I)=UD(I)
         DO 39 I=2,ND
 39      DD(I)=UD(I)-UD(I-1)
         DO 40 I=1,ND
         IF(KEY(I)) 11,12,13
 12      R(I)=RT(I)-(YT(I)-DD(I))*S(I)
         SP(I)=S(I)
         IF (DD(I).GE.YC(I).AND.DD(I).LT.YT(I)) GO TO 40
         IF(DD(I).LT.YC(I)) GO TO 15
         KEY(I)=1
         SP(I)=0.0
         R(I)=RT(I)
         GO TO 40
 13      IF(UV(I).GT.0.) GO TO 40
         KEY(I)=0
         SP(I)=S(I)
         YT(I)=DD(I)
         YC(I)=DD(I)-2.0*RT(I)/S(I)
         R(I)=RT(I)-(YT(I)-DD(I))*S(I)
         GO TO 40
 11      IF(UV(I).LT.0.) GO TO 40
         KEY(I)=0
         SP(I)=S(I)
         YC(I)=DD(I)
         YT(I)=DD(I)+2.*RT(I)/S(I)
         R(I)=RT(I)-(YT(I)-DD(I))*S(I)
         GO TO 40
 15      KEY(I)=-1
         R(I)=-RT(I)
         SP(I)=0.0
 40      CONTINUE
         DO 50 I=1,ND
         X(I,ND1)=F(I+1)*(-SM(I,I))-R(I)+R(I+1)
         DO 45 J=1,ND
         X(I,ND1)=X(I,ND1)-SC(I,J)*UV(J)
 45      X(I,J)=SM(I,J)
 50      CONTINUE
         CALL SOLVE (ND,X)
         DO 60 I=1,ND
         UA(I)=X(I,ND1)
 60      WRITE(6,250) T,T,UD(I),UV(I),UA(I)
 90      CONTINUE
170      FORMAT(1H1,6X,'THE RESPONSE IS',/,5X,'CORD.',6X,'TIME',9X,
     1   'DISPL.',6X,'VELOC.',11X,'ACC.'/)
250      FORMAT(T10,F10.3,3F15.4)
         STOP
         END

         SUBROUTINE SOLVE (N,A)
         IMPLICIT REAL * 8 (A-H,O-Z)
         DIMENSION          A(30,30)
         M=1
         EPS=1.0E-10
         NPLUSM=N+M
         DET=1.0
         DO 9 K=1,N
         DET=DET*A(K,K)
         IF(DABS(A(K,K)).GT.EPS) GO TO 5
         WRITE(6,202)




         GO TO 99
 5       KP1=K+1
         DO 6 J=KP1,NPLUSM
 6       A(K,J)=A(K,J)/A(K,K)
         A(K,K)=1.
         DO 9 I=1,N
         IF (I.EQ.K.OR.A(I,K).EQ.0.) GO TO 9
         DO 8 J=KP1,NPLUSM
 8       A(I,J)=A(I,J)-A(I,K)*A(K,J)
         A(I,K)=0.D0A
 9       CONTINUE
202      FORMAT(37H0SMALL PIVOT -MATRIX MAY BE SINGULAR )
 99      RETURN
         END
```

Index